The Deep River Coalfield

The Deep River Coalfield

Two Hundred Years of Mining in Chatham County, North Carolina

JAMES H. CHAPMAN

McFarland & Company, Inc., Publishers
Jefferson, North Carolina

LIBRARY OF CONGRESS CATALOGUING-IN-PUBLICATION DATA

Names: Chapman, James H., 1952– author.
Title: The Deep River Coalfield : two hundred years of mining in Chatham County, North Carolina / James H. Chapman.
Description: Jefferson, North Carolina : McFarland & Company, Inc., Publishers, 2017 | Includes bibliographical references and index.
Identifiers: LCCN 2017025656 | ISBN 9781476668987 (softcover : acid free paper) ∞
Subjects: LCSH: Coal mines and mining—North Carolina—Chatham County—History. | Deep River Watershed Region (N.C.)—History.
Classification: LCC TN805.N8 C45 2017 | DDC 622/.3340975659—dc23
LC record available at https://lccn.loc.gov/2017025656

BRITISH LIBRARY CATALOGUING DATA ARE AVAILABLE

ISBN (print) 978-1-4766-6898-7
ISBN (ebook) 978-1-4766-2902-5

© 2017 James H. Chapman. All rights reserved

No part of this book may be reproduced or transmitted in any form or by any means, electronic or mechanical, including photocopying or recording, or by any information storage and retrieval system, without permission in writing from the publisher.

Front cover: two miners take a break from their rescue work (Ben Dixon MacNeill Photograph Collection #P0078, North Carolina Collection Photographic Archives, Wilson Library, University of North Carolina at Chapel Hill)

Printed in the United States of America

McFarland & Company, Inc., Publishers
Box 611, Jefferson, North Carolina 28640
www.mcfarlandpub.com

To the miners and workers who labored and perished in the mines at Deep River. They lived near the mines, worked in the mines and, suddenly, died in the mines. Their lives intermingled with a deep, dark world, sometimes harsh and unforgiving, but always where they returned.

They were miners like so many others in so many places. They were men and boys. Their story has been largely forgotten. It is time to be told.

"Its people have only to put forth their energy and enterprise, to stand with the first States in this Republic."
—Ebenezer Emmons, 1852, State Geologist of North Carolina, writing of coal mining at Deep River

*"Her mineral wealth is boundless,
In copper, iron and gold,
Her coal is exhaustless,
Her riches undeveloped,
can never be told."*
—"Old Chatham," by L.J.J., 1879

Table of Contents

Acknowledgments ix
Preface 1
Introduction 3

1. Geological Aspects of the Deep River Basin 7
2. Early History of the Deep River Coalfield 12
3. Geology as a Science Stimulates Geological Surveys 19
4. The Antebellum Period 36
5. The Deep River Coalfield at the Time of the Civil War 48
6. Post–Civil War Considerations 60
7. The 1880s Through 1900 78
8. The 1895 Mine Explosion at Cumnock 94
9. The 1900 Explosion at Cumnock 109
10. Coal Production in North Carolina, 1900–1925 115
11. Tragedy Strikes Coal Glen: May 27, 1925 127
12. The Nature and Causes of Mine Gases 138
13. The State of Coal Mining After 1925 142
14. New Challenges for Coal Mining 145

Conclusion 155
Appendix A: The Dan River Coalfield 157
Appendix B: Jackson's Survey of the Deep River Coalfield 159
Appendix C: Daddow-Bannon Map of the Deep River Coalfield 162

Table of Contents

Appendix D: Transcript of Inquest Concerning the December 19, 1895, Disaster at Cumnock — 164

Appendix E: Individuals Killed at Farmville Mine, 1925 — 168

Appendix F: Total Tonnage Mined Yearly at Deep River — 172

Appendix G: Company Records — 174

Appendix H: Active Coal Mining Companies, 1867–1896 — 176

Chapter Notes — 178

Bibliography — 217

Index — 225

Acknowledgments

Robert G. Anthony, Jr., curator of the North Carolina Collection at the University of North Carolina at Chapel Hill, suggested the idea for this book some years back because of my interest in coal mining and North Carolina history. Bob's encouragement throughout the project kept me on track to complete the work. The helpful staff of the North Carolina Collection made available to me the Ben Dixon MacNeill photograph collection, which contains images of scenes in the wake of the May 27, 1925, explosion at the Farmville Mine of the Carolina Coal Company at Coal Glen, and responded in kind to my requests for copies of photographs and printed material. The staff at the Special Collections and Archives Department of the Z. Smith Reynolds Library at Wake Forest University, Winston-Salem, North Carolina, located and allowed me to use a rare copy of Hamilton Fulton's *Report of Sundry Surveys* (1819). The staff at the State Archives of North Carolina at Raleigh assisted me with vintage newspapers and vital records in their collection. Greg Williams, of General Timber, Inc., gave permission to visit the old Farmville Mine workings on his property at Coal Glen. The Mine Safety and Health Administration of the U.S. Department of Labor provided a transcript of the findings of the 1925 Farmville Mine inquest as well as answering questions about mine safety. The folks at J.R. Moore & Son general store in Gulf took time to answer my questions and made arrangements for me to take photographs of items in their personal collection. Their business dates to 1935 and is a throwback to the small retail stores that served small communities before the advent of shopping malls and online stores and where the locals congregated for talk and fellowship. The Kentucky Geological Survey at the University of Kentucky, Lexington, gave me permission to reproduce illustrations from their website. The U.S. Bureau of Mines was a great source for photographs, including one of the several mine safety cars maintained by the bureau and dispatched to locations where rescue and recovery efforts were urgently needed. Randall Rigsbee, editor of the *Chatham Record*, put out a call to readers who had knowledge of the mining operations at Deep River. His interest in the topic includes several articles he has written about the history of the area. Michael Hill, of the Historical Research Branch of the North Carolina Office of Archives and History, expressed an early interest in the book and provided encouragement. My wife, Cynthia, stuck by me every phase of the project and was my biggest supporter.

Even in a scholarly study, one must not forget the human side of the story which

Acknowledgments

affected real people in a real way. Sandra Cameron took time to talk with me about her mother, Margaret Whicker, who was out chopping cotton the morning of the explosion at Coal Glen in 1925. A girl of seven at the time, she never forgot the resonating sound followed by a plume of smoke which billowed up into the sky, an experience that remained with her all her life and one she shared with others until her passing many years later. W.B. "Dub" Mason, Jr., of Cumnock, told me of his father, W.B. Mason, Sr., who worked at Coal Glen and helped bring out bodies after the explosion. Dub allowed me to see and copy the letter his father received from the coal company in appreciation of his service during that most difficult time. Mike Tysor wrote to me about his two grandfathers, Hiram and Joe, who were farming only a few miles away on that fateful spring morning. Joe worked at the Farmville Mine but his older brother Hiram insisted that Joe stay home to help plant corn. Whether the young men heard the explosion or learned of it soon after, Joe's decision to pick up a hoe instead of a shovel most likely kept him from joining the others on the casualty list. Fate, coincidence, and paradox, however one wants to call it, all came into play in the many manifestations of coal mining at Deep River.

Preface

In April 1864, a family of five picked up their lives and sailed from the coalfields of South Wales to Canada's shores, then proceeded south to the anthracite coalfields of Northeast Pennsylvania, where they settled in the mining town of Plymouth. Soon, they met up with the two remaining members of the family, who only a few years earlier had arrived on American soil. All the men in the family pursued mining and some eventually spread out to other areas to continue their chosen profession. My great-grandfather was one of the members of that family who followed that vocation with an unspoken understanding of the perils it brought.

Some years later he and two brothers ventured to West Virginia where they continued to pursue coal mining, one of them becoming an inspector of mines for the state. Unfortunately, his tenure was shortened when a runaway coal car jumped the tracks and pinned him against a wall of coal. Six days later he succumbed to his injuries, leaving a grieving widow and seven children. He was the last member of his immediate family to be involved in coal, and, ironically, the only one to die as the result of an accident.

I tell this story because mining is typically family-oriented and my family was no different from the rest. Like so many others, the men in the family worked in the mines and the women relied on their wages to pay the rent and buy the food that sustained them. The loss of pay of even one family member could mean losing one's home and livelihood. The small mining communities along Deep River included unmarried men, but families accounted for most of those who settled in the mine patches. The individuals who settled Egypt, Gulf, and Farmville had the same hopes and dreams as others, namely to make a better life for themselves and their families or simply to have steady work and wages despite a highly dangerous occupation.

The tragedies that struck Egypt and Farmville in the late nineteenth and early twentieth centuries affected everyone in the community. Men of dissimilar backgrounds worked alongside each other where differences in race, ethnicity, and creed had no place if there was to be any possibility of survival.[1] The miners who entered the workings at the beginning of their shift relied on the fire boss to check the surroundings for poisonous gasses. The miner who lit the fuse of the squib that detonated the black-powder charge that brought down the coal at the face had to ensure that all safety precautions had been followed before setting off that chain of events. Miners' laborers, or "butties," were responsible for the back-breaking job of loading the coal into the cars while ensuring that the miner received every penny due him. Rock, bone, slate, and shale counted for nothing toward a miner's wages, and if his wages suffered so did his helper's. The laborer might aspire to become a miner but it took hard work and a long apprenticeship to learn the trade, and there was no guarantee that a roof cave-in, explosion, or strike would not take that opportunity away. Everything and everybody in the community was connected and had a purpose both below and above ground. Survival often meant respect for that connection. Those that severed it often found themselves having to cope with the hardest of tragedies. Families, at least, benefitted from the solace of each other though multiple losses could never be completely salved. Families stayed on if their financial circumstances allowed it or they left the area and relied on the largesse of others before contemplating their next step into the future.

Though coal mining along Deep River, North Carolina, has a long history, total production does not begin to compare to that of her sister states Virginia and Tennessee. Kentucky, Pennsylvania, West Virginia, Maryland, Ohio, and Indiana became the leaders in total tonnage before the completion of the transcontinental railroad in 1869 opened the western coalfields in Colorado and Wyoming. A long line of railroad tycoons, some unscrupulous and others savvy, appeared on the horizon and staked themselves to the mineral rights underneath their feet prior to extracting their valuable fuel source from the ground.

During Reconstruction, businessmen and speculators flocked to North Carolina at the behest of testimonials and slick promotions to exploit the coalfields along Deep River with the purpose of stoking the boilers of ships and locomotives as well as the local iron forges that dotted the landscape and beyond. The outlook for industrial development throughout the South and the rest of the country depended on a steady supply of coal and the ability to mine and transport it. Large corporations that built their success by shipping coal by way of canals and railroads to markets outside their region soon discovered that mining coal had two advantages: providing fuel for their own needs and transporting surpluses to a growing market hungry for cheap, readily accessible fuel.

Entrepreneurs, legislators, and scientists who took interest in the coalfields along Deep River reasoned that the presence of several rich seams discovered and worked on a small scale and an ability to reach markets by means of river navigation and railroads meant huge profits both domestically and commercially. However, the enormous cost overruns associated with the challenged internal improvements movement to provide an unimpeded route to markets kept North Carolina at bay compared to other states. Funds dried up in the course of surveying, dredging, and paying wages. The problematic geology of the Deep River coalfield with its numerous fault lines and narrow seams interfered with mining efforts, which also drove up costs. All through the nearly two-hundred-year history of coal mining at Deep River, capitalists and speculators invested as much energy as anyone else in pursuing an enterprise they believed, and were told, would mean great prosperity, only to succumb to the vicissitudes of others and the forces of nature.

Introduction

On Thursday morning, May 28, 1925, the *Raleigh News and Observer* reported to some 31,000 readers that on the previous day, May 27, three successive explosions of gas ripped through the Farmville Mine of the Carolina Coal Company at Coal Glen in Chatham County, North Carolina, leaving forty miners dead in the aftermath.[1] The May 31 Sunday edition of the same newspaper revised the victim count to fifty-three dead and proclaimed that with the rescue work finished, the mine "will go on" despite the loss of half of the male population in this small mining community. The news of the event summarized not only the state's worst mining and industrial accidents on record but also reflected an enthusiasm that for no better reason gave hope for an industry that could still mean wealth, jobs, and success. The terrible tragedy that struck Coal Glen on an otherwise normal work day was the culmination of trials that had their origins in a history of over 150 years, and despite the frustration and tragedy throughout that time, overcoming unfulfilled plans and sounding the charge still held sway despite the day's calamitous event.

Although North Carolina never experienced a coal-mining rush as was accorded gold in late- eighteenth and early nineteenth-century Rowan, Cabarrus, and Mecklenburg counties, North Carolina boasted a long legacy of coal mining, beginning in the 1770s to the demise of the last serious efforts at coal mining in the Deep River coalfield some fifty years after the 1925 disaster at Coal Glen.[2] The purpose of this study is to provide a brief history of attempts at coal mining in three areas within close proximity of each other near Deep River: the Carolina Coal Company at Coal Glen in Chatham County; the Cumnock Coal and Mining Company (formerly the Egypt Coal Company), just across Deep River in Lee County; and, the Gulf property in southwest Chatham County where the nearby Horton mine operated on a smaller scale along with an iron works owned by John Wilcox. The devastating explosion at the Farmville Mine in Coal Glen represents a culmination of events leading to the tragedy and its aftermath, which sealed the fate for any further significant coal-mining efforts in the state.[3] Coal mining in North Carolina merits a reexamination from the standpoint of its attempts to emerge as a progressive, modern state after its dormancy in the early part of the nineteenth century up to and beyond the cessation of hostilities during the Civil War. Moreover, the lives of the fifty-three men lost inside the Farmville Mine should remind us of a terrible tragedy often relegated to a footnote in the pages of North Carolina histories.

Introduction

Once navigable by commercial traffic, Deep River is beset by shallow water and clogged with fallen trees and branches, though in some places it can accommodate recreational paddling and canoeing (photograph by the author).

Introduction

The Deep River runs through several counties—Chatham, Lee, and Moore—in central North Carolina, creating the natural north-south border between Chatham and Lee counties before uniting with the Haw River to form the Cape Fear River, whose terminus is the Atlantic Ocean near Wilmington, the state's largest port city. In a much localized area comprising these three counties, coal mining emerged along the river bank in time for North Carolina to make bold proclamations about itself as an industrial and commercial contender in the nation's marketplace. But perhaps the boldness of these types of statements belied a more serious issue beneath the surface. This work focuses mostly on the Deep River coalfield, which largely spreads over the three counties. The Dan River coalfield, which is found in Stokes, Rockingham, and Madison counties, never held promise for successful mining to the extent of the Deep River field and is not examined as closely in this study, though efforts to assess its value are not without their merits.[4]

1

Geological Aspects of the Deep River Basin

The two major Triassic basins exposed in the Carolinas containing measurable amounts of coal are the Deep River and Dan River basins. The Dan River basin, which extends northerly into Virginia under the name Danville basin, is bordered on the west by the Chatham fault zone. The Deep River basin occurs mostly in North Carolina but extends just across the state line into South Carolina. Within the Deep River basin, two coal beds, the Cumnock and the Gulf, contain the most valuable and accessible deposits of coal.[1] The coal-bearing part of the Deep River basin is about thirty-five miles long and five miles wide.[2] The deposits containing coal of the most important economic value are the Cumnock coal "after the Triassic sedimentary formation in which it occurs."[3] This area is further characterized by an association of shale and sandstones belonging to the Newark group, which includes the red sandstones of the Connecticut Valley and Massachusetts as well as the red sandstones and shale of Virginia. Coal of this type can lend itself to breakage and splintering when extracted by the most common means available at the time, namely pick axe and explosives, but the seams along Deep River proved to be an exception. Although this coal had commercial value, efforts to mine it could render any expectation of profitability challenging. In contrast, most of Pennsylvania's coal, both anthracite and bituminous, was formed during the much earlier Carboniferous period of the Paleozoic era more than 250 million years ago without most of the geological obstacles inherent in the Deep River coalfield. This period was characterized by lush tropical fern forests and peat bogs, which, from increased pressure combined with heat from the earth's core, transformed the peat into lignite and ultimately soft, or bituminous, coal. Additional, powerful upheavals in the earth in what is now Northeast Pennsylvania resulted in the formation of mountains lined with anthracite coal of unusually high carbon content.[4]

Coal is generally classified and ranked according to its carbon content and ability to burn cleanly at high temperatures. All coal begins as peat, which matures, first, into lignite, or brown coal; then, bituminous; and, lastly, anthracite. Lignite, which consists mostly of compressed peat, is soft and ranks lowest on the scale due to its low carbon content, resulting in lower heating capability. However, history has shown that use of lignite provided a viable fuel source in areas where there was an absence of higher-quality coals. Bituminous coal has a much higher heating value than lignite and is valued for having more varied applications, such as heating and steam production. Anthracite, which is formed from bituminous coal, has undergone extreme temperatures and pressures over millions more years, resulting in the hardest and most compacted of coals with generally the highest carbon content. Each main classification of coal—lignite, bituminous, and anthracite—can be further categorized according to its B.T.U. or British thermal unit, which is the amount of heat required to raise one pound of water by one degree Fahrenheit.[5]

Bituminous coal is typically equal in parts of carbon content and volatile matter, but superior grades result in excellent coking qualities and large volumes of gas upon distillation. Coke, a by-product of coal when baked at high temperatures in the absence of oxygen to remove the volatile matter, heats at a higher temperature than coal and was preferred in early iron-making enterprises in areas where anthracite was absent. Most of Deep River coal falls into the bituminous category with some varieties approaching semi-anthracite. Deep River coal performed favorably in tests with its Pennsylvania, Virginia, and West Virginia counterparts and in some areas where iron-ore deposits existed parallel to a coal seam, it proved to be an excellent coking coal and had a ready market in iron production. In Stokes, Rockingham, and

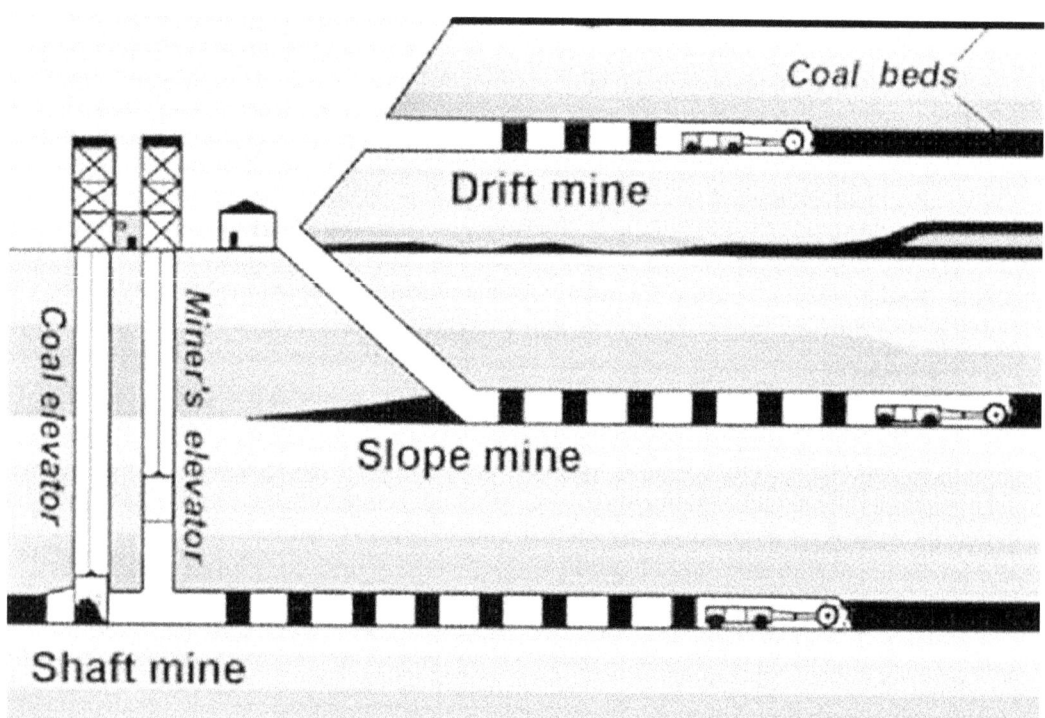

Three types of mines most often seen in coal mining: shaft, slope and drift (courtesy Stephen Greb and the Kentucky Geological Survey at the University of Kentucky).

1. Geological Aspects of the Deep River Basin

Madison counties in the Dan River coalfield, a variety of anthracite was mined but not on any extensive scale. Virginia bituminous coal, such as that mined at the Midlothian collieries near Richmond, provided the main source for Virginia's ordnance manufacturing during the Civil War.[6]

Anthracite, of which most of the world's supply is found in Northeast Pennsylvania, generally has the highest percentage of carbon content, reaching as much as 90 percent, and ranks highly in heating value. Anthracite is characterized by its hardness and sheen, giving it the name "stone coal" or "hard coal." Although more difficult to ignite than the softer bituminous varieties, its heating capacity and ability to burn cleanly, makes it a highly valued coal in both domestic and industrial applications. Anthracite's successful uses in industry, such as iron foundries and blacksmiths' forges, helped pave the way for America's industrial revolution. Producing little smoke and ash, it became an ideal coal for domestic use. Anthracite is typically more expensive per ton than the other varieties of coal but the most economical because of its ability to burn hotter for longer periods of time. Deposits of a semi-anthracite, or steam variety, whose heating qualities and reduced volatile matter places it between bituminous and anthracite, is mined in the valley coalfields of western Virginia near

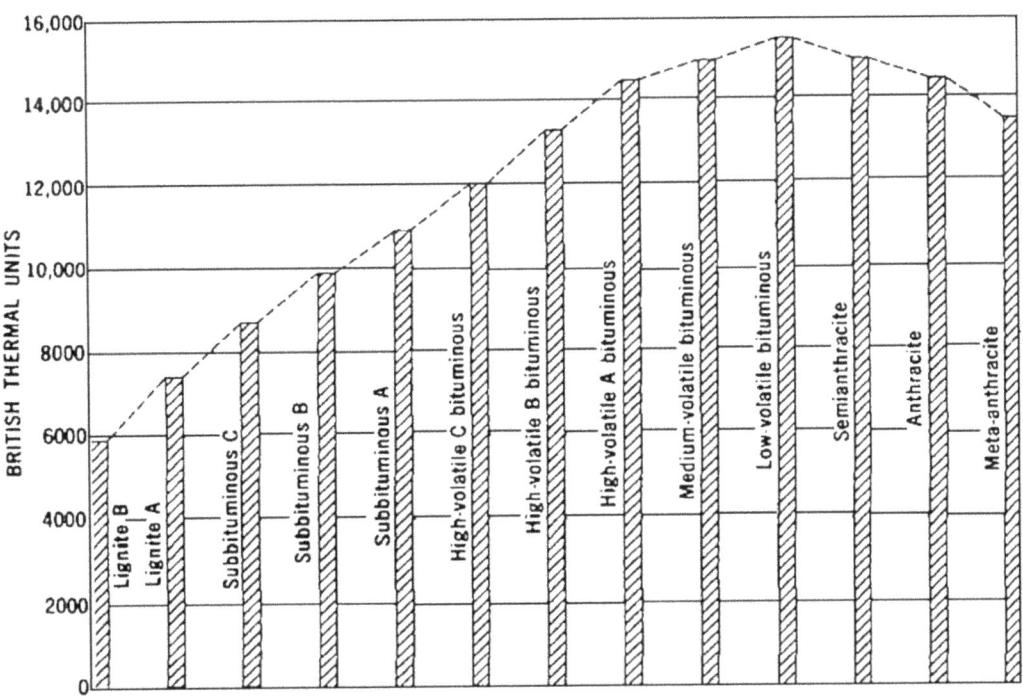

Graph showing the relative heat value and carbon content of various types of coal (Paul Averitt, *Coal Resources of the United States, January 1, 1974. Geological Survey Bulletin 1412*, Washington: U.S. Government Printing Office, 1975).

present-day Merrimac.[7] In the early years of steam transportation, railway engines and steamships relied on this type of coal because of its ability to provide a quick boiling point with lower emissions than many bituminous varieties.

Based on a study performed in 1922 by geologists Marius R. Campbell and Kent W. Kimball, the following table shows a comparative analysis of North Carolina coal and coals from other regions. Note by the data in the ten categories that Deep River coal compared favorably with coals of its same type. Later, geologists and entrepreneurs in an attempt to induce investment in the state's natural resources would push for Deep River coal as among the best in the country. At the time, coal was the major fuel for industrial and domestic uses. Other fuel sources, such as natural gas, oil, and electricity, would come along later though coal long remained the preferred fuel source for most Americans.

Comparison of Different Bituminous Coal Types by Location

Mine Name	Moisture	Volatile Matter	Fixed Carbon	Ash	Sulfur	Hydrogen	Carbon	Nitrogen	Oxygen	Heat Value (B.T.U.s)
Cumnock Mine, (composite of four samples)	2.5	32.0	57.3	8.2	2.0	5.3	75.5	2.0	7.0	13,620
		32.9	58.7	8.4	2.1	5.1	77.5	2.0	4.9	13,970
		35.9	64.1	—	2.3	5.6	84.6	2.2	5.3	15,260
Carolina Mine (composite of two samples)	1.8	29.0	40.2	29.0	2.9	4.4	57.2	1.7	4.8	10,450
		29.5	40.9	29.6	3.0	4.2	58.3	1.7	3.2	10,650
		41.9	58.1	—	4.3	6.0	82.7	1.7	4.6	15,120
Pocahontas field (WV)	1.7	16.5	76.0	5.8	.5	4.3	84.2	1.1	4.1	14,490
New River field (WV)	2.1	21.6	71.9	4.4	1.5	4.7	83.1	1.6	4.7	14,540
Toms Creek (VA)	2.2	32.4	59.2	6.2	.6	5.2	79.7	1.6	6.7	14,130
	2.2	32.4	59.2	6.2	.6	5.2	79.7	1.6	6.7	14,130
Big Stone Gap field (VA)	2.2	33.7	56.4	7.7	.8	5.1	77.2	1.5	7.7	13,790
Morgan County (TN)	2.1	37.2	54.7	6.0	2.7	5.5	76.5	1.8	7.5	13,880

Source: Campbell and Kimball, *The Deep River Coal Field of North Carolina.*

While anthracite does not exert as many B.T.U.s as some types of upper-level bituminous coals, its much longer burning capacity was ideal for ships, whose dependence on a ready fuel supply was critical when far from their home base. While semi-anthracite coal has about equal carbon content as anthracite, it is sometimes considered a more superior steam coal than anthracite as demonstrated when stoked into the boiler of a steamship, it could generate a faster rate of steam, resulting in an

1. *Geological Aspects of the Deep River Basin*

almost immediate surge in speed. The Pocahontas coalfields in Virginia and West Virginia, as well as the coalfields in South Wales, contain rich semi-anthracite seams. In 1851, after a series of exhaustive trials, the Royal Navy found Welsh steam coal to be the most desirable for its vessels.[8] Welsh steam coal was also sought by Southern shipping companies because of its high B.T.U. output, its smokeless quality, and low ash content.

Underground mines can be categorized into three main types: shaft, slope, and drift. Shaft mines are characterized by their limited surface but are sunk vertically at a considerable depth for finding coal seams, pumping out standing water, ventilating workings, and providing ingress and egress by hoisting and lowering men and equipment. Representing the deepest type of mine, a shaft mine typically contains at least two openings, one for haulage, such as coal cars, and the other for miners and other mine employees, and in some instances shaft mines contain a dedicated opening for ventilation. The Egypt Coal Mine in Lee County is an example of a shaft mine, which reached over 400 feet in depth. Slope mines are inclined passages that intersect a coal seam at an angle. An example of a slope mine is the Carolina Coal Company Mine at Coal Glen in Chatham County. Drift mines follow the vein of coal horizontally and do not intersect the coal seam as shaft and slope mines do. They require the least amount of expenditure and manpower for a mining company to access the coal seam. Some individuals who had a rudimentary knowledge of mining fundamentals dug small "bootleg" mines of the drift type usually outside the purview of the mining companies.

2

Early History of the Deep River Coalfield

During the colonial era, when Deep River coal was utilized at local forges, outcroppings of coal—the part of a bed or layer that is visible at the surface—provided a small industry mostly involving iron making and domestic use.[1] This early consumption would eventually translate into fabrication of iron to supply the American cause during the Revolutionary War, and the nearby coal deposits had a critical role in its manufacture. Reporting at its session on April 24, 1776, the North Carolina Provincial Congress authorized Thomas Person, Martin Pfifer, and Ambrose Ramsay to make arrangements with John Wilcox for firing his furnace and iron works at Deep River for the purpose of "casting pieces of ordnance, shot, and other warlike implements" to furnish the American troops.[2] The arrangement also called for the employment of smelters, wood cutters, colliers (i.e., miners), and other workers necessary for the production of weapons. In a letter dated July 3, 1776, to the Council of Safety, James Milles expressed his views about potential iron production and casting when he mentioned riding out to M' John Wilcox's forge and bloomery on Deep River in Chatham County. A furnace, which existed about ten miles from the forge, was sufficiently large to accommodate the casting of cannon of at least one ton. Milles observed that near the forge was a supply of pit coal with the potential to reach an extensive seam and produce greater quantities on a regular basis.[3] However, in a letter dated June 6, 1777, Ramsay, Mial Scurlock, and John Birdsong complained to Governor Richard Caswell that "Mr. Wilcox now waits on your Excellency and the Honorable Council of State to have something done with respect to the Iron Works in this County (Chatham)."[4] On a visit to the foundry, the three men noted that the furnace had not blown and that there was not a sufficient supply of coal, ore, and limestone. Later that year, on December 22, the House of Commons approved the acquisition of "unappropriated lands" adjoining the Iron Works in Chatham Co. for supplying the works with "ore, coal, timber, and stone."[5] Regardless of the availability of new fuel sources, historian Robert B. Gordon noted that problems such as inclement weather and procurement of qualified farriers to sustain a successful furnace hindered the foundry's overall production.[6]

In the post–Revolutionary War era, Americans saw themselves as part of a nation still defining itself, witnessing from afar the growth of the Industrial Revolution in

2. Early History of the Deep River Coalfield

Europe in the late eighteenth-century and its expanding transportation network that beckoned entrepreneurs and capitalists to invest and share in the wealth. The new republic yearned for greater economic growth and began seeking ways to improve its own natural resources so that the path to new prosperity might be realized. In a new nation where regions were separated geographically without the benefit of a shared commerce and industry, leaders at both the federal and state levels saw the advantage of establishing an inter-regional network to foster communication and promote business growth. State legislatures soon began to recognize the economic benefits of internal improvements and passed legislation accordingly to promulgate the benefits of public works. At this time, North Carolina, along with neighboring states, embarked on an effort to dredge waterways, construct roads and highways, and build canals and railroads to link the interiors of their states with the coast so that goods and commodities, including coal, could be conveyed on ocean-going vessels and stimulate commerce not only with other states but Europe and other foreign markets as well. Historian Alan D. Watson observed that "North Carolina must emulate 'her sister States, in the struggle for wealth and distinction' or be left to retrogression and decay."[7]

As early as 1784, a push for internal improvements in North Carolina came in the form of Governor Alexander Martin's message delivered to the General Assembly in which he stated that

> the trade and navigation of this Country is of lasting consequence, and require your immediate interposition and patronage. It is necessary our rivers be rendered more navigable, our roads opened and supported, by which the industrious planter may have his produce carried to market with more ease and convenience. Thereby more merchants of opulence would be induced to settle in the State and open new resources of industry among our Inhabitants whose labors being fully compensated daily additions would be making to their respective wealth, in proportion to which the revenue of the State would be also increased.[8]

Soon thereafter bills began to appear before the House calling for clearing and opening navigation of rivers and ports. Legislation supporting the pilotage and facilitating the navigation of the Cape Fear River was read for the first time, passed, and forwarded to the Senate for its vote.[9]

By 1790, bills put before the legislature called for creating and maintaining accessible roads, tasking the County Courts of Pleas and Quarter-Sessions to lay out public roads, establish and settle ferries, build bridges, and clear inland navigation.[10] In 1792, the General Assembly passed an act to establish a company for the purpose of facilitating navigation of the Cape Fear River from Fayetteville to the confluence of the Deep and Haw Rivers. Acknowledging that "navigation is the life and main spring of commerce" and that to make possible safe and easy passage on the waters would greatly benefit the inhabitants of the western part of the state, the General Assembly was poised to undertake public works projects to accomplish those goals.[11] Further, the General Assembly appointed a nine-member commission "for regulating the pilotage and navigation of the Cape Fear River."[12] North Carolina resolved to promote its own economic agenda to open up the state for trade by way of improving its infrastructure, particularly, rivers, harbors, inlets, and roads. The promise of creating new markets within the state itself, instead of seeking trading partners with neighboring South Carolina and Virginia, meant that revenue would remain in North Carolina.[13] By 1796, the General Assembly had passed additional legislation to improve navigation on the Roanoke River, the Deep and Haw rivers, as well as on the Catawba, Tar, Yadkin, and Pee Dee rivers. An even more ambitious action on the part of the General Assembly called for "the cutting of a navigable canal from Roanoke river ... near the town of Plymouth, to Pango [sic] river."[14]

On the national level, Tench Coxe, a Pennsylvanian who served as Assistant Secretary of the Treasury during Washington's administration, set about writing an economic account of the United States early in the nation's history, published as *A View of the United States of America, in a Series of Papers, Written at Various Times, between the Years 1787 and 1794*. Taking stock of a wide variety of topics from civil and religious liberties to general improvement, Coxe tasked himself with assessing and describing the present condition of the new country with the acumen of a businessman. He was one of the first of a growing number of economists and lay scientists who saw the importance of identifying and utilizing the natural resources of the United States. He also had the foresight to consider coal a staple of industry and enterprise. Speaking of the collieries on the James River, in Virginia, Coxe noted they "will not only abundantly supply the extensive territory watered by the rivers of the Chesapeak [sic]" but supply the fuel of ships delivering goods to "all the sea-ports of the United States."[15] His view that the country should rely less on foreign coal, calling it a "very losing commodity," focused on the resources in the interior of the country which is "plentifully supplied by nature with this fossil." However, transporting coal to market was as much a challenge as "winning" it at the face, as the reality of undertaking internal improvements would demonstrate.[16]

Other than a few local merchants advertising the sale of Liverpool-imported coal, no further mention of the Deep River coalfield appeared until 1809, when Raleigh's *North-Carolina Star* printed a short article describing a coal mine in Chatham County near Deep River which was reported to be inexhaustible but "in a part of the country, where, at present, no advantages can be derived from it."[17] A letter from a writer signed "Star Gazer" appeared in the newspaper the following month in response to the aforementioned article, but, on this occasion, the writer cited a fragment of an old Philadelphia almanac in his possession in which "a gentleman in Bohemia" claimed that mineral coal broken or ground contained excellent manure-like qualities, superior to Plaster of Paris, "in promoting the vegetation of grasses, and especially clover." His recommendation to the editor that publication of the idea might induce farmers in the area to experiment and profit by coal fertilization appeared to have some merit given the nearby supply and accessibility to outcroppings.[18] In July 1811, the same newspaper printed a letter to the editor referring to "a bed of coal on Deep River in this county [Chatham] that is easily raised and of an excellent quality." The writer

2. Early History of the Deep River Coalfield

attested that three European blacksmiths of his acquaintance had used it in their forges and "pronounce it as good coal as they ever used in their respective countries, and each of them preferred using it to charcoal."[19] One hundred years later, the idea of tapping into oil shale for use as a fertilizer still held interest though it was not pursued at any great length. Still, use of coal outside its traditional function as a fuel source was already beginning to be investigated and considered.

Having a dependable transportation system in place was a critical factor in reaching the state's natural resources, especially its mineral wealth, and opening up trade routes to ship the product to market. The call for internal improvements in order to remain competitive in the marketplace quickly fueled a nationwide movement, as state governments rallied to support such projects to further their economic interests. As early as 1795 the General Assembly passed legislation to construct a canal network for the purpose of diverting trade from Virginia and South Carolina ports to those within the state which could then link up with ocean ports to ship their cargo to more lucrative markets. However, it was Archibald Debow Murphey, lawyer, educator, and state senator from Orange County, and one of the state's leading advocates of internal improvements, who, in 1815, as chairman of the senate's committee on internal improvements, prompted the General Assembly to subscribe to stock in the Roanoke and Cape Fear Navigation Company and "authorized state-funded surveys of the Tar, Neuse, and Yadkin Rivers."[20] Subscriptions of stock would soon present themselves when other navigation companies, who were anxious to partner with the state, would invest their capital as a joint venture and assume part of the risk.

In 1816, the legislature's Committee on Inland Navigation proposed an interconnected waterway system within the state for the purpose of diverting trade from South Carolina and Virginia to the ports of North Carolina which, in turn, provided an outlet to the sea.[21] Murphey's view that internal improvements were "among the most important objects of state policy to improve the navigation of the Yadkin and its waters" meant that if valuable commodities, such as lumber and naval stores, were to reach their markets in a timely manner, waterways must be unimpeded for vessels to pass in order to deliver their goods. On December 9, 1816, he submitted a report of his findings to the North Carolina legislature outlining his observations and recommendations on inland navigation.[22] Writing enthusiastically, and with a flourish of state pride, Murphey stated:

> let us devote our labours [sic] to the honor and glory of the state in which we live, by establishing and giving effect to a system of policy which shall develope [sic] her physical resources, draw forth her moral and intellectual energies, give facilities to her industry, and encouragement to her enterprise.[23]

Murphey had visions of North Carolina establishing a commercial center by either taking advantage of natural geography or subscribing to a plan of building projects to facilitate the creation of access points along the coast. To fund his idea, Murphey looked to the state legislature to incorporate a company "with suitable capital, and subscribing on behalf of the state such part of that capital as the public funds will admit."[24] Murphey's report, which appeared in a serialized format in some local newspapers, promised real solutions, although untested, with cost estimates to improve navigation of rivers, inlets, and waterways as well as the construction of turnpike roads to link cities and towns with embarkation points on rivers and canals for shipping cargo to the coast and then on to further destinations. The report influenced the legislature sufficiently to charter the Lumber River Canal Company for the purpose of "opening a communication" between the Yadkin and Cape Fear rivers.[25] Soon thereafter, bulk commodities, such as coal, had a more favorable outlet to markets compared to the more costly option of overland transportation by wagon and cart.

In 1819, the legislature enacted a law for the creation of a fund for internal improvements and to "establish a board for the management thereof."[26] The Board of Internal Improvements, appointed by the General Assembly, was instructed to have "sundry surveys" made and to employ a principal engineer for the state and one or more surveyors to provide recommendations for proceeding with internal-navigation projects.[27] Hamilton Fulton, the newly appointed state engineer, was given a series of instructions by the committee containing a review of intended improvements contemplated by the General Assembly. These instructions mandated (1) visiting rivers, (2) giving instructions to navigation companies, and (3) examining a proposed route for opening communication by water.[28] Based on Fulton's findings the legislature passed an act to facilitate the navigation of the Lumber River in the south-central part of the state in the coastal plain. Timothy Griffin, Neil Buie, and Malcolm McNiel were appointed to oversee the efforts and provide the Board of Internal Improvements "a statement of their acts, contracts and proceedings."[29] Based on his observations, Fulton concluded that navigational difficulties could be overcome, pointing out that improvements between the Haw River to Rockingham Co., and Deep River to Gulf, in Chatham County, could result in accessing the coal deposits in those areas to meet the demand of the domestic as well as commercial markets.[30]

An article appearing in 1819 in the *American Farmer* included a letter from a member of the North Carolina Catawba Navigation Company addressed to a gentleman in Camden, South Carolina, claiming that progress was being made by the company within the boundary of North Carolina, and if approved by the citizens of South Carolina, especially those contiguous to the waterways, would see "boats running on its waters for more than 150 miles above the dividing line of the two states." Successful removal of impediments would mean that goods from the counties of Mecklenburg, Lincoln, Iredell, and Burke could easily flow into favorable outlets such as Camden and Charleston. The company's contractor, a Mr. [Turner] Abernathy, was engaged in cutting a canal around the shoals of Mountain Island in present Gaston County at considerable expense and that with the necessary locks in place would soon extend to the south fork of the Catawba River.[31]

In April 1823, a correspondent, signed "Walter Raleigh," addressed an open letter in the *Weekly Raleigh Register* to Charles Fisher, secretary of the Rowan Agricultural

2. Early History of the Deep River Coalfield

Nineteenth-century woodcut engraving of coal arks approaching weigh station on the Lehigh Canal. Draft animals were needed to navigate and pull the arks containing heavy loads of cargo. Compared to the nation's overall canal miles, North Carolina's share was minimal ("Down Among the Coal Mines—Weighing the Cargoes in the Weigh-Lock on the Lehigh Canal," by Paul Frenzeny, *Harper's Weekly*, February 1873).

Society, concerning the abundance of North Carolina coal and its untapped potential. Citing dwindling supplies of wood for fuel, the author observed that the more populated areas of the state were beginning to feel the effects of its scarcity and high cost, and predicted that coal was destined to become "an article in much greater demand in our country than it is at present."[32] Raleigh's perceptive observation that "subterranean supplies for fuel" eventually will become more common especially given that in the large manufacturing areas demand for wood had led to the destruction of the surrounding forests sent an encouraging message that North Carolina as well as surrounding states that "contain it [coal] in abundance will find it a considerable and valuable possession." Noting the beginnings of a burgeoning coal trade in Virginia with its proximity to three coalfields, the writer was confident that North Carolina would be "among the states which will be so fortunate as to embrace a full supply of this article within our own territory." However, in meeting the demand of regions desirous of a viable coal trade, internal improvements, especially to river outlets, were needed to effect trade and commerce. Predicting that North Carolina's role in the coal trade would be significant despite once displaying "but little cold indifference," Raleigh reminded readers that a state with proximity and an ample supply of resources enabled its inhabitants "to excel in arts and manufactures." Using Pittsburg, Pennsylvania, as an example of a region surrounded by a large supply of coal to fuel its many factories and industrial plants together with an outlet to transport it to local markets and beyond, Raleigh exhorted North Carolina to seize the opportunity in order to participate in the "grand march of improvement," as in this case what coal afforded.[33]

After legislation to implement internal improvements passed both houses of the General Assembly, the financial burden became a controversial issue and one that was not always settled by the state. The state's commitment to internal improvements also included finding means to provide an outlet to markets previously unattainable. To help fund projects, the state sold stock to investors and speculators and by the early nineteenth century North Carolina began to be seen as a region for growth and trade, with some entrepreneurs taking on the majority of the burden of funding their own internal-improvement projects. Sponsored by state and private enterprise, men of science arrived in North Carolina to assess its potential resources and ultimately encourage investment with promising dividends. Some of these men offering up their knowledge in the course of pursuing economic growth were serving as faculty members at the University of North Carolina at Chapel Hill. Internal improvements would ultimately prove to be a controversial issue in antebellum North Carolina as paying for them became a thorny issue particularly when an area's problematic geography, or political infighting, could drive up costs for the work, sometimes halting a project dead in its tracks.

The number of miles allocated to canal building in North Carolina was sparse compared to the overall miles of canals in the United States on account of the state's failure to obtain sufficient private investments, which proved to be the bane of future efforts. Each region's challenge was to help finance and complete internal improvements in a timely manner so that an inter-connected network of canals, roads, and turnpikes could be realized not only to facilitate trade within the state but as part of a national communication network envisioned by the Founding Fathers and reiterated by men of industry and vision such as Tench Coxe. However, location and responsibility of maintenance continued to nag state leaders and businessmen, who sometimes looked after their own interests to the detriment of an extensive network benefitting all concerned. It would be left to future investors and entrepreneurs backed by surveys and studies conducted by scientists, geologists, and engineers to articulate their vision and promise profitable results to state officials whose investment in these projects continued to be crucial for their success. The state's role as investor became a partnership early corporations sought and developed to push their agenda forward notwithstanding sectional and partisan differences. The alliance of government with business was critical in overcoming financial setbacks, limited resources, and natural impediments for trade to progress within the state and beyond, and as will be seen later in this study, the railroad held the most promise.

3

Geology as a Science Stimulates Geological Surveys

Almost commensurate with the attention to internal improvements, and as part of an economic and scientific awakening, North Carolina became a center of interest for scientists and teachers, who were employed to use their specialized knowledge to assess locations and conditions that might forecast favorable investment opportunities. According to historian Michael S. Smith, transportation and agriculture motivated the state's early efforts to complete a geological survey in order for investors and, in some instances, the state, to evaluate the extent of their region's holdings as well as consider viable transportation outlets to and from those locations and ultimately to awaiting markets.[1] After the end of the gold rush, new markets opened up for absentee investors looking for other opportunities, such as agriculture and the mineral industry, to make and sometimes remake their wealth. To that end, this new breed of investor utilized and relied on the reports of academicians and scientists, who represented a new branch of science called geology.[2] Fueled by the promise of quick profits with borrowed money, entrepreneurs from New York and New England organized themselves into corporations expecting to realize a return on the investment supplied by the shareholders.

Early scientists, such as William MacClure (1763–1840) and Horace Henry Haydn (1769–1844), came to geology from different disciplines. MacClure, born in Ayr, Scotland, worked for American importers in New York City and after making his fortune in the mercantile business devoted time to his main interests, science and geology. McClure advocated a rigorous application of the scientific method to objects and phenomena in order to extract new knowledge and understanding. Haydn, a renowned dentist of his era, pioneered early geological and botanical studies, publishing what many authorities recognized as the earliest work on geology by an American. The mineral Hadenite was named in his honor.[3] As a result of these early efforts, future scientists with more ambitious aims established themselves in universities with even greater opportunities to study and teach in their specialized fields and publish their findings, even securing positions in the employ of newly formed corporations whose directors and shareholders relied on these reports and surveys to make informed business decisions regarding investments in internal improvements and infrastructure.

One of the earliest treatises to deal with mining in general in Europe and the United States was a work by the American geologist and explorer, William Hypolitus Keating, who is best known as a member of Stephen Long's expedition to the Great Lakes region in 1823.[4] Taking to task those who had misconstrued and underrated the "art of mining," Keating affirmed the need to understand mining in terms of its "nature and object," while not subscribing to "false and disastrous speculation" for economic gain. For Keating, the art of mining was part of a world view that empowered and enriched a nation by combining with the study of natural science to make it one of the most important branches of public economy.[5] Citing England, France, and Germany as models of a proper and successful mining culture, the author addressed the problems and drawbacks of mining in the United States and which course should be taken to overcome them.[6] Keating's belief that individuals must become acquainted with the country's natural resources, and invoke mature reflection and deliberation to make informed decisions, could we then ensure "the advantageous introduction of the art of mining."[7]

Early State Geological Surveys

According to historian Charles S. Sydnor, geological surveys were undertaken at public expense in all but two Southern states. Following graduation in 1813 with a B.A. degree from Yale College, Denison Olmstead, a native of Connecticut, spent the next four years teaching school and studying theology. In November 1817, he joined the faculty of the University of North Carolina, becoming professor of chemistry and geology, a newly created position within the university, which was pressing forward with the new study of science as part of its curriculum.[8] In 1820, Olmstead published the first scientific account of coal in the Deep River field, making the following observations:

> An extensive secondary formation has lately been discovered near us. On the road between this place [Chapel Hill] and Raleigh, traveling eastward, we come to it four miles from the college; but at another point it has been discovered within two miles of us. It is a sandstone formation.... It was natural to look for coal here and I have for some time directed the attention of my pupils and of stonecutters to this object. Two or three days since one of the latter brought me a handful of coal, found in this range, on Deep River, in Chatham County, about 20 miles south of this place. The coal is highly bituminous, and burns with a very clear and bright flame. It is reported that a sufficient quantity has already been found to afford an ample supply for the blacksmiths in the neighborhood.[9]

In 1821, Olmstead presented to the North Carolina Board of Internal Improvements a plan for "a geological and mineralogical survey to be made of the state of North-Carolina."[10] The Board read and approved the study to the legislature, but took no notice of the matter until two years later when in 1823 the proposition was renewed and authorized by act and ratified by the General Assembly.[11] Professor Olmstead,

3. Geology as a Science Stimulates Geological Surveys

fitting the criteria of competent skill and knowledge of science, was appointed to begin the survey and report his findings to the State Board of Agriculture, which published the results "for the benefit of the public, as provided by the sixth section of the act of the last General Assembly, entitled 'An act to promote Agriculture and family Domestic Manufactures within this State.'"[12] The legislature appropriated the amount of $250 a year for four years to help fund scientific projects.[13] Using his vacation time away from his university teaching responsibilities, Olmstead later submitted two reports on the state's geology before returning to Yale in 1825 and his work continued by his former classmate and colleague, Elisha Mitchell.[14]

Elisha Mitchell

A graduate of Yale in the same class as Denison Olmstead, Elisha Mitchell spent three years in various educational capacities before his appointment in 1816 as tutor at the university. During that time, Mitchell studied for the ministry at Andover Theological Seminary and in 1817 received a license to preach as well as an appointment to a professorship in mathematics and natural history at the

Top: Prof. Denison Olmstead (1791–1859). He was one of the early scientists to champion the study of geology and mineralogy (*North Carolina University Magazine*, December 1860). *Bottom:* The Rev. Elisha Mitchell (1793–1857). An ordained Presbyterian minister, he is best known for his geological studies of North Carolina and for whom Mt. Mitchell in Yancey Co. was named (Rt. Rev. James H. Otey, Hon. David L. Swain, et al., *A Memoir of the Rev. Elisha Mitchell: Together with the Tributes of Respect to His Memory, by Various Public Meetings and Literary Associates, and the Addresses Delivered at the Re-interment of His Remains,* Raleigh: J.M. Henderson, 1858).

University of North Carolina.[15] The following January, Mitchell began his teaching duties at the university and in 1821 was ordained a minister by the Presbytery of Orange County. Upon Olmstead's return to New Haven four years later, Mitchell assumed the duties of his colleague's professorship and began a more extensive study of the geology of the surrounding area, conducting field trips and excursions with his students in tow. While continuing to write on the occasional theological topic, he contributed his scientific findings to Silliman's *American Journal of Science* and in 1825 published *Papers on Agricultural Subjects ... Published by Order of the Board of Agriculture of North Carolina, Vol. I.* Completing Olmstead's report on geology to the Board of Agriculture, Mitchell, in November 1827, published *Report on the Geology of North Carolina Conducted under the Direction of the Board of Agriculture, Part III.*[16] Mitchell would later be credited with measuring the height of the mountain named for him, Mount Mitchell, later succumbing to injuries sustained in an accidental fall in 1857.

Lemuel Williams

Lemuel Williams, of Massachusetts, visiting the state on behalf of northern capitalists, had been in Chatham County for the purpose of analyzing and reporting his findings on the value of the Deep River coal deposits. In an address given in the senate chamber at Raleigh, on January 14, 1851, Williams stated that successful mining of coal depended on the following elements: extent and thickness of the beds, quality of the coal, and facilities and cheapness of transportation to tide water and then to a market.[17] In determining the extent of the bed, Williams calculated the area of the coal-bearing field to be one hundred and five square miles, less than the area Walter R. Johnson estimated in his report. In determining thickness of the beds, Williams based his estimation of six million tons to the square mile on one vein of coal at six feet thick underlying the field. He noted three varieties of coal in his report—highly bituminous, semi-bituminous, and anthracite—that he enthusiastically claimed are "unsurpassable; in variety unequalled by any location in the United States."[18] Williams continued to state his case that the profitability of coal would exceed the mining of precious metals, including gold and silver, as in Great Britain where the total value of coal contributed to that nation's supremacy as a manufacturing, commercial, and military power which can be traced to its coal production and trade.[19]

Williams's solution for transportation was slack-water navigation especially considering that "improvement of Cape Fear and Deep Rivers affords a cheaper transit to the ocean than any other improvement."[20] In comparing roundtrip carriage with other mining and transportation companies in the mid–Atlantic states, some of which could take up to fourteen days to deliver their cargo and return, Williams projected that a round trip journey from the mines on Deep River to Wilmington required only four days[21] A means for transporting the coal was of primary concern and the initial plan of the use of slack-water navigation met with enthusiasm. However, the project never developed into a viable solution, as necessary improvements made by the Cape Fear and Deep River Navigation Company were stymied by labor stoppages and freshets, which damaged the company's locks and dams. The company sought financial relief from the legislature, which increased its capitalization to $350,000, but, again, weather and other factors, curtailed work and enthusiasm for the project. Williams believed that anthracite coal from the workings could be shipped from Wilmington to markets in New York and New England, without causing undue competition with the anthracite mines of Pennsylvania. However, Williams considered that high fees and taxes for transporting Deep River anthracite coal to New York might render it untenable to compete with Pennsylvania coal. His vision to see the Deep River region "within a very few years, inhabited by a dense population, and adorned with flourishing villages," included increased commerce and traffic for Wilmington, leading to favorable comparisons to Philadelphia.[22]

Walter R. Johnson

Walter Rogers Johnson, born June 21, 1794, in Leominster, Massachusetts, was a distinguished geologist and chemist, who from 1839 to 1843 taught chemistry and physics at the University of Pennsylvania. One of the most prolific authors on geological subjects, he next began, under the authority of Congress, investigations into the different types of coal and their value.[23] In October 1849, while on a tour of middle and western North Carolina, he stopped at the Deep River coal region, and with his friend and colleague, Elisha Mitchell, spent several days examining the outcroppings and mine workings of the field along their journey.[24] After the North Carolina General Assembly passed legislation to incorporate the Deep River Mining and Transportation Company, of Albany, New York, the company engaged the services of Johnson to conduct an analysis of the mineral content on their property in Chatham County. The published report appeared in 1851 addressed to the board members of the company.[25] Johnson represented one in a procession of geologists and scientists hired by private concerns to inspect their property and make determinations as to value and feasibility of its development. Taking into account that the outcroppings in relation to the water table traversed through the coalfield reached only twenty-seven feet, he recommended that mining operations exist below water level, with sufficient pumping machinery in place to effect adequate drainage.[26]

Focusing on the Farmersville property, and Pit No. 5 specifically, Johnson measured the thickness of the seam and determined it to be seven feet nine inches. A statement made by an individual who had taken measurements farther towards the dip indicated that the seam measured eight feet eight inches.[27] His analysis of the coal included samples of eight different specimens from three open pits at Farmersville. Johnson used the outcroppings of coal at or near the surface for his analysis although outcroppings did not always represent the true quality of the coal especially in light of weather conditions and surface influences that could render the outcroppings less favorable.[28] Still, Johnson was convinced that at greater depths the coal would prove to be more satisfactory. In his next series of experiments, Johnson used forty samples of Farmersville coal rendered into a finely powdered state to determine the quantities of carbon and hydrogen.[29] Comparing the Farmersville coal to that used in American and British trials, Johnson observed that the coal's "brisk and brilliant combustion" made it ideal for parlor grates and smithing.[30] In terms of the costs associated with mining and transportation, Johnson recommended an appropriate location where the mine would be clear of water and where separation of impurities and loading could all take place. The proximity of the coal beds within less than half a mile to the river favored slack water navigation from the mines to the ocean via Fayetteville. Based on his assessment, Johnson arrived at a cost from forty- to forty-five cents per gross ton, and the delivery into barges no more than fifteen cents a ton. He noted that a portion of the Wilcox property contained a farmhouse and various other structures to accommodate miners and other individuals associated with the mine.[31]

Johnson's report identified and assessed coal deposits in relation to the properties owned by the Deep River Mining and Transportation Company at Farmersville on Deep River. His map showed mining operations in progress, and his accompanying study was the first of its kind to note water levels in proximity to coal seams, and his recommendations for mining at or below levels alerted mine owners to potential risks regarding safety, cost, and transportation. His observation that the field contained high-grade anthracite coal beds was a boon in terms of reaching other markets, such as Boston, as it could be mined and transported much cheaper than any of the Pennsylvania anthracites.[32] The board of directors of the company quickly praised the report and invested additional capital to meet the challenges of the necessary navigational improvements. Again, banking on the field reports of professional geologists to make informed decisions, mining companies were better prepared to predict costs and determine transportation routes to and from market.

With a subscription of 10,000 shares of stock at $100 per share, the company wasted no time in making improvements to navigation on Deep River to facilitate transportation of their coal to Fayetteville and from that point to tide water in Wilmington.[33] In an article appearing in the *Fayetteville Observer* in December 1851, as a follow-up to Johnson's report, the board of directors of the Deep River Mining and Transportation Company submitted an assessment of economical considerations concerning their possessions at Farmersville. In it they identified three key elements for determining value of coal: quality, quantity, and cheapness with which it could be mined, including facility and cost for delivery to market.[34] The directors noted that in comparison to the cost of improvements made during the construction of the Lehigh Navigation at $4,500,000, the entire cost of the Cape Fear and Deep River improvements did not amount to $400,000. In consideration of the costs of delivering coal to New York markets, and the return on investment, to those at Smithville or Fayetteville, the board recommended that the best market for their coal was Smithville, as "it is directly in the track of the steamers that ply from New York to

3. Geology as a Science Stimulates Geological Surveys

Charleston, Savannah, New Orleans, [and] Texas. No other company can compete with your company in southern markets."[35] However, the board's recommendation to use steam power to draw coal barges and draft animals to pull coal arks on the canals depended on improvements to the waterways before any transportation issues could be resolved, and, as will be shown later with the advent of the railroad, canals were beginning to be seen in a less favorable light.

Thomas L. Clingman

One of the state's more colorful and accomplished citizens was Thomas Lanier Clingman—political leader, states' rights advocate, Confederate general, and scientist. Born in 1812 in what is now Yadkin County, North Carolina, Clingman graduated in 1832 first in his class at the University of North Carolina. Upon graduation, Clingman read law and soon thereafter entered politics, culminating in his 1840 election to the U.S. House of Representatives and in 1860

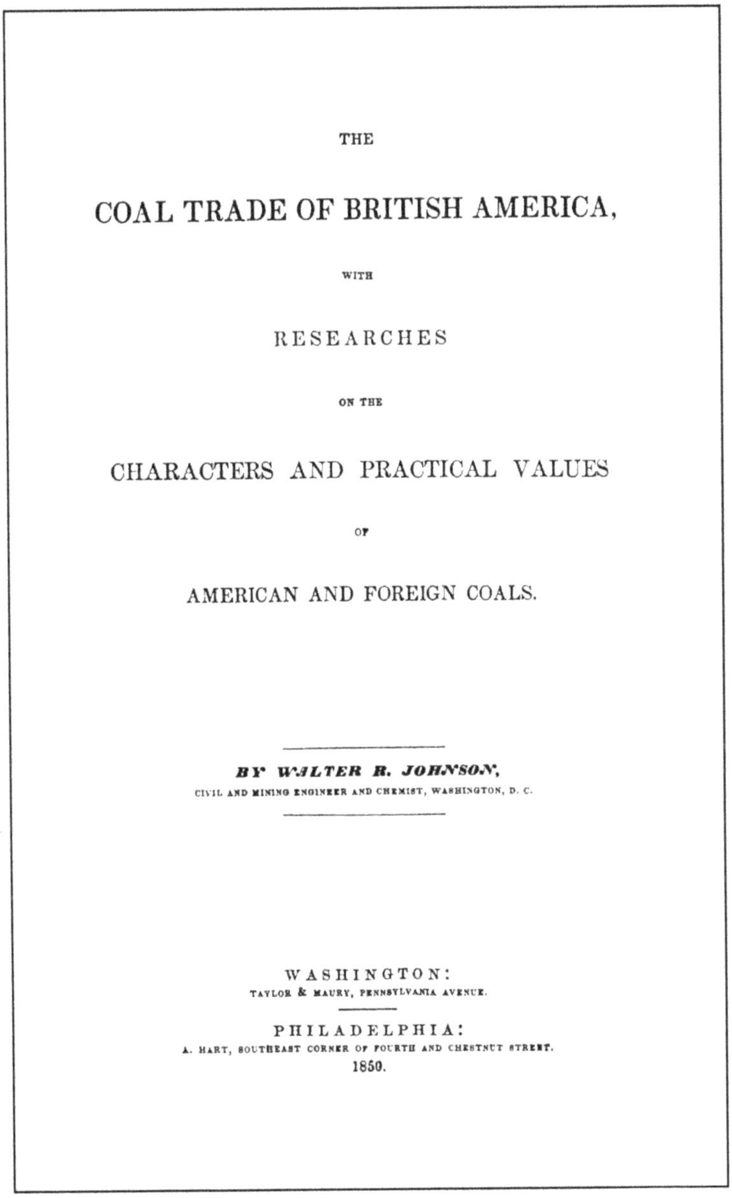

THE

COAL TRADE OF BRITISH AMERICA,

WITH

RESEARCHES

ON THE

CHARACTERS AND PRACTICAL VALUES

OF

AMERICAN AND FOREIGN COALS.

BY WALTER R. JOHNSON,
CIVIL AND MINING ENGINEER AND CHEMIST, WASHINGTON, D. C.

WASHINGTON:
TAYLOR & MAURY, PENNSYLVANIA AVENUE.
PHILADELPHIA:
A. HART, SOUTHEAST CORNER OF FOURTH AND CHESTNUT STREET.
1850.

Walter R. Johnson's *The Coal Trade of British America* contained a chapter on one of the first published studies of Deep River coal. He was also a friend of Elisha Mitchell (Walter R. Johnson, *The Coal Trade of British America, with Researches on the Characters and Practical Values of American and Foreign Coals*, 1850).

election to the U.S. Senate.³⁶ As a native son of North Carolina, with a desire to promote the resources of the western region, Clingman authored an article published in *DeBow's Review* in 1858 giving a general accounting of the state and its resources, with a review of minerals and mining interests. Clingman noted the favorable abundance of iron ore deposits at Deep River with the presence of the particularly rare black band ore, which lay in close proximity to the coal seams. He pointed out that although the coal measures were not as extensive as those of some of the other states, they were inexhaustible and of the best qualities for fuel, making of gas, and the manufacture of iron. He further stated that the iron ore was in greatest abundance between the coal seams and that consideration be given to providing proper outlets and internal improvements so that upon completion of these works "iron can be made and transported to Wales, and sold at as cheap a rate as that for which the Welsh manufacturers now afford the article."³⁷ Clingman was an early advocate of domestically produced iron as an answer to the monopoly England and Wales enjoyed over their American counterparts. His belief that domestic iron could compete with the more famous "branded" types manufactured in Great Britain was already being realized by the Scrantons in Northeast Pennsylvania.³⁸

Thomas Lanier Clingman (1812–1897). His early analysis of North Carolina geology and mineralogy led him to advocate domestically produced iron over foreign imports (courtesy Library of Congress).

Ebenezer Emmons

Ebenezer Emmons, geologist, educator, and physician, was born in western Massachusetts in 1799. Showing an early interest in natural science, he was introduced to the study of geology by Amos Eaton while a student at Williams College in Massachusetts. In 1824, Emmons enrolled in the Rennselaer Polytechnic Institute at Troy, New York, where he continued his geological studies. He graduated from the institute in 1826 and published his *Manual of Mineralogy and Geology*, a handbook for students. Between 1826 and 1828, Emmons studied medicine at Berkshire Medical College

3. Geology as a Science Stimulates Geological Surveys

and became a practicing physician while still conducting research and teaching in the field of his first allegiance, geology. His big break as a geologist came in 1836, when the state of New York appointed him one of four geologists to undertake a new state geological survey. Locked in a heated controversy with New York State Geologist James Hall regarding the age of a system of stratified rocks Emmons called the Taconic system, he was eventually banned from the practice of geology in New York State, and, with the loss of a subsequent lawsuit charging Hall with slander, left New York.[39]

On January 24, 1851, the state legislature passed an act to make a geological, mineralogical, and agricultural survey of the state, thus resuming the work that had been defunded in 1827. Governor David Settle Reid, in a somewhat controversial move based on his reputation in New York, hired Emmons as its chief geologist. In October 1851, the *Weekly Raleigh Register* printed the news of the governor's appointment of Emmons along with the text of a glowing letter from former New York Governor William L. Marcy praising Emmons for his part in preparing the New York geological and agricultural survey together with his impressive output of scientific literature.[40]

Emmons would eventually conduct three important scientific surveys of the state: 1852, with a focus on the soils and agriculture of the lower counties and the coalfields of Rockingham, Stokes, Chatham, and Moore counties; 1856, which concerned itself more with the geology in the midland counties—an expansion of his 1852 survey on the same topic—which would form the basis for more serious attempts at developing the coal lands along Deep River; and 1858, which examined the agriculture of the eastern counties.[41] Emmons's knowledge base and proven competence helped make the surveys the definitive scientific works on the topic for many years.

On November 22, 1852, Governor Reid submitted Emmons's first report to the general assembly. Emmons noted that the state could be divided into two great districts: agricultural and mining. The former was predominantly agriculture while

Ebenezer Emmons (1799–1863). Before arriving in North Carolina, he worked extensively on New York's geological survey. His surveys of North Carolina in 1852, 1856 and 1858 gave the most thorough assessments of the geology and agriculture at that time (*Appleton's Popular Science Monthly*, January 1896).

the latter was both mining and agriculture, which was equally productive with other districts. In the mining community, Emmons observed, workers lived cheaply and could subsist on the sale of coal as well as agriculture. His able assistant,

Dr. Spencer "Spence" McClanahan, of Chatham County, reported that he had discovered an abundance of marl, a deposit consisting of clay and calcium carbonate used in the manufacture of fertilizer and as a neutralizing agent in acidic soil. McClenahan also had discovered that extending along the shoreline of the Cape Fear and Neuse rivers, marl was accessible to boats and when river navigation was permissible could be picked up at a small cost. He also observed that boats on the Cape Fear River transporting Deep River coal to its destination might return with marl, which "may be used in Chatham and Moore counties as a fertilizer."[42] During his investigations, McClanahan discovered anthracite coal on the Dan River as well as lignite—an inferior type of coal—eighteen miles above Fayetteville on the Cape Fear River.[43] Neither of those deposits would be worked to any considerable extent because of narrow seams and flooding.

Emmons's report included, among other things, an analysis of the quality of Deep River coal.[44] He noted two varieties of coal at Deep River: semi-bituminous and bituminous. Semi-bituminous coal was an inferior type of bituminous coal, averaging between fifteen to twenty per cent of volatile matter.[45] Building on previously encouraging reports of Deep River coal, Emmons confirmed that it was clean of smut and "adapted to all the purposes, for which the bituminous coals are specially employed."[46] He continued to state that the best quality coal was found at Hornesville and Farmville, both of which were in the same vicinity. The Taylor mine; the Gulf, or Horton mine; and the Murchison mines also offered good prospects for mining quality coal, based on the role the Hornesville mine played on the American side during the American Revolution. But, for Emmons, shipping the coal to market was as large a consideration as removing the coal from the ground. Emmons maintained that had the potentials of the Horton mine played out after the war, the region would have seen an increased population as well as enterprise. Emmons trusted that the forthcoming improvements to the land and waterways portended a home market for Deep River coal.[47] Regarding the cost of transporting the coal to market, Emmons, taking a page from Malthus, equated population increases with diminished resources, such as wood, where it once flourished in the area but now was non-sustainable. For Emmons, calculating the enormous amount of coal underground to be mined and transported, its proximity to iron deposits and transportation routes, outweighed any other considerations for investment of labor and capital in the Deep River coal lands.[48]

Working tirelessly with his son and McClanahan, Emmons completed his report and on November 27, 1852, the *Raleigh Register* issued a statement that the governor was in communication to introduce the report of Professor Emmons, "with a proposition on the part of the House of Commons to print 3000 copies of the same."[49] However, demand soon exceeded supply, and the legislature authorized an additional run of the report. The following March, *The Raleigh Register* printed a more

3. Geology as a Science Stimulates Geological Surveys

circumspect article about Emmons's geological survey, noting that his observations pertaining to agriculture and geology already had been stated and examined mostly by Elisha Mitchell, and that Emmons "has to a remarkable degree endorsed the conclusions made so long ago ... and has reaped the benefits executed by his predecessor."[50] However, the newspaper did acknowledge the thoroughness of Emmons's report, which should serve notice that active, continued efforts to develop those resources should be the current focus. Despite the paper's critical review, Emmons had gotten off to an impressive start especially given that he was assuming another scientist's work that had been inactive for the last twenty-five years. While Emmons had managed to provide a report using material amassed by his predecessors, he successfully incorporated their findings with his own to make for a more thorough study than what had already been presented.

In late 1856, Emmons published his second installment of the North Carolina geological survey, encompassing the geology of the midland, or piedmont, counties of the state.[51] In the preface to his survey, Emmons noted that opportunities for inspecting mines and mineral deposits had been greater than in the first survey of 1852, particularly those located in the midland counties, adding that the coalfields on Deep River had been re-examined and found to have more potential for use than previously estimated. Indeed, Emmons devoted eleven chapters to the Deep River and Dan River coalfields in his latest study.[52] His survey also included observations and analyses of mineral deposits in the region, including iron, gold, silver, copper, manganese, and graphite, but his analysis of the Deep River coalfield demanded most of his attention and generated the most interest. Both surveys contained more data and analyses than any one person before him provided, and gave extra value for businessmen poised to invest in the coalfield notwithstanding the surveys they paid private contractors to conduct on behalf of their company's directors and shareholders. In discussing the overall quantity and quality of the Deep River coalfield, Emmons justified his previous analysis of 1852 by referencing the deep shaft at Egypt and its findings, which he used to refute Elisha Mitchell's accusation of a substandard and incomplete assessment of the area.[53] Examining coal deposits at various locations at Deep River, including Egypt, Farmersville and Wilcox's, Emmons found that Farmersville coal tested favorably and was well adapted to the use of smiths as well as domestic use. It was also cheaper than charcoal and more plentiful.[54] For the coal seams at Egypt, which were already in a state of active mining, Emmons's favorable assessment that "no analysis shows a better composition for all the purposes for which coal is employed," did much to attract the attention of capitalists and speculators. However, coal was not the only object of Emmons's interest in the coalfield. The presence of iron ore held promise on two levels: its high quality content, and its close proximity to the coal seams. Emmons believed that where the iron-ore deposits were near the surface of the earth, only a small investment was further needed to mine and remove the coal at the same time.[55]

In April 1854, Raleigh's *Southern Weekly Post* contained a letter that Emmons had sent to Governor Reid regarding the coal and iron fields of Chatham County. While on a return trip from visiting Anson and Stanly counties for the purpose of researching his geological survey, Emmons traveled through Randolph, Davidson, and Chatham counties, areas he claimed to not have already visited, and was so impressed with the coal and iron measures, particularly in Chatham County, that his enthusiasm prompted a reconsideration of the resources on Deep River, culminating in a recommendation to the governor to build a national foundry there. For Emmons, what he once considered a "local project," had become one of national importance. He proposed that the bituminous coal found in the area could be used as coke for reducing and smelting iron ore. In his assessment of the iron ore, Emmons declared that it is "abundant and excellent."[56] The anticipated production of copper nearby only furthered Emmons's argument that the Deep River area was the ideal place for the foundry. As to transporting the product to market, Emmons recommended building thirty miles of railroad track at "some point ten miles west of Raleigh, and ending at or near the Gulf on Deep River."[57] The completion of the line meant bringing Deep River access "within twelve hours of Norfolk, passing through Raleigh, Weldon, and Portsmouth."[58] The favorable climate and variety of agriculture only helped to convince Emmons further of the

Ebenezer Emmons's first survey of the geology of North Carolina was published in 1852, titled *Report of Professor Emmons on His Geological Survey of North Carolina*, but it was his 1856 study of the midland counties, pictured here, which contained a thorough analysis of the Deep River coalfield (author's collection).

3. Geology as a Science Stimulates Geological Surveys

advantages of developing the Deep River area, but in the final analysis North Carolina was eliminated from consideration and the foundry was built elsewhere.

Emmons-Mitchell Controversy

In a series of letters appearing between 1852 and 1854 in *The Raleigh Register* and the *Greensborough Patriot*, Elisha Mitchell sparked a prolonged debate questioning the results of Emmons's and McClanahan's findings in their 1852 state geological survey of the Deep River coalfield. In a July 2, 1853, letter to the *Greensborough Patriot*, Mitchell claimed that Emmons, who was not new to controversy, had overestimated the size of the Deep River field at forty-three square miles as stated in the geological survey of 1852. Mitchell took particular exception to the large investments of capital entrepreneurs had spent toward efforts extracting and transporting the coal from the seams which might not guarantee a return on investment. In one instance, Mitchell pointed out that a charter had been granted to a company with a capital of $500,000 to construct a railroad line leading from Fayetteville to the coalfields for the purpose of transporting the product from mine to market. Mitchell believed that investing such large amounts of money should have hard science to back it up which he believed Emmons did not satisfactorily provide.[59] Taking up the pen again in a letter appearing in the October 19, 1853, edition of *The Raleigh Register*, Mitchell stated that "Dr. Emmons has nothing of the nature of proof, to offer, that the coal of Chatham extends to any particular distance from the outcrop."[60] The quarrel would continue to play out in the public eye until 1856, when Emmons's second geological report was published and took into consideration the recent drillings near Egypt which Emmons concluded had confirmed his earlier findings.[61]

Charles D. Wilkes, U.S.N.

In compliance with an April 13, 1858, resolution of the U.S. Senate to examine the coal, iron, and timber resources of the Deep River region of North Carolina, Secretary of the Navy Isaac Toucey ordered a four-person commission, led by Captain Charles Denby Wilkes, U.S.N., to explore the coal and iron-ore region at Deep River for the purpose of establishing machines and workshops for the construction of engines and boilers for naval vessels.[62] Wilkes was charged with submitting a written report of his findings to the Navy Department whereby the secretary would forward his recommendation to the Senate. The Navy Department already maintained four yards—in Boston, New York, Philadelphia, and Norfolk—and the promise held by the nearby coal and iron ore deposits in one geographic area of the state seemed to bode well for adding another facility in North Carolina. In his report to the secretary, Wilkes noted and credited the previous work of Denison Olmstead and Ebenezer Emmons, who examined the sandstone formation at Deep River where the coal measures were found. The close proximity of the coal deposits to the beds of iron ore was of particular interest and value not only to the Navy Department but the state of North Carolina as well for commercial development.[63] Though Wilkes's main concern was with the accessibility of the iron ore, which was found between the coal seams, he believed that the nearby coal deposits could be mined concurrently for use in the reduction of the iron ore.

Wilkes focused his attention on the shaft sunk by the Governor's Creek Coal and Iron Company, at Egypt, which reached a depth of 460 feet below the surface including the lower coal seam.[64] Examining samples of coal taken from the Egypt shaft, Wilkes found it to be light and highly bituminous, "yielding a shining and very porous coke and purplish ash, an excellent coal for making gas or burning."[65] Speaking of the timber in the area, Wilkes believed that a sufficient supply of the middle size was abundant and serviceable for naval purposes.[66] Wilkes noted improvements in water power along the many creeks flowing into Deep River and that slack water navigation was unimpeded. The prospects of building mills for grinding cereals and sawing timber relied on the repair of existing structures as well as the construction of new ones all pointing favorably to a regional naval site.

Regarding logistics and transportation of coal to market, Wilkes pointed out that the slack water navigation of the Cape Fear and Deep rivers was achieved by nineteen dams and locks from Jones' Falls on the Cape Fear near Fayetteville to Evans' Bridge on Deep River, resulting in a pool reaching as far as the Woomble branch of Deep River, a total distance of some ninety-eight miles. It was Wilkes's belief that barges carrying between 100 to 120 tons, drawing six feet of water, could reach the upper part of the area, passing near the coal and iron deposits. The outcropping of coal on this route was approximately 100 feet above water level and could be extracted without mine flooding, thus facilitating faster mining efforts and shipping. However, Wilkes was critical in his assessment of attempts to construct the dams and locks, noting that had construction been completed in a timely manner the area would already have been a

Metamorphic slates.   Section of coal field, Deep river.   Metamorphic slates.

1. Conglomerate; 2. Red sandstone; 3. Black slates; 4. Dark sandstone, fire clay; 5. Shales and iron ore; 6. Coal; 7. Argillaceous slates.

Cross section of the Deep River coalfield drawn by Capt. Charles Wilkes, USN. Note the close proximity of the iron ore bed to the coal seam (Charles Wilkes, USN, *Report of the Secretary of the Navy, Communicating the Report of Officers Appointed by Him to Make Examination of the Iron, Coal and Timber of the Deep River Country in the State of North Carolina, Required by a Resolution of the Senate,* **1859).**

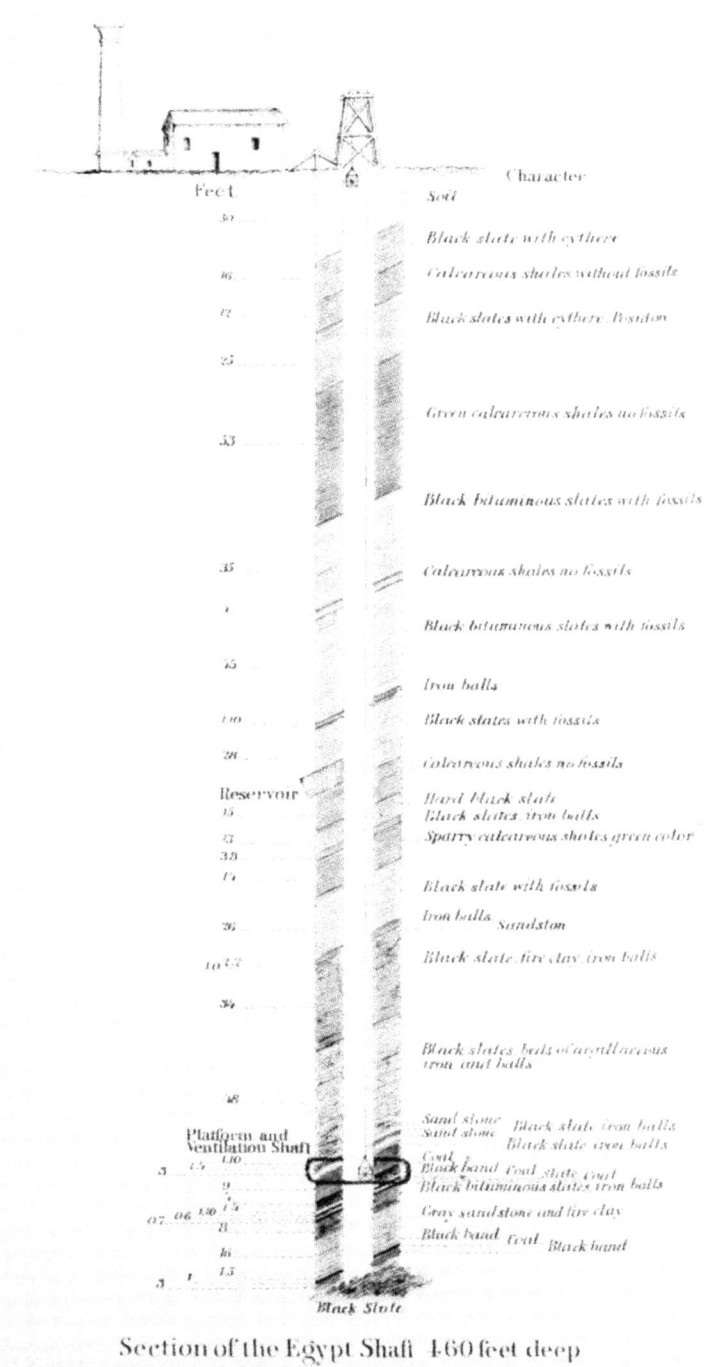

Section of the Egypt shaft drawn by Capt. John Wilkes, USN. By that time, the shaft already had reached over 400 feet (Charles Wilkes, USN, *Report of the Secretary of the Navy, Communicating the Report of Officers Appointed by Him to Make Examination of the Iron, Coal and Timber of the Deep River Country in the State of North Carolina, Required by a Resolution of the Senate*, 1859).

significant producer of coal and iron ore. Squandering of state funds and substandard construction had resulted in "inefficient dams and insecure locks," justifying the legislature's reluctance to let go of its purse strings for future funding endeavors.[67]

Wilkes gave credit to mining engineer William H. McClane for his persistence opening the Egypt shaft which confirmed the large quantities of coal discovered by Ebenezer Emmons in his earlier 1852 survey. With an eye still toward slack water navigation as the most readily means for transportation of coal, Wilkes's concern that during times of drought an insufficient amount of water would be available to float a steamer drawing more than eighteen inches of water was borne out by his associate's misfortune of having to rely on a small steamer during that condition to transport him from Fayetteville to Wilmington, a journey of 100 miles requiring thirty hours to complete.[68] However, the limited number of months available for ships to ply their trade during winter, along with the high cost of improvements and maintenance, added to the woes of slack water prospects. Railroad construction, already well underway in the region, continued to attract proponents, who witnessed firsthand the continued hemorrhaging of state funds directed at repairing slack water infrastructure. Navigation continued to hold interest among some sections of the state through the end of the antebellum period.[69]

Fayetteville, in addition to its slack water navigation, was pursuing construction of a railroad directly to the coalfields with the intention of transporting coal to where the line connected with the North Carolina Central Railroad near High Point.[70] Wilkes was careful to point out that railroad traffic would not compete with slack-water navigation as there would be an abundance of traffic for both. At the end of 1860, trains were running to the coalfields but the cost of purchasing rolling stock and equipment to build shops, constructing wharves along the Cape Fear River in Fayetteville to ship the coal, and constructing a bridge across Deep River all required additional funding from the state, which was reluctant to do so.[71] Wilkes's comments about the significance of the railroads included a network of roads connecting with each other with the expectation that would put the coalfields "into direct communication of a few hours with Norfolk."[72]

DEEP RIVER COAL.

BITUMINOUS COAL of the best quality can be had at the works at Egypt, at a reasonable price by the Ton.
                                      WM. McCLANE,
                                           Mining Engineer.
May 21, 1856.   6-tf

Coal advertisement appearing in the *Fayetteville Observer*. William McClane also served as agent for the New York interests who owned the Egypt Mine (*Fayetteville Observer*, May 21, 1856).

3. Geology as a Science Stimulates Geological Surveys

In conclusion, Wilkes's displeasure with the fact that the North Carolina Central Railroad Company used iron from other states in its manufacture of machinery for maintaining the roads and the discovery of substandard iron foundries in the area were not enough deterrents to discourage his recommendation of constructing a national foundry in the Deep River district. In Wilkes's mind, the sum total of the abundant coal and iron ore deposits, existing timber and crop lands, favorable weather, and the close proximity to Washington made the area ready for such a project. However, it would be left to Wilkes's superior, Isaac Toucey, Secretary of the Navy, to make the final decision.[73]

4.

The Antebellum Period

The first statistic for coal production in North Carolina appeared in the United States Census for 1840, when seventy-five bushels, or three tons, were mined in Chatham County, with only one person employed either as a miner or property owner.[1] The Egypt Mine, or Cumnock, as it was later to be called, had its beginnings in 1830, when Peter Evans, owner of a plantation in the northward bend of Deep River dug a shaft near the village of La Grange. Evans renamed the village Egypt after witnessing a twelve-wagon caravan of locals he called "Israelites" arriving at his plantation during a drought to purchase corn.[2] In 1851, Brooks Harris and L.J. Houghton acquired the property from Evans for about $15,000, and Harris bought out Houghton's interest and sank a shaft eventually reaching 460 feet into the Cumnock coal bed.[3] The property changed hands several times until 1854, when the Governor's Creek Steam Transportation and Mining Company took it over and operated the mine during the Civil War.

From 1851 to 1856, the General Assembly incorporated no less than six coal mining companies. The Pittsboro Mining and Transportation Company, chartered on January 28, 1851, included as its directors Henry B. Hewitt, Horace White, and C.W. Goddard.[4] The Deep River Mining and Manufacturing Company, chartered on January 28, 1851, comprised John Taylor, Lemuel Williams, and Henry B. Hewitt.[5] Williams, who had addressed the Senate chamber earlier on January 14, 1851, on the merits of the Deep River coalfield, had an interest in the area from the standpoint of a geologist and investor, sometimes acting on behalf of other investors who relied on his concise reports for investment purposes. Hewitt also served as a director of the Pittsboro Mining and Transportation Company (1851), the Lagrange Mining and Transportation Company (1851), and the Chatham Mining and Transportation Company (1851). In 1852, the legislature passed an act to incorporate the Hillsboro Coal Mining and Transportation Company, which was to subscribe to 10,000 shares of stock at $100 each, and represented an ambitious project with the company investing a considerable amount of money. The year 1855 saw the incorporation of the McIver Coal Mining Company on February 14, and both the Clarendon Coal Field Company and the Gulf Coal Mining Company on February 16 with John H. Haughton and Lawrence J. Haughton serving as directors, who were also involved with the Cape Fear and Deep River Navigation Company. Another incorporation of note was the Gulf and Deep River Iron Manufacturing Company.[6]

4. The Antebellum Period

Coal mining as a business enterprise requires a great deal of capital even at the outset before mining operations can begin. The company must purchase or lease land, survey the property, analyze its mineral content, construct buildings for storage of coal and equipment, purchase rights-of-way, hire laborers, build accommodations for miners, gain access to sources of timber for mine props, and pay for transportation of coal to market. In its early history, mine equipment may consist of at least a hand windlass or boring machine to drill a sizable hole through the bedrock if a shaft was required to reach the coal seam. Often when a mine opening flooded, expensive pumping engines were required to drain the standing water away from the mine, or, simply, the mine shut down its operations until the water receded significantly to restart efforts. Mines required a directed flow of fresh and return air to be circulated throughout the workings to continuously supply workers and diffuse dangerous gases.[7] The cost of transporting coal to storage areas or market could bankrupt a company if rates suddenly spiked and the mine found itself at the mercy of the transportation company.

As early as 1850, the Deep River Mining and Transportation Company had begun purchasing land in Chatham County for their coal operations. In November, the *Fayetteville Observer* reported that Thomas Parish sold his plantation and coal mine on Deep River for $7,500 to a "northern gentleman, or company, whose intention was to proceed at once to working the mine, with a larger force."[8] The outlay made by the unidentified capitalist, the article noted, meant that the speedy completion of the works by the company would enable the purchaser, and other owners of coal mines on Deep River, to ship coal to Fayetteville and on to Wilmington for transport by ocean vessels to outside markets. As part of its overall business plan, the company bought property in Wilmington for the purpose of storing, loading, and unloading coal from its mines on Deep River. The *Wilmington Tri-Weekly Commercial* reported the purchase, by an unidentified party, of a site at Wilmington where the Phoenix Steam Saw Mill formerly stood.[9] The depot, or coal yard, could provide easier accessibility to the coal scheduled to be shipped from Wilmington to points beyond the area or for retail pockets serving a local market.

In 1853, Charles Thomas Jackson (1805–1880) of Massachusetts, a physician with expertise in geology and chemistry, was hired by New York businessman and speculator F.W. Camman to conduct a survey of the coal-bearing land in various locations at Deep River particularly near Gulf.[10] Camman, already involved in the Capps Hill Gold Mining Company, which was chartered by the state of North Carolina in February 1855 and had workings in Mecklenburg County, represented a new type of entrepreneur who approached business with an analytical point of view, employing experts of similar thinking to survey and examine potential mineral-bearing lands in order to evaluate their economic as well as commercial value. Using data already compiled from previous state and privately funded surveys, these experts had a basis for their work, which tended to focus more on specific areas or regions in order to provide their employers with the latest information available. Although the legislature supported new business and provided some capitalization, especially for internal improvements and surveys, the more wealthy businessmen who had formed partnerships and even corporations benefitted greatly from their own experts who reported directly to them.

Jacksons's report, which was reprinted in the *Greensborough Patriot* and various other local newspapers, focused on the coal mines owned by J. & L. Haughton which supplied the local demand for coal at smiths' forges.[11] Early in his investigation, Jackson noted that only small amounts of coal had been thus far extracted and that open cuts and pits served as access points to extract coal from the outcroppings, some of which were buried deeper into the ground than normal. African-Americans sent to the area to work as miners, hewed out the coal with an adze or mattock, making for a labor-intensive effort without benefit of machinery to move the heavy weight of coal, which could be attached to rock and require further separation. Even by these primitive methods, Jackson was able to examine the coal "over a considerable area of ground, in numerous pits along the line of outcrop." Jackson discovered that the aggregate thickness of the workable beds was six feet, which represented greater access to the coal seam and a potentially greater amount of coal although no one as yet could predict that the seam would dwindle to nothing or remain workable especially without bore holes to properly assess the extent and value of the coal.[12]

Jackson mostly agreed with the scientists before him who had submitted favorable reports on the Deep River coalfield and added that the coal was good for gas making for lighting and to be "of excellent quality for steam-engine furnaces."[13] Witnessing first hand demonstrations of the coal's applications in the forges of local smiths, Jackson was impressed with its ability to bond steel in an axe and further noted that its welding was "perfectly effected." Confirming Emmons's findings that 9,800 tons of coal could be taken from every acre and that a square mile could yield up to 6,272,000 tons, he called for the immediate erection of a steam engine and "all the machinery for pumping water and for raising the coal."[14] Jackson further added that the mine shaft needed to descend great depths to access the larger seams which often meant additional expenditures for ventilation and haulage. Also, noting that the near completion of slack-water navigation of the Deep River eventually meant shipments of coal traveling to the coast, via Fayetteville, new markets could be tapped and exploited, and coal delivered to New York would be preferred to that from Newcastle, England, and at a greater profit.[15] Jackson's report to Camman took into consideration mostly those qualities of the coal he examined in terms of a scientific and economic approach.

An article appearing in the *Weekly Raleigh Register* in 1855, reported that the Committee on Internal Improvements found that William H. McClane, a stockholder in the Deep River Mining and Transportation Company, had advised the newspaper that the company already had expended $140,000 in land and machinery, and with the expectation that river improvements were soon to be completed, the goal was set at 350,000 tons of coal the first year and, with other mining companies looking for

4. The Antebellum Period

ways to transport their product to market, annual shipping figures would increase from 400,000 to 500,000 tons. However, there was a stipulation that should the state legislature refuse to grant funds to improve river navigation, the Hillsboro Mining and Transportation Company would abandon its efforts to mine coal at Deep River.[16] Meanwhile, the coal property at Gulf was being advertised for development and habitation. On February 16, 1855, the General Assembly ratified an act to incorporate the Gulf Coal Mining Company, with John H. Haughton and Lawrence J. Haughton as directors.[17] Though organized on a smaller scale than the Egypt Coal Company, the firm did provide a need in supplying coal for local uses after the end of the Civil War.

In his report to the directors of the Deep River Mining and Transportation Company, William Rogers Johnson's observations and analysis centered on the geology of the coal lands as well as projected expenditures and income the company would need to consider. Agriculture and woodlands factored into the overall costs as well since the report made provisions for supplying new mining communities with produce as well as timber for structures and mine props.[18] In the construction of a mining town the company assumed a great risk by the scope and expense of the project which could deplete capital already allocated to mining operations or transportation. Johnson based his report on findings he made by sampling coal from three open pits within the overall Farmersville area. Using eight different samples of coal taken from the pits, he next formed a mixture of forty specimens to obtain an average specimen of the whole. Johnson's conclusions eventually presented to the board of directors that Deep River coal was equal in thickness to the Pittsburgh seam and not inferior to the main seam at New Castle were encouraging along with its quality "well adapted to domestic use, to metallurgic arts, and to steam navigation."[19]

Commenting on the successful drilling of a coal deposit at Egypt, the February 28, 1856, issue of the *Asheville News* reported that it was "a seam of exceedingly rich bituminous coal."[20] The article noted that the operations were supervised by William H. McClane, mining engineer, who represented the interests of New York speculators in a business venture incorporated as the Governor's Creek Coal and Iron Manufacturing and Transportation Company.[21] Describing the seam as one mile extending across Deep River from the outcrop at Farmersville, the projection that even larger deposits could be mined at Egypt seemed possible. It appeared that McClane delivered up one ton of coal from the Egypt shaft to use in a steam engine at the railroad office, finding that two bushels equaled one cord of pine. The projection that in two months the mine could ship 300 tons daily held promise if only a means of transportation were available and, as a result, interest shifted to the completion of a rail line connecting the coalfield with Fayetteville. Amidst the planning for increased production, an explosion of fire-damp killed five men working at the Egypt shaft.[22]

In late 1858, William Gammell, a Baltimore speculator, addressed a letter to the *Weekly Standard*, of Raleigh looking to manufacture iron at Deep River.[23] Gammell, like some others before him, immediately focused on the benefits of having coal deposits near iron ore seams. Using Scottish technology and production standards,

Gammell believed that the manufacture of pig iron could be a great deal cheaper if made at facilities on Deep River.[24] Noting that Scotch pig iron was the least expensive of its type produced in the world, the proximity of the coal to the ore was more advantageous than importing it from Scotland and other iron-producing countries. Businesses could now compete favorably with Europe and New York, whose cost per ton was three times the price of iron made on the Tyler property at Egypt on Deep River. Having been satisfied and encouraged by the results of the work thus far, Gammell engaged to loan $26,000 to the proprietors to help boost future development. His estimated investment of $150,000 to buy mineral lands, sink the pits, and erect two furnaces based on the Scottish principle would pay for itself in a short time. Surmising that two furnaces could produce 400 to 500 tons of iron a week, or 100,000 tons per year, the savings from using local facilities would far outweigh the costs of New York and Boston iron.[25]

> **FOR SALE,**
>
> THE GRANGE LAND, formerly the property of A. McBryde, sen., dec'd, in Chatham county, on the waters of Deep River and Indian Creek, containing 1,000 ACRES. There are few Tracts of its dimensions on the River so compact in its form or of more general fertility and variety of soil. Corn, Cotton, Wheat, Tobacco, &c. can be advantageously cultivated, each in its proper soil. There are numerous Springs convenient on the land, with a Mineral Spring near the house, that has been successfully used as a medical agent. A large two-story DWELLING, with a good Barn and Stables; about 200 acres good low grounds, and 50 or 60 equally as good, uncleared; 200 acres of second low grounds, the most of which is uncleared; 40 or 50 acres of fine meadow land, &c. A vein of stone coal passes through it, which may be used. The Land is very well timbered, and a healthy situation. Any person wishing to purchase a valuable Plantation, will please call and examine it, or address
>
> ARCH'D McBRYDE,
> or W. M. McBRYDE,
> Tyson's P. O., Moore County, N. C.
> April 3, 1844.   1-5w

An early reference to anthracite coal is this notice of land for sale at Deep River in Moore County where a vein of "stone coal" ran through the property. Anthracite, or stone coal, was found in one area of Moore County though its production was low compared to the bituminous variety (*Fayetteville Weekly Observer*, April 24, 1844).

In an article appearing in November 1860 in the *New York Times*, an unnamed correspondent reported that he had been investigating "the improvement of the navigation of the Deep River" as well as the mineral wealth. In the improvement project nineteen locks were constructed with a corresponding number of levels for slack water navigation varying in depth from eight to twenty feet.[26] Perhaps one the most lucrative prospects offered in the Deep River coalfield was the extraction of oil from coal, iron ore, and slate rock. In an article appearing in 1861 in the *Mining Journal and Journal of Geology, Mineralogy, Metallurgy, Chemistry, and the Arts*, an unnamed correspondent of the *New York Times* gave a detailed report of the processes involved in the preparation and production of oil by this method.[27] Citing the Deep River Coal and Iron Company as the forerunner in the enterprise, iron ore was extracted from

4. The Antebellum Period

the mines and subjected to baking or roasting to drive off the bitumen, similar to coking, after which the iron ore was calcined, or roasted, in preparation for the blast furnace. The ore, reduced to one-half during the process, meant that two tons of calcined ore produced one ton of pig metal. More than twenty gallons of pure oil could be had from each ton of ore, giving an advantage in cost and transportation over the black band ores of Scotland. The slate stone produced about thirty-nine gallons to the ton, with a maximum yield at forty-five gallons. The iron stone yielded an average of twenty-five gallons of oil to the ton.[28] This forward thinking added yet another advantage for mining Deep River coal and iron.

According to the *Times* correspondent, the Deep River Coal and Iron Company was extracting oil from iron ore and the coal deposits at Egypt, adjoining those of the Deep River Company, and was a similar quality and that the company would confine its operations to the mining of coal for local and New York markets.[29] The cost of coal delivered to New York, including commission, dockage, and storage, was $4.46 a ton. At the current rate it sold for $8.00 a ton for the purpose of making gas. The yield of gas was from 10- to 11,000 feet per ton, leaving a residual of about 1,600 pounds of fine coke, which was used by silversmiths and other metallurgical workers. The famed Newcastle coal—the benchmark for high-quality coal—yielded between 9- and 10,000 feet of gas a ton, but from 1,000- to 1,200 pounds of rough coke. The results paved the way for a consistently superior product at a competitive price.[30] With plans to expand the operations and add new machinery at considerable cost, the Deep River Company was poised to profit from their enterprise. The *Times* correspondent was confident that operations would continue even amidst concerns relating to "the talk of secession and disruption."[31] However, murmurs of war

Notice to contractors in 1859 soliciting bids for work on the Western Railroad. Completion of the stretch of track to the coalfields would be completed in September 1863 (*Fayetteville Observer*, February 10, 1859).

and the eventual shelling of Fort Sumter in April 1861 put most industrial projects in the state on hold with the exception that coal mining and iron production continued in support of the Southern war effort.

The chartering of the Western Railroad Company in 1852 to build a forty-three-mile connection from Fayetteville to the Deep River coalfield largely met with the approval of the pro-coal faction, who believed that an inroad to the coalfield would not only benefit their section but the state as well. Finally, there was a serious attempt to mine and transport coal on Deep River with consideration for cheaper haulage. In addition to state support, the railroad enjoyed financial backing from Northern interests, who also saw advantages of tapping the coal deposits.[32] However, the overall effort to build the line languished for nearly six years due to construction delays brought about by dissension, contract disputes, and obtaining rights of way, leaving only twelve miles of track completed by 1859 of the estimated forty-three miles total.[33] State Senator John T. Gilmer, representing Cumberland and Harnett counties, delivered a speech December 2, 1858, calling on his fellow senators advocating passage of a bill to aid in the construction and equipment of the Western Railroad from Fayetteville to the coalfields on Deep River.[34] Gilmer noted that machinery to extract the mined coal from the ground was in place and all that was needed was a transportation outlet from the coalfields to a place of embarkation for shipping the product to market. Noting that industry and transportation came to rely more on coal than ever before, Gilmer urged that we "cannot too highly estimate the great value of these coalfields."[35] That the state had already invested as much as $40,000 in the commissioning of geological surveys helped make the argument that the timing was right for completing the rail line. Though passage of the bill was a cause célèbre, it was the commitment and backing of key individuals, like Gilmer, who pushed legislation forward so that work could continue toward the completion of the Western Railroad link to the Deep River coalfield.

The Coming of the Railroad

According to historian Alan D. Watson, the appearance of the railroad represented the most profound development in internal improvements within the state.[36] During the legislative session of 1831–1832, the General Assembly witnessed a new enthusiasm for transportation in the form of railroads, chartering the Cape Fear and Yadkin Railroad Company and the Tarborough and Hamilton Railroad Company.[37] The incorporation of these roads was soon followed by the chartering of others by the legislature, including the Fayetteville Railroad Company and the Fayetteville and Western Railroad, among others.[38] One of the legislature's most important actions during its session was to incorporate the Cape Fear and Yadkin Rail Road Company "for the purpose of effecting a communication by a rail road and a canal or canals, from the town of Wilmington through or by the town of Fayetteville to the Yadkin river."[39] With a capital stock of $2,000,000, sold in shares of $100 each, the appointed commissioners on May 8, 1832, opened the books for subscriptions.[40] However, after much publicity and determination to succeed in establishing the railroad, the following May the *Fayetteville Weekly Observer* printed a story first appearing in the *People's Press and Wilmington Advertiser*, announcing that plans to connect the seaboard with the mountains by rail had been abandoned and that the money deposited by subscribers on their shares would be returned on a pro-rated basis.[41] With the failure of the Cape Fear and Yadkin Rail Road Company, Fayetteville had to find alternative methods for securing viable transportation outlets both east and west particularly if its goal included reaching the Deep River coalfield.[42] In 1834, Fayetteville experimented with an early prototype of a railroad based on the British wagon-way system, which utilized wooden rails with a rounded center fastened to the ground for pulling horse-drawn cars capable of transporting up to thirty passengers or four tons of freight.[43] In some instances wooden rails were replaced with more durable iron track, leading to further consideration and development of a steam-powered railroad.

Perhaps going hand-in-hand with the push for internal improvements, a call for an examination of North Carolina's metallic and mineral wealth portended independence and "unrivalled wealth" for the state. The key to unlocking these natural resources resided with enterprise, which held that the citizenry must put forth vision and effort such as that shown by Thomas W. Clegg, who, according to the April 3, 1839, issue of the *Fayetteville Weekly Observer*, had extracted a specimen of coal from his property at Egypt in Chatham County. The sample tested favorably in a forge at Fayetteville, prompting encouraging reports to further determine the extent of the coalfield and its wealth. The newspaper coverage enthusiastically proclaimed that such discoveries could enable the state to greater self-sufficiency and "lead to her independence of her neighbors."[44] Self-sufficiency meant that the state could enjoy the profits of mineral extraction without their going into the coffers of nearby states such as Virginia, who already had a significant hold on coal mining in the region. However, successful mining ventures required opening up routes to markets using the most cost-effective and efficient means available, and by that time the railroad was starting to win over business for their transportation needs.

By the 1840s, efforts to promote internal improvements held great promise when attention was being directed to the railroads as the preferred method to move goods and material, including a plan to open up the Deep River coalfield. In 1840, the aggregate length of railroads in North Carolina was two-hundred fifty miles while the aggregate length of canals was a paltry thirteen- and one half miles.[45] However, navigation never lost its place in the design of a transportation network, and during its 1848–49 session, the legislature chartered the Cape River and Deep River Navigation Company with the purpose of improving the Cape Fear River above Fayetteville to the coal deposits in Moore and Chatham counties.[46] Recognizing the importance of improved navigation on the Cape Fear and Deep rivers to access the rich, natural resources of the area, such as the coal and iron deposits, as well as naval stores, local

Fayetteville businessmen engaged William Beverhant Thompson, civil engineer, to conduct a survey of the Cape Fear and Deep rivers extending as far as Hancock's Mill in Moore County.[47] Investors were keen to put their case before the legislature particularly when they could submit a report by an eminent engineer to back their claims. Thompson's recommendation that implementing a system of locks, dams, and canals, to effect slack water navigation took on even greater meaning when compared with the scope of the Chesapeake and Ohio Canal for the purpose of reaching an Atlantic port. His appraisal of the Deep River coalfield took into account similarities to the "far famed bituminous coal of Alleghany County, Maryland," which, by way of the recent expansion of the Chesapeake and Ohio Canal, could be transported to seaport as well as "lay open to the inhabitants of the Atlantic States, the mineral treasures of the Alleghany country."[48] For North Carolina, whether by navigation or railroad, she was poised to no longer witness "the diversion of her products to the building up of the Commercial Cities in the States on her North and South."[49]

Responding to the state of Virginia's application to the North Carolina legislature for a charter to build a southerly line through the state, the legislature chartered the North Carolina Railroad Company on January 27, 1849, for the purpose of building such a line as the Richmond and Danville Railroad required. Afterward, the Richmond and Danville Railroad sought to implement a physical connection with the North Carolina Railroad at Greensboro, and after the Richmond and Danville surveyed a line through Milton, North Carolina, did not respond to the efforts and it was not until after completion of the main line to Danville in 1856, that efforts were made to obtain a North Carolina charter for the extension. Assuming the risk of building a western extension of their line, the Richmond and Danville Railroad prompted the Virginia legislature to pass an act on April 7, 1858, to authorize "connections between the Richmond and Danville R.R. and the North Carolina Central Railroads."[50] The act also granted permission to the Richmond and Danville Railroad, or any North Carolina corporation "which might be chartered for that purpose to build a railroad from Danville to Greensboro, to connect the Richmond and Danville Railroad and the North Carolina Railroad."[51] In response, the North Carolina legislature chartered the Dan River Coalfield Railroad Company with the intention of constructing a railroad from a place on the Virginia line near Danville to the coalfields of the Dan River and further stipulating that the road "shall not run within twenty miles of the North-Carolina railroad."[52] Although the plan to connect Danville with the Dan River coalfield was still alive at the beginning of the Civil War, military priorities of the Confederate government dictated that the Piedmont Railroad was the preferred connection between the Richmond and Danville Railroad at Danville and the North Carolina Railroad at Greensboro. The Richmond and Danville consequently dropped the Dan River Coalfield Railroad Company from consideration, quashing further interest in accessing the Dan River mineral deposits.[53]

Pressured by Fayetteville interests that claimed improvements were cutting off their trade, the legislature, on December 24, 1852, incorporated the Western Railroad

4. The Antebellum Period

Company to "construct a railroad from the town of Fayetteville to some point in the coal region, in the county of Moore, or in the county of Chatham."[54] The action of the legislature elicited positive response, as noted in the newspaper coverage of the event. Fayetteville's *North-Carolinian* in January 1853 claimed that the addition of a railroad to tap the Deep River coalfield would increase Fayetteville property values by fifty percent and serve as a basis for a network of railroads carrying freight throughout the region and beyond.[55] However, in February, the same paper warned that speed was imperative in constructing the railroad as interests in important commercial cities, such as Charleston, Wilmington, and Norfolk, may instead exploit the Deep River field, "passing us by and leaving us off their routes."[56] Clearly, the Western Railroad became a rallying point for increased commerce and wealth not only for Fayetteville but the entire North Carolina–South Carolina–Virginia region.

During its 1854–55 session, the legislature saw through passage of an act to authorize the Cape Fear and Deep River Navigation Company to issue bonds. In an assessment of the type of work needed to be completed to ensure unimpeded navigation, Charles F. Fisher, Chairman of the Committee on Internal Improvements, in 1855, issued a report of the company, noting the recommendations of its chief engineer, E.A. Douglass which called for the construction of a sufficient number of locks and dams along with widening and lengthening the canal at Buck-Horn on the Cape Fear River.[57] Singling out Douglass for his success in all previously contracted improvements as a slack-water engineer, particularly his work as principal engineer with the Lehigh Coal and Navigation Company, Fisher recommended that the General Assembly spare no expense meeting his recommendations.[58] Looking at a source of revenue to help fund the improvement projects, Fisher, placed the greatest prospects on the coal deposits "near the banks of the Deep River."[59] He cited the observations of William H. McClane, mining engineer, who believed the coalfields to be extensive, consisting of 900,000,000 tons of coal waiting to be mined and shipped to markets.[60] In a report of the stockholders of the Cape Fear and Deep River Navigation Company to the General Assembly, the recommendation was for the state to invest a much smaller amount of capital in improvements, in this case $600,000, compared to the $11,000,000 required by the Lehigh and Delaware improvement with greater return on investment.[61] A follow-up report, this time submitted from the president and directors of the Cape Fear and Deep River Navigation Company to the General Assembly, put Deep River coal at the top of its priorities which had the backing of scientists such as Johnson, Emmons, and Jackson as well as the "actual experiment of the most experienced miners from the North and from Wales."[62]

An unnamed interested party writing to the *North-Carolinian* in January 1853 proposed solutions for the eastern and western extensions of the railroad, the eastern solution calling for linking Fayetteville with Beaufort harbor as the best opportunity for a ready and global market, while a western plan advocated connecting Fayetteville with the North Carolina Central Railroad at Greensboro or a point nearby, passing through the rich Yadkin Valley, and on to points farther west.[63] Optimistic reports submitted by experts proclaiming the vastness and value of the coalfields made a route to the area crucial for supporters of the project especially during the years leading up to the Civil War, sectional differences and agendas aside.

Other railroad projects ensued when the General Assembly, during its 1854–1855 session, approved funds in the amount of $1,000,000 to issue stock in the Atlantic and North Carolina Railroad, and $4,000,000 for the Western North Carolina Railroad, which were eastern and western expansions of the North Carolina Railroad. While river improvements had met with some success, as on the Roanoke and Cape Fear, and plank road improvements met with some promise, all attention was ready to be focused on the new railroads for their ability to connect sections of the state with areas outside the limitations of road and water travel. As in the case of the Western North Carolina Railroad Company, it meant faster travel and greater distances "beyond the Blue Ridge."[64] Soon a railroad construction boom would mean interconnecting multiple lines into a network of travel and transportation heretofore unseen in the state.

It would take nearly three decades, along with the findings of professional geologists, until North Carolina was poised to undertake an ambitious railroad-building project that it could begin to see the potential of its seemingly endless coal deposits.[65] During the same 1854–1855 session, the General Assembly, on February 14, 1855, enacted legislation to incorporate the Chatham Railroad Company for the purpose of establishing a transportation link between Raleigh, or some point on the North Carolina Railroad west of Raleigh, "at or near the Coal-fields in the County of Chatham."[66] The Cheraw and Coalfield Railroad Company, among others, also sought a connection to the coalfields, but as Alan D. Watson noted, the failure of these efforts "stemmed immediately from the expense of the projects."[67] Attempts to raise the necessary capital from the state and individual entrepreneurs proved unsuccessful in many instances, putting interest in coal markets temporarily on hold. The completion of the Western Railroad's link from Fayetteville to the Deep River coalfields seemed to hold the most promise for reaching the coal lands and mineral deposits, but as will be shown, that promise did not immediately materialize or arrive without issues.

In 1856, the announcement of a rich bituminous coal seam at Egypt in Chatham County, prompted the company to send its steamer, the *John H. Haughton*, to Haywood, at the confluence of the Deep and Haw Rivers, for a cargo of coal though the only haulage that had transpired was the shipping of coal to Fayetteville by way of wagon. Actually, the *Haughton*, which was on its maiden voyage on Deep River, had left Lockville with flour, cotton, turpentine, and rosin, only to be held up for several weeks at Buckhorn Canal on account of the low water levels of the river.[68] Due to fluctuations in demand for coal and in water levels on the Deep and Cape Fear rivers to accommodate ships with a considerable displacement, the company found itself in the next year's legislative session without financial support. Coupled, with the state's failure to link the Cape Fear River, the state's only direct outlet to the Atlantic Ocean, with the Yadkin and its markets, undermined the completion of future canal-building projects, practically sounding the death knell for any further interest in canal transportation and coal haulage.[69]

In an article concerning developments on the upper Cape Fear River, the *Fayetteville Weekly Observer,* on July 21, 1856, announced to its readers that with the completion of the "locks, dams, sluices, and canals from Cross Creek to Lockville," a channel for trade and commerce had opened up in the area.[70] Reporting on his excursion from Fayetteville to Haywood taken on the steamer *Brothers*, the writer in a detailed manner described the scenery along his journey, giving way to a description "useful as well as ornamental."[71] Plying his trade as a journalist with regard to factual reportage, he next commented on the "solid and substantial locks and dams upon the river" and his observations that operations should have commenced at Cross Creek and extended their improvements upwards, he took some comfort in noting that problems like misaligned locks and dams were being corrected by the company with the use of the *Haughton* and six flats transporting rocks from quarries to the weakest points.[72]

According to the writer, the only obstacle that kept coal from reaching market was at Lockville, a point where a "fall of thirty-five feet in one-and-a-half miles" had to be corrected. The Egypt coal mine, about eleven miles higher up the river past Lockville, was a place of disembarkation where he descended the mine shaft and noted that the timbering and sheathing were "all strong and substantial."[73] William H. McClane and his foreman were on hand to whom the writer gave ample credit for sinking the shaft 482 feet, installing pumps to rid the workings of excess water, and excavating the coal approximately one-hundred yards from the shaft. As with others conversant on the subject, the writer noted that lack of improved transportation was the greatest obstacle to successful trade, again all the more crucial for a railroad network to supplant navigation.

5

The Deep River Coalfield at the Time of the Civil War

Mineral resources have long held sway in a nation's capacity to wage war. According to historian Robert C. Whisonant, at the outset of the Civil War some ninety percent of manufacturing existed in the North as well as a majority of the coal and mineral deposits.[1] Coal represented an important staple of both the Union's and Confederacy's war effort as it was needed to fuel railroads and steamships as well as foundries and other important industries critical for the manufacture of weapons, steam locomotives, and iron plating for ships. For the Confederacy, a reliable source of coal for domestic purposes became a significant issue when the North cut off supplies to Southern states early in the war. Mining operations at Deep River included three locations, all near the river. Gulf, on the southern border of Chatham County, west of Egypt; Egypt, on the south side of Deep River; and, Farmersville, east of Egypt, on the north side of Deep River. The distance between Gulf and Farmersville was four miles. The Deep River Coal and Iron Company was listed in the U.S. Federal Census Non-Population Schedule, 1860, showing an investment of over $136,000 in real estate and property located at Goldston, Chatham County. The company's operation included iron manufacturing and coal-oil production but it is not known to what extent their business was domestic, commercial, or military.[2]

In 1861, the North could boast five anthracite railroads serving its numerous coal mines in northeast Pennsylvania.[3] Local smiths who relied on wood, which was necessary for charcoal production, found that coal resources were more desirable unlike the former, which was becoming in short supply due to extensive timbering. A large part of the North's advantage in coal mining had to do with its significant deposits of anthracite, of which it possessed three-fourths of the world's supply and the South virtually none.[4] With an extensive network of roads, canals, and railroads already in place to transport coal, the Union enjoyed access to a wide transportation outlet that reached port facilities in New York, Philadelphia, and Baltimore. Indeed, at the beginning of hostilities, the North claimed 22,000 miles of rail compared to the South's 9,000 with more facilities for making rails and locomotives.[5] All considered, it was small wonder that the northeast corridor was beginning to earn itself the reputation as a world-class center for heavy industry. The following tables show the amount of coal mined by the South and North during the time of the Civil War. North

5. The Deep River Coalfield at the Time of the Civil War

Carolina had the third highest totals behind Virginia and Tennessee. Some Southern states such as Arkansas maintained some coal mining but in negligible amounts.

Coal Production of Southern States, 1861–65

Year	Alabama	Arkansas	Georgia	North Carolina	Tennessee	Texas	Virginia
1861	5,000	2,000	2,500	15,000	150,000	1,000	94,697
1862	18,000	2,200	3,500	30,000	140,000	1,000	115,495
1863	18,000	2,500	6,000	30,000	100,000	1,000	112,068
1864	20,000	2,700	10,000	25,000	100,000	1,000	111,472
1865	8,000	3,000	10,000	20,000	100,000	1,000	73,730
Totals	69,000	12,400	32,000	120,000	590,000	5,000	685,378

Coal Production of Northern States, 1861–65

Year	Illinois	Indiana	Iowa	Maryland	Ohio	Pennsylvania Anthracite	Pennsylvania Bituminous	West* Virginia
1861	948,800	56,000	52,500	287,073	1,855,300	10,245,156	4,562,000	—
1862	1,038,200	67,200	56,700	346,201	1,890,400	10,106,435	4,995,600	—
1863	1,198,300	79,900	60,000	877,313	1,923,500	12,267,446	5,332,600	276,489
1864	1,315,868	92,500	65,200	755,764	1,952,500	13,027,168	6,051,600	313,398
1865	1,518,100	103,900	69,574	1,025,208	1,891,288	12,076,996	6,372,900	484,215
Totals	6,019,268	399,500	303,974	3,291,559	9,512,988	57,733,201	27,314,700	1,074,102

Source: Howard N. Eavenson, *The First Century and a Quarter of American Coal Industry* (Pittsburgh, PA: Privately Published, 1942)

*Production commencing after statehood in 1863

Virginia's coal and iron ore deposits represented the largest in the Confederacy, making the state a crucial source for its manufacturing ability. Much of Virginia's coal production came from the Midlothian and Clover Hill companies while a variety of Midlothian coal called Black Heath was used exclusively in Richmond's cannon foundries.[6] Within close proximity to the coalfields, Virginia's iron deposits proved invaluable in the manufacture of much-needed weaponry, as both resources went toward keeping the Tredegar Iron Works running which represented the South's largest and most important center for iron and steel production. In addition to its numerous bituminous fields, Virginia contained significant deposits of semi-anthracite coal which were mined by the Price Mountain Coal Company near Blacksburg, and produced a cleaner-burning coal than its bituminous counterpart. Still, Virginia struggled with transportation issues such as its inability to keep pace with the growing demand for iron necessary to repair and construct new railroad lines.[7]

At the outset of the war, the Confederacy's supply of gunpowder and niter, or saltpeter, had dwindled to significantly low levels, necessitating the importation of these vital resources in order to supply its forces. With the Union blockade of Southern ports in place and demonstrating increasing success, it became imperative for the South to find new outlets to compensate for those deficits, and a search began for supplying the need locally. Accorded a high degree of mineral diversity, Virginia could boast large deposits of coal, and its limestone-rich caverns became a key source for obtaining niter, a crucial ingredient in the manufacture of gunpowder. Requiring an agency for the express purpose of locating and processing niter, as well as needed mineral sources, the Confederate government in 1862 created the Nitre and Mining Corps to meet the need. The department also was charged with disbursing iron, copper, lead, zinc, and coal to areas having the greatest need within the Confederacy. On April 11, 1863, the Confederate Congress ratified legislation to create the Nitre and Mining Bureau for the purpose of creating an "organization of a corps of officers for the working of the nitre caves, etc."[8] Under the direction of Brigadier General Josiah Gorgas, the bureau became an important department for meeting the need of gunpowder manufacturing facilities requiring a steady supply of niter. In June 1864, legislation authorized recruitment of additional personnel, including chemists and assistants, many of whom had training in geology.[9] Addressing the need for a reliable supply of coal, the bureau "opened new coalfields in North Carolina and Alabama, and coordinated the flow of mineral fuel to Confederate naval stations along the coast."[10] The Confederate Navy drew its coal supplies from three points: Richmond, the Egypt mines in North Carolina, and the mines near Montevallo, Alabama. Several months' supplies of coal had been stockpiled at Richmond and Wilmington but transportation problems only allowed quantities sufficient for immediate needs. It would take until September 1863 before a railroad link to the Deep River coalfield was completed, allowing for larger amounts of coal to be transported to Fayetteville and then on to Wilmington.

In early 1862, John Manning, Jr., lawyer, congressman, and former delegate to the Secession Convention in Raleigh, advocated that there should be no delay in building a railroad connection between the coalfields at Deep River and the North Carolina Railroad. Referencing the sinking of the shaft at Egypt, the slopes at Farmersville and on the Taylor property, the pits at the Gulf, Murchison's, and Foushee's, Manning

> **NORTH CAROLINA COAL.**
>
> THE SUBSCRIBER BEGS LEAVE TO ANNOUNCE to the citizens of Raleigh that he has made arrangements to keep on hand a constant supply of Coal from Egypt, in Chatham county.
>
> Persons wishing to have Coal delivered at their residence can have it by making early application.
>
> P. FERRELL,
> Wilmington street.
>
> August 28, 1861. 78—tf.

At the beginning of the Civil War, local merchants stocked supplies of North Carolina coal in anticipation of the dwindling amounts caused by the war effort.

5. The Deep River Coalfield at the Time of the Civil War

believed that the coal and iron deposits were integral to the military effort.[11] His focus was not coal entirely as he advanced efforts to mining the iron ore deposits in juxtaposition to the coal beds. The shaft at Egypt, with a depth of 464 feet, he observed, contained seams of alternating layers of coal and black band iron ore, and that mining efforts could bring up both simultaneously. In keeping with the reports of others before him, such as Emmons, Manning pointed out that for lack of thirty miles of railroad, the area could be mined and its coal and iron rendered usable for the Confederate war effort, including the manufacture of agricultural implements. Looking to transportation as an important factor in the prosecution of the war, Manning believed that the railroad held the most promise for supplying iron and promoting development in the region, noting that the Confederate government had awarded two large contracts with companies for the purpose of manufacturing shell, shot, and pig iron. Manning's pride for his state, particularly in regard to coal and iron production made him one in a long line of advocates pushing for exploitation of these natural resources and like the many before him he recognized the need for a ready outlet to ship coal, iron, and manufactured goods to market.[12]

The first real foray into constructing a rail line to the Deep River coalfield came when the General Assembly enacted legislation for the purpose of constructing a railroad line between Deep River at or near the coalfields in Chatham County through the county to the city of Raleigh or some point west of the city of Raleigh and the formation of a corporation with a capital stock of $800,000. An act to charter the Chatham Railroad Company already had been ratified by the General Assembly on February 14, 1855.[13] On February 23, 1861, the General Assembly passed an amendment to change the wording in the original legislation "to connect with the North Carolina Rail Road at Raleigh, or some point west of Raleigh not exceeding twelve miles.[14] An accord was signed on November 25, 1862, constituting the articles of agreement between the North Carolina Railroad Company and the Chatham Railroad Company "in consideration of mutual benefit and advantage to be derived therefrom."[15] The Chatham Railroad Company agreed to construct at Cary, "a commodious depot, sufficiently large for the deposit of coal and iron which may be brought on the said Chatham Railroad, to be carried thence forward on the North Carolina Railroad, and that said depot shall be located conveniently for such transportation."[16] The agreement also provided for the installation of switches in which the loaded cars could pass from the Chatham Railroad "to and upon and along the North Carolina Railroad."[17]

On February 15, 1861, legislation for the creation of a second iteration of the Chatham Railroad Company passed "for the purpose of building and constructing a railroad ... from the coalfields in the county of Chatham through said county to Raleigh, or some point west of Raleigh not exceeding twelve miles."[18] On January 30, 1862, the General Assembly enacted an amendment to the previous legislation to read "to connect with the North Carolina Railroad at Raleigh, or some point west of Raleigh not exceeding twelve miles."[19] In October, the *Weekly Standard* announced that the directors of the Chatham Railroad intended to locate their railroad so that it reached the Deep River coalfields.[20] The newspaper stated that Messrs. Washington, Hoyt, and Murdoch were manufacturing iron at their furnaces on Deep River and that a large quantity of coal had been mined and lying at the mouth of the pits ready to be transported to its final destination. There was urgency for the Chatham Railroad to complete its connection in order that the coal and iron reached the markets of the Confederacy. As was pointed out previously, the expense of hauling one ton of coal by wagon from Lockville to Raleigh was a resounding eleven dollars and fifty cents. In Raleigh, a ton of coal sold for eighteen dollars whereas if the road were completed it would sell for between five and six dollars without the crippling carriage charge. However, grading for the railroad came to a halt and the line was not completed until after the Civil War, leaving the Western Railroad poised as the best option to complete its line to Egypt.

Kemp Plummer Battle (1831–1919), lawyer, businessman, and educator, served as state treasurer and later president of the University of North Carolina at Chapel Hill. During the Civil War, as president of the Chatham Railroad Company, he was in favor of the railroad transporting coal from the Deep River mines to Confederate munitions factories.[21] Noting that authorities of the Raleigh and Gaston Railroad, as well as some citizens of Raleigh, were looking to the railroad as a contingency plan when Union troops might be in a position to disrupt railroad transportation, Battle purchased stock in the company and advocated in the state convention financial assistance for the enterprise. An ordinance passed by the convention authorized the railroad to subscribe for $200,000 stock in the Chatham Railroad Company, and the City of Raleigh, $50,000. Battle was elected president and secured the services of Col. Elwood Morris, who was instrumental in building the Baltimore and Ohio Railroad, to oversee the project. The charter required the terminus of the road to be Raleigh and the coalfields of Chatham County. The Cape Fear and Deep River Navigation Company had already put in place a water route from Lockville to the coalfields. Battle was tasked to locate and grade the road from Raleigh to Lockville, a mile above the junction of the Haw and Deep rivers. Morris believed that he could secure a line from Walnut Creek into Raleigh shorter and less expensive than the one that ran by Cary, but if the North Carolina Railroad Company would allow the Chatham Railroad Company to make improvements along the right-of-way then that route would be best, resulting in expedient transport of coal and iron for use at company shops and foundries. However, Confederate authorities took little interest in the project and as a result the plan for the necessary grading languished and was not completed in a timely manner.[22]

One individual who foresaw the benefits of Deep River coal in support of the Confederate cause was Charles Beatty Mallett (1816–1872), merchant and businessman. Born in Eutaw, North Carolina, near Fayetteville, Mallett attended Kenyon College in Ohio, where he studied science and engineering. After graduation, he returned to North Carolina, involving himself in the family's textile business and ownership

5. The Deep River Coalfield at the Time of the Civil War

of the Mallett Cotton Mill. Later, his interest in and contributions to the railroad industry led to his presidency of the Western Railroad Company between 1855 and 1865.[23] During his tenure as president of the railroad, particularly in its formative years, Mallett saw through legislation in February 1861 to extend the line to the Deep River coalfield, which called for two divisions of the railroad: an eastern division to extend from the Cape Fear to the west bank of Deep River near Egypt in Chatham County and a western division to extend from the west bank of Deep River to a point on the North Carolina Railroad to be determined by the company's president and directors.[24] By June 1861, a connection was completed to Evander McIver's Depot at Big Buffalo Creek, which represented the terminus of the Western Railroad. The Fayetteville and Coalfields Railroad was running within three miles of the Egypt property but would not be completed until May or June 1862; meanwhile, some fifty to sixty tons of coal per day were laboriously carted over three miles of dirt roads to the railroad, which transported the coal to Fayetteville and then loaded on to barges bound for Wilmington.[25] On May 10, 1862, the state legislature passed an act for the Western Railroad Company to extend and complete its road from the coalfields to the North Carolina Railroad with $100,000 in continued state funding contingent on the railroad completing the line in ten-mile increments.[26] In early 1863, W.T. Horne, of Chatham County, advertised the sale of his land at Hornesville, near Farmersville, an area whose coal deposits Ebenezer Emmons called "of a decided character."[27] In early May, the *Fayetteville Weekly Observer* noted that the Western Railroad had cars running within one-half mile of the Egypt shaft, cutting out nearly three miles of poorly maintained road.[28] Beginning May 27, freight could be received and discharged at Egypt station, and the station known as McIver's Depot was discontinued.[29]

Commensurate with his attention to the Western Railroad, Mallett partnered with James Browne, a coal merchant from Charleston, South Carolina, to enter into a business arrangement with the Confederate government for the purpose of supplying coal, iron, nails, and other materials critical to the South's war effort. Coal and other freight were to be shipped by rail via the Western Railroad from Egypt to Fayetteville and then by barge or steamer to Wilmington, where it would be used to fuel locomotive and schooner boilers or for further distribution outside the area. Mallett and Browne took into account coal for domestic use as previous outlets for meeting this demand were either shut down or diverted to needed areas within the Confederacy. The company's first recorded sale during the war was June 13, 1861, when it shipped coal to the Deep River Coal and Iron Company in the amount of $395.66.[30] In August, Browne shipped 30,000 tons of coal from the Egypt mine to Charleston and Columbia. A specimen of about fifteen to twenty tons arrived by train in which Browne exclaimed that it was the "most beautiful coal he ever saw."[31] Together with coal from the Richmond basin and the Chattanooga and Alabama fields, North Carolina coal was doing its part to fuel Confederate industries and steam transportation.

In March 1864, the Mallett and Browne Coal Company of Deep River shipped from their mine at Egypt thirty-eight tons of coal valued at $1,570 to the Nitre and Mining Bureau, which, because of a need for its own operations, required additional shipments for use by the navy.[32] On March 24, 1864, Mallett and Browne, in turn, shipped forty-seven tons to the War Department.[33] The company's contribution to the Confederate war effort was significant, as it supplied coal to industries as diverse as munitions factories, railroads,

Egypt Coal Mine.

THE undersigned were, at the November Term of the Confederate Court, District of North Carolina, appointed Managers of the Egypt Coal Mine property, and have entered into copartnership for the purpose of mining and selling Coal, and solicit orders for the same in any desired quantity. Orders for any amount can be supplied on short notice. The Coal from this property is undoubtedly the best in the Confederate States. Applications may be made to Chas B Mallett, Fayetteville, N C, or James Browne, Charleston, S. C.

CHARLES B. MALLETT
JAMES BROWNE.

Fayetteville, Jan'y 20, 1863.   96tf

In addition to supplying coal to the Confederacy, Mallett and Browne solicited customers to purchase coal for domestic use as seen in this 1863 advertisement that appeared in the *Fayetteville Semi-Weekly Observer*.

paper mills, and blockade runners. In July 1864, Mallet & Browne shipped eighty-five tons of coal valued at $8,500 to fuel the blockade runner *Florie*, which was owned by the state of Georgia and counted among her investors Gov. Joseph Brown and Col. C.A.L. Lamar. She made several successful runs to Wilmington before refugees reported her destroyed September 10, 1864, after running ashore in Wilmington harbor.[34] Seven tons of coals valued at seven-hundred dollars were shipped to meet the needs of the Bath Paper Mill, located at Bath, Aiken County, South Carolina, which manufactured paper for use in Confederate currency, ledgers, tablets, and receipts.[35] Totals of sales of coal to the Nashville & Chattanooga Rail Road, the East Tennessee and Georgia Rail Road, and the Georgia Rail Road attested to the importance and reliance on North Carolina coal these railroads came to expect in the throes of war.[36] Indeed, Mallett and Browne's sales of coal for July 1864 reached a notable $32,990.[37]

The Egypt mine was worked regularly during this time. Coal was loaded onto barges and taken down Deep River, which was navigable by locks, to Wilmington to supply blockade runners. However, on November 10, 1863, an explosion of fire-damp, or methane gas, at the mine resulted in the deaths of six men, five of whom were conscripts assigned to work in the mines to avoid military service. Initial reports of the explosion stated that firedamp was the cause but one worker removed from the mine stated that no firedamp was present and that blasting powder most likely was the source of the explosion. One of the consequences of the explosion and wreckage of the mine workings was that the citizenry of the area were now cut off from obtaining coal for heat and other domestic purposes.[38] Of even greater consequence to the war effort was that the mine supplied all foundries in the state and as a result of the damage done to the mine other sources of coal had to be located.[39]

Mining operations at Egypt eventually resumed when the Confederate Court, during its November 1863 session, appointed Charles B. Mallett and James Browne

5. The Deep River Coalfield at the Time of the Civil War

as managers of the Egypt coal mine property. Advertisements in local newspapers solicited orders for coal in any quantity but in actuality, production levels were not as high as seen before the explosion.[40] Deep River coal at Egypt furnished the Fayetteville arsenal with coke and foundry iron, two critical elements for the manufacture of arms.[41] Additionally, the Confederate government relied on this coal for its own foundries and workshops. In February 1865, when Federal troops were poised on the outskirts of Fayetteville to attack the city, Lt. Col. F.L. Childs, in a dispatch to Brig. Gen. Josiah Gorgas, pleaded with the general of the importance of the city having cotton factories, machinery, naval ordnance, and coal and iron "of Deep River country."[42] On reaching Fayetteville in March 1865, Sherman assigned the First Regiment of Michigan Engineers to dismantle and demolish the Confederate arsenal there. However, before he could do so, Confederate authorities had shipped out by rail some of the arsenal's machinery and stored it in the Egypt mine, where it was hidden from view of Union troops.[43]

In January 1863, the *Fayetteville Weekly Observer* contained an article originally appearing in the *Raleigh Standard* concerning a letter dated December 12, 1862, addressed to Kemp Battle from Capt. Thomas R. Sharp, Superintendent of the Confederate States Locomotive Shop in Raleigh, attesting to the superior quality of Deep River coal. Captain Sharp's statement was further corroborated by the superintendents of the Raleigh and Gaston Railroad Shops in Raleigh and of the N.C. Railroad Shops at Company Shops, a community near present day Burlington. In his letter to Battle, Sharp stated "In answer to your inquiry as to how the coal furnished by the Egypt pits compares with other coal ... the Deep River coal is far superior to any that I have ever used before." He continued to praise the coal for its blacksmithing qualities and affirmed it was more valuable than the Clover Hill, Virginia, coal previously used for that same objective. For foundry purposes as a coking coal it succeeded very well in that application, according to Sharp.[44]

Refueling Confederate steam ships often meant short, clandestine trips to areas where coal could be quickly transferred to awaiting vessels. Thus, it was important for the Confederate Navy to have safe, dependable places for fueling. The Confederacy's challenge of finding and shipping coal also had a great deal to do with the Union blockade of Southern ports which was part of the Union's "Anaconda Plan" to squeeze and isolate the South from its suppliers. As Union efforts frustrated Confederate attempts to reach ports safely, the Confederate Navy found sources of fuel wherever it could, mainly Virginia and North Carolina. However, North Carolina's mostly bituminous coal had mixed reviews, aside from Captain Sharp's endorsement, that although it could be mined and put to use immediately in forges and steam engines, its qualities, for some uses, did not compare favorably to those of anthracite, Welsh steam coal, or even Virginia coal. In one instance, involving the Confederate blockade runner the *Advance*, its chief engineer, James Maglenn, related how the ship was ultimately captured:

> We left Wilmington about 9 September 1864. Trying to escape the attention of Union vessels, smoke, sparks, and flame from the stack had to be kept down. This was difficult to do as we had

used our last shovel full of good coal [Welsh steam]. Our black smoke was giving us away. Some of the Union fleet were following it and began to chase us. We were using Chatham, or Egypt coal, which was very inferior, in fact nothing but slate or the croppings of the mine.[45]

Such incidents often meant that vessels hugged the coastline, leaving them open to capture or sinking. Confederate blockade runners would sometimes take on superior Welsh steam coal when docked at English ports for supplies required to sustain the war effort and meet demand on the home front. However, the British exercised caution when supplying coal to the Southern cause as doing so could be construed by the Union as a hostile action or infringement of neutrality laws, posing dire consequences for the Union's war effort to suppress the rebellion. The case of the U.S.S. *San Jacinto*'s seizure of the R.M.S. *Trent* in 1861, resulting in the removal of two Confederate diplomats bound for England and France to solicit diplomatic recognition and military aid, had already led to strained relations between the two countries, and was still resonating in the minds of many people.

The use of North Carolina–produced coal for the Confederate war effort has been a controversial topic. The strong reaction over its use, which erupted in late 1864 when North Carolina Governor Zebulon B. Vance alleged that coal from the C.S.S. *Advance* had been appropriated for use by the navy department to fuel the C.S.S. *Tallahassee*, leaving the *Advance* with inferior quantities of North Carolina coal, resulted in her capture by the blockader U.S.S. *Santiago de Cuba*. Answering Vance's accusation in a letter to Flag Officer R.F. Pinkney, commander of naval forces at Wilmington, J.A. Willard, Naval Coal Agent, stated that based on evidence presented to him by the Messrs. Power, Law & Co., the agents and part-owners of the *Advance*, no coal was impressed by the Confederate States government to supply other vessels, a statement Secretary of the Navy Stephen R. Mallory confirmed and forwarded to President Jefferson Davis. From that point forward it was recommended that ships using North Carolina coal do so at night in order to avoid detection by the enemy's navy as a result of heavy smoke emissions.[46] Any anthracite coal captured or impressed by the Confederate navy had limitations in its use in steam-generating boilers. Lacking the technology to adapt its steamships to anthracite, the Confederate navy found itself confronting situations as a dispatch dated July 2, 1864, revealed when a Capt. Morris, C.S. Navy, commanding the *C.S.S. Florida*, reported that in reference to the anthracite coal he captured as part of the cargo on board the Union bark *Greenland*, "it was of no use for our furnace."[47] The navy continued to rely on the Virginia and North Carolina coals supplied by the Nitre and Mining Bureau as well as any other amounts it could muster. Although Deep River coal experienced mixed reviews, it did provide an adequate solution as a needed fuel source but never on par with the Union's vast stockpiles of far superior anthracite. As the war dragged on and precious supplies of coal for the Confederacy began to noticeably dwindle, it became imperative that vessels, railroads, and industries keep operating on any coal available to meet military objectives.

One entrepreneur and champion of his adopted state was Jonathan McGee Heck,

5. The Deep River Coalfield at the Time of the Civil War

C.S.S. *Advance* after her capture on September 10, 1864, as a Confederate blockade runner by USS *Santiago de Cuba* (James Barnes, *The Photographic History of the Civil War in Ten Volumes; Volume Six, the Navies*, New York: Review of Reviews Co., 1911).

a Virginian by birth, and an early proponent of the coal and iron resources at Deep River. Early in the war, Heck, had been commissioned a colonel of the Thirty-first Regiment of Virginia Volunteers. An astute entrepreneur in securing materiel for the Confederacy to manufacture into war implements, he received authority from the Confederate Adjutant and Inspector General's Office for his firm Heck, Brodie & Co. to provide employment for sixty Union prisoners of war located at Salisbury prison in North Carolina and Danville prison in Virginia[48] The Raleigh Bayonet Factory's close proximity to the iron and coal regions along Deep River provided much-needed resources for manufacturing these essential implements. The additional "enemy manpower" was critical in keeping the factory open and making up for a declining workforce when Southern men left their jobs to join the Confederate military or were conscripted into service. As part of the arrangement, prisoners were expected to sign statements to the effect that they volunteered to work for Heck, Brodie & Co. and not cajoled into serving by Confederate authorities.[49] Some Union captives found this type of work more agreeable than sitting out the war in a prison often rife with disease and hearing the desperate cries of the wounded and sick for whom they could do so very little.

Capt. William L. Brodie and Lieut. Reese W. Butler, who were already commanding Company A, City Battalion, North Carolina Reserves, in December 1864 organized with Heck the Gorgas Mining & Manufacturing Company at Gorgas, North Carolina, for the purpose of mining coal and producing metal work of all types but mostly for the war effort.[50] The firm was to operate at Gorgas, once Nathaniel Clegg's mills, four miles above Lockville. A call for one-hundred African-American laborers, twenty-five carpenters, and fifteen blacksmiths to work at their factories soon followed.[51] The company remained in business after the war, appearing in the 1866-'67 edition of *Branson and Farrar's North Carolina Business Directory*, with J.M. Heck, president, until the company and its contents were sold at auction. In April 1862, the firm Mendenhall, Jones, and Gardner purchased and retooled a grist mill on the east bank of Deep River for the production of rifles and rifle barrels for the Confederacy but relied on water power instead of coal in the manufacturing process. The abundant water power supplied on Deep River illustrated the importance of proximity to resources in consideration of the location of a factory.[52]

When it became more evident in the early spring of 1862 that rapidly advancing Federal troops were poised to take Norfolk, Virginia, the government relocated to Charlotte all ordnance and other stores that could be saved from the Gosport Navy Yard at Portsmouth. Although situated considerably inland, the Charlotte Navy Yard had access to a rail transportation network that provided ingress to seaboard cities and its interior location helped protect it from additional Union incursions along the coast.[53] By May, machines for the manufacture of heavy ordnance guns, projectiles, and other equipment critical to the South's war effort were in place, including a steam hammer rescued from the Pensacola Navy Yard which could forge the heaviest shafting used on shipboard or the largest frigate's anchor. Two coke ovens on the premises supplied the yard with coal coke, which smelted longer than ordinary coal and was more suitable for firing blast furnaces in the manufacture of pig iron.[54] The navy yard and Charleston came to rely on meager supplies of coal trickling in from Richmond; however, to help compensate for the loss, the Nitre and Mining Bureau increased to 290 tons monthly the output of the Egypt mines allotted for the navy, to be delivered at Fayetteville to meet demand within the state, including steamers and workshops. The increasingly difficult task in obtaining transportation by river to Wilmington then by railroad to Charlotte and Charleston made deliveries uncertain at best particularly late in the war.[55] The capture of Wilmington in February 1865, signaling the Union's occupation of all of North Carolina's main ports, resulted in the continuing loss of supplies from abroad critical to the Confederacy's survival both on the front lines and on the home front.

One of North Carolina's most important manufacturing facilities during the war was the Endor Iron Works located at Lockville near Egypt on Deep River in current Lee County,[56] built by John and Donald McRae and John W.R. Dix, all from Wilmington. Realizing a need to produce pig iron on an industrial scale for the Confederate war effort, the McRae brothers were attracted to the area by the close proximity of the coal and iron-ore deposits needed for manufacturing weaponry, rails, ship parts, and other equipment deemed essential.[57] In 1862, the company built a blast furnace for the purpose of manufacturing iron castings, pig iron, railroad car wheels, and iron plate, among other critical components for weapons and accessories needed to

5. The Deep River Coalfield at the Time of the Civil War

sustain the South's war effort.[58] The furnace required a reliable fuel source and company management entered into a business relationship with local coal suppliers. An entry in Endor's business ledger for May 22, 1864, showed that the company paid the Farmville Coal Company $422 for 500 bushels of coke and 700 bushels of coal, and in July another $1,150 for coal.[59] To supplement dwindling coal supplies, the Endor furnace continued to use wood for making charcoal as a fuel source for its smelting furnace and on June 9, 1864, purchased 392 cords of wood from local supplier Thomas Smith.[60] From January 1 to September 27, 1864, the Endor facility spent $11,541 on coal for its iron-manufacturing operations.[61] After the conclusion of the war, the firm Clegg, Downer & Co. purchased the Endor Iron Works, located at the terminus of the Western and Coal field Railroad, consisting of a blast furnace, rolling mill, and foundry. An advertisement in local newspapers of the time solicited business orders of all kinds specializing in wrought iron, pig iron, and mill work.[62] The Lockville Mining and Manufacturing Company in June 1866 assumed ownership of the Endor Iron Works, with John A. Smith to work the foundry on shares and the company to furnish wood and iron.[63]

6

Post–Civil War Considerations

After the end of hostilities, the financial strain on the South was considerable as state treasuries found themselves bankrupt or encumbered with substantial debt. State governments had to look at ways to shore up damaged and destroyed infrastructure while finding the means to pay for it. When the public treasurer's report appeared in January 1866, North Carolina state treasurer Kemp P. Battle minced no words when it came to the sizeable debt North Carolina had incurred during the war, noting that "accumulated investments have been swept away." In addition to the need for repairing infrastructure, states had to confront the restoration of transportation outlets, especially the miles of destroyed and neglected railroad track necessary for moving people and freight to their destinations. Not to be deterred by the financial encumbrance to taxpayers, Battle spoke of the state as "a country of very great resources," noting the "fertile soil, genial climate … and minerals and metals" all contributing to the state's destiny as one of the "great centres [sic] of civilization and trade."[1] The state's ante-war debt, including interest, by October 1 had reached $13,033,000 (approximately $188,000,000 by 2015 inflation-adjusted standards). Battle's involvement in the railroad and local industries before the war made him an especially apt spokesman for pointing out the benefits of the state and encouraging businesses to invest heavily not only for their benefit but the state's as well.

Fortunately for North Carolina a means to mending its shattered economy came from the unlikely aid of its former enemy. With the help of Northern speculators, the state began to realize that a healthy return on investment was in the offing when advertising the merits of their own resources, appealing to interested parties near and far who were looking to make quick money and have a stake in the state's newly found identity. No areas of the country were off limits, especially the South, which lay claim to being rehabilitated and reconstructed, and, most importantly, again open for business. Indeed, as early as July 1865, the newly organized North Carolina Land and Real Estate Agency was "formed in Raleigh for the purpose of buying and selling the surplus lands in North Carolina" with special attention to developing the agricultural and mineral wealth of the state.[2] With an eye toward the improvement of their state and region, erstwhile promoters Kemp P. Battle and Jonathan M. Heck enthusiastically welcomed "immigrants" who would invest in land and cash in on its resources. The Washington correspondent of the *New York Times* noted that the "prosperity of the state can only be restored by fostering emigration, and by such a

6. Post–Civil War Considerations

hearty submission to the Union and the laws as will assure capitalists that their investments are safe beyond a doubt."[3] Battle and Heck's aggressive advertising campaign featuring numerous tracts of land for sale offered something for everyone who would only invest their "energy, skill, and industry," together with commitment.[4] State legislatures began to charter new companies for the purpose of extolling their area's wealth and in the case of the North Carolina Land Agency, providing opportunities to both southern and northern investors. As bellicosity and states' rights gave way to cooperation and the promise of universal prosperity between the two sections of the country, investing in a new South, a resurrected South of untapped wealth and potential, seemed a start in the right direction for all willing to take a chance.

In May 1866, the *Fayetteville News* printed an advertisement announcing that the Egypt Mine had reopened. After reducing its operations in 1864 because of the war, a new owner of the mine, Robert Paton, or Payton, was announced. The advertisement claimed that the mine produced a superior grade of bituminous suitable for grates and blacksmith's use. Here, coal was considered for its suitability in both domestic and indus-

---

[No. 33.]

1000 ACRES OF LAND ON DEEP RIVER FOR SALE!

IN THE VICINITY OF THE GREAT COAL FIELDS—NEAR THE CELEBRATED EGYPT PROPERTY!

WE ARE AUTHORIZED TO SELL, AT reasonable rates, a most valuable Tract of Land on Deep River, that wonderful region of Mines and water power, containing about

ONE THOUSAND ACRES!

This Tract is situated on the South side of Deep River, and bounded *entirely* on the North by that stream. It is near the celebrated

EGYPT PROPERTY,

only a small body of Land separating.

It has never been explored for Coal, but its contiguity to the aforesaid property renders certain the existence of that mineral in great abundance, and of Iron also. Indeed, the present owner has a letter from Prof. Emmons, in which he states that, in his opinion, *there is no doubt but that it contains both!*

Of this Tract from

150 TO 200 ACRES ARE CLEARED,

and the remainder WELL TIMBERED. The greater part of the Land is fine Tobacco Land. the growth being Oak, Dogwood and principally Pine. Superior Tobacco has already been raised upon part of it, and the entire soil is pronounced, by experienced Planters, to be the best of soils for that Plant.

The proximity of the Tract to the River gives it the most eminent advantages, whether reference be had to the cultivation of Tobacco or the almost certain development of

COAL.

The River is NAVIGABLE from a short distance above to Locksville. And the property is also near the line of the proposed extension of the Chatham and Cheraw and Coal Fields Railroads.

For further particulars enquire of the

NORTH CAROLINA LAND AGENCY.

Raleigh, July 8, 1865.   1—tf.

Soon after the end of hostilities between North and South companies such as the North Carolina Land Agency were poised to sell prime real estate to speculators who sought high returns on their investments. Inducements on a grand scale claimed that 1,000 acres of land adjacent to the Egypt property would yield 9,800,000 tons of coal (*North Carolina Advertiser*, July 29, 1865).

trial uses.[5] However, the optimistic news of the commencement of mining at Egypt was short lived, as an article appearing in October in the *Raleigh Weekly Sentinel* reported news of the mine's closing and discontinued operations because of the high freight rates imposed by the Western Railroad, representing another setback for postwar coal operations at Deep River.[6] Not to be outdone, a brief letter from the president of the Egypt Coal Mines to the president of the Western Railroad, appearing in the November 8, 1866, issue of the *Wilmington Journal*, denied the closing. Indeed, *Branson's North Carolina Business Directory* listed the Egypt Mining Company, with Robert Payton as engineer, in its 1867–68 edition.[7]

In May 1867 an article appearing in *The Wilmington Daily Dispatch* as an open letter from state geologist W.C. Kerr to Governor Jonathan Worth concerned the overall disrepair of the infrastructure of the Deep River coal region.[8] Having spent a week touring the Deep River area, Kerr reported that "stagnation of business and suspension of work" had led to waste, apathy, and decay, and that extensive works planned, and in some instances implemented, at the Gulf, Egypt, Endor, and Ore Hill were now abandoned.[9] Conditions especially significant for Kerr's attention included the absence of coal mining and iron making, pits full of water, extinguished furnaces, dilapidated buildings, disrepair of the dams and locks on Deep River, and the unfinished state of the Chatham Railroad. Kerr next reiterated the significance of the natural resources in the area, and that making a concerted effort to restore transportation links and completing existing ones would make those resources more accessible to enterprises such as manufacturing.[10] He observed that improvements in farming could be realized by tapping existing lime and phosphorous deposits as soil conditioners. Discovery of copper and gold deposits that had received only scant attention, along with Kerr's recommendation that the area be surveyed and studied, could lead to an extensive area having great value in its mineral content.[11]

Newspapers and other publications beckoned speculators and potential property owners to develop tracts of land in North Carolina. On July 8, 1868, one such newspaper, *The North Carolina Advertiser*, published under the auspices of the North Carolina Land Agency, advertised land of all types for sale in North Carolina on reasonable terms. Pointing out that land ownership and traditional farming methods were now undergoing a significant transformation since the end of the war, the weekly newspaper encouraged local property owners to sell off or lease portions of their land to attract speculators from outside the region. This "tenantry system" would allow land owners to keep their holdings while delegating maintenance and farming to lessees, who employed their own workers and paid the property owner a stipulated rent for the use of the land. The agency sought to encourage the purchase of properties based on their abundance of natural resources and close proximity to transportation centers. No longer was coal the only resource touted at Deep River, but timber, domestic crops, cash crops, and water power also were given their due. So enthusiastic was the newspaper's claims that it printed 10,000 copies for general distribution. Among the properties listed for sale was land on Deep River adjacent to the Egypt coal-mining operations. As seen in the company's prospectus, a healthy infusion of new money would go far to develop the area and benefit everyone associated with such an enterprise.[12]

The North Carolina Land Company, incorporated by the legislature on February 8, 1869, to develop the resources of the state, had the authority "to buy and sell, lease, mortgage, or otherwise to convey land ... [and] to engage in any species of agricultural or manufacturing enterprise ... [and] own and manage steam or other vessels"[13] for transporting freight into the state as well as carrying out any other purposes associated with the company. With offices in Raleigh and New York, the company's board in April 1864 published its *A Guide to Capitalist and Emigrants*, which beckoned to the agriculturalist, the vintner and fruit grower, as well as the miner and manufacturer "to invest their funds in this State,"[14] Relying on the expertise of gentlemen familiar with the state, the guide provided a sufficient "amount of information precisely suited to the wishes of all classes of persons who desired to seek a home, in one of the most highly favored portions of the earth."[15] With George Little as president and K.W. Best as secretary, the company consisted of "gentlemen of the States of New York and North Carolina" whose mission was to encourage relocation of Northern and European immigrants to the area, offering inducements to stay and invest. The overall message conveyed that there was something for everyone to be had if they only made the commitment to come and put down roots as many had already done and continuing to do.[16]

The authors of *A Guide to Capitalists and Emigrants* noted that in the middle or central belt of the state "minerals of great variety and value were found in this section of North Carolina."[17] Though every county in the state was considered in terms of its agricultural and industrial potential, the coalfields of Chatham County, in particular, were described as very extensive and valuable, and of good quality. Located near the coal deposits were iron mines "of great but undeveloped value" that added extra regard for the area. The completion of the Chatham Railroad needed only to be completed to link to the other roads at Raleigh "thence to the Ocean by the harbors of Norfolk and Beaufort and ports of Wilmington and New Berne [sic]."[18] Noting the state's progress in terms of completing a number of internal improvements, the authors offered assurances to investors and speculators that whatever concerns they may have would be addressed and a great return on investment realized. As Governor W.W. Holden stated in a letter to the North Carolina Land Company, "No State on the American continent can present greater inducements to immigrants than North Carolina."[19]

Entrepreneurs soon began to place advertisements in local North Carolina newspapers continuing to praise the value of the natural resources the state still possessed at Deep River, namely the coal and iron ore deposits. The old adage "energy and enterprise" represented a new sort of business attitude aimed at greater efficiency and purpose. Building projects, such as the repair and construction of railroads, required skilled labor and well-informed business decisions, such as railroad companies' banding together to accomplish the same goal, or the organization of "the right men," to manufacture iron, while providing employment for the displaced individuals whose mechanical proficiencies were needed to meet demand. Railroads were critical for the state to be competitive with other markets and North Carolina's coal and iron resources represented a significant advantage for achieving that goal.[20]

Branson's North Carolina Business Directory for 1867–8, showed three businesses concerned with coal mining and iron manufacturing in Chatham County: Egypt Mining Company, Egypt, Robert Payton, engineer; Foundry and Machine Shop, Lockville, Silas Burns; and Iron Manufacturing Company, Egypt, Clegg, Dennis & Co.[21] A meeting of the stockholders for the Egypt Mining Company was called January 5, 1869, for the purpose of electing directors for the current year.[22] In early 1870, Raleigh's *Daily Standard* announced that the coal mines at Egypt were to be worked again on a large scale. John Atkins, of Philadelphia, arrived to take charge of the operations, which had been rendered into a dilapidated condition since the end of the Civil War.[23] The same newspaper announced on June 3 that the Egypt Company was to be sold by John Kessler for $10,000 at public sale.[24] By 1871, developments at the Egypt mine were receiving more exposure in the newspapers of the day. The *Raleigh*

Publications such as the *Guide to Capitalists and Emigrants*, published in 1869, helped entice speculators and entrepreneurs for the purpose of investing capital in various enterprises in the state. Some states even had their own immigration agent working on behalf of the state.

6. Post–Civil War Considerations

Daily Telegram announced that machinery to pump the standing water from the mine was being erected with the resumption of mining scheduled within two or three months.[25] By March 1872, advertisements were appearing in local newspapers, such as the *Raleigh News*, which announced that coal was being sold at six dollars per ton based on 2,240 pounds. H.H. Potter, coal agent, stated that a ton of coal cost no more than one cord and a half of wood and would go twice as far.[26] The outlook for the coal-producing region at Deep River looked encouraging when Wilmington's *Daily Journal* reported that the Egypt Mine was being worked extensively by the Governor Creek Coal Company, producing 100 tons daily with the ability to double the quantity could ready transportation and customers be found.[27]

Experts were in agreement that the Dan River coalfield contained coal of little or no commercial value, leading to no serious attempts at mining in the region especially with the demise of prospects for a coalfield railroad. However, in 1869, the state legislature passed an act authorizing the Dan River Coalfield Railroad Company to construct and extend their road to Germanton in Stokes County and "thence to some point on the Western North Carolina Rail Road, at or near Statesville."[28] The act also approved construction of branches with the Richmond and Danville Railroad and the North Carolina Rail Road provided that the gauge of the Dan River Coalfield Railroad was the same as that of the North Carolina Railroad.[29] Despite consideration as a possible carrier to and from the Dan River coalfield, the Dan River Coalfield Railroad represented a number of business concerns that never made inroads into what was once regarded as an accessible and profitable property laden with a promising mineral field. However, this field is important from the standpoint of early investigations into the area by noted scientists such as Olmstead, Emmons, McLanahan, Kerr, Hale, Holmes, and Stone. Emmons went so far as to claim that coal deposits found at the Dan River field did not "differ materially from those of Deep River."[30] J.A. Holmes, who at the time was state geologist, in response to R.W. Stone's examination, conducted much later than Emmons's, revealed that the Dan River field possessed similar characteristics with the sedimentary rocks of the Richmond Basin's coalfield in Virginia, and that the Dan River field's stratigraphy was similar to that of the Deep River field.[31] Also the significance of the Dan River field's location in Stokes, Rockingham, and Madison counties near the Virginia border offered prospects for commerce and trade between the two states as well as providing for an outlet to western markets.

Although the Dan River Coalfield Railroad never met the objective for which it was chartered, in 1868 a new charter was approved in an ordinance of the State Constitutional Convention with the same provision as the first, namely, "for the purpose of constructing a Rail Road from some point on the Virginia line near the town of Danville, in Virginia, to the Coalfields of Dan River."[32]

Taking into account the need for repairing existing infrastructures for most industries in the state, finding the most motivated individuals to accept the risk, and the time required to complete the task, the Egypt Coal Mine was not ready to reopen until machinery was erected to pump out water that had collected in the shaft over the years with the view that mining would commence in two to three months.[33] With the completion of the Howe Truss bridge over the Haw River, the *Wilmington Journal* called attention to the progress being made with the extension of the Chatham Railroad to Lockville, nine miles from Pittsboro, and that a connection to Jonesboro, where it intersects the Western Railroad, was imminent.[34] *The Wilmington Daily Journal* added that the Endor Iron Works near Egypt were being relocated to nearby Lockville and that the company was clearing out the river of obstructions between Buckhorn and the Gulf as well as repairing the roads and networks so as to facilitate transportation of coal from the Gulf and Egypt to Lockville along with iron ore from Buckhill.[35] With commitments such as these beginning to take place, the expectation that coal and iron-ore mining at Deep River would soon resume on a significant scale became more real as well as signifying a source of optimism in a state looking to increase interest from investors and speculators.

One Fayetteville newspaper, *The Eagle*, described in its November 30, 1871, edition an excursion to Fayetteville with a stop at the Egypt coal mining operations on the Chatham and Western railroads. Though the writer of the article remained anonymous, he described visiting the mine workings at Egypt, where the shaft had reached a depth of 500 feet with a production rate of eighty tons per day.[36] The writer noted that the Egypt Coal Company offered to send free of charge coal to municipalities for the purpose of testing its quality in their gas works, with the expectation that they would find the coal superior to all others and decide on Egypt coal as their primary fuel source.

In May 1872, the *Greensboro Patriot* reported that the Governor's Creek Coal Company was currently working the Egypt shaft with a daily output of 100 tons and an expectation of doubling the amount provided that adequate transportation and salability could be obtained.[37] Sales of Egypt coal to gas companies had been favorable and the company was encouraged to expand its customer base to include the larger markets outside the area with the help of the railroads. Coal sales to these markets meant that North Carolina coal would be competing with the best coals already established in the East and Midwest and that the quality of the coal would speak for itself. In late 1872, an announcement in Raleigh's *Daily News* confirmed that the Egypt coal property had been reopened and was being worked by Alexander J. Derbyshire, of Philadelphia; however, records are sparse regarding developments at the Egypt mine under Derbyshire's ownership.[38] We do know that Derbyshire was a successful businessman involved in railroads before coming to North Carolina. He was once director of the Pennsylvania Central Railroad and the Mine Hill Railroad, and served as president of the Little Schuylkill Railroad Company. He also served on the board of the Pennsylvania Railroad as an officer. Derbyshire and his cousin John established A.J. Derbyshire & Co., a successful flour business in Philadelphia, and the company took interest in the development of the railroads and mining, locally and out of state.[39] Both the Pennsylvania Central and the Mine Hill railways serviced the anthracite coal region in Northeast Pennsylvania and the Mine Hill Railroad eventually became part

6. Post–Civil War Considerations

of the Philadelphia and Reading Railroad Company, which was one of the largest of its type in Pennsylvania and the rest of the country.

The *Raleigh Daily Sentinel*, in October 1874, announced that operations at the Egypt coal mine had been shut down for lack of transportation to more extensive markets outside the area. The construction of the Raleigh & Augusta Air Line Railroad, which was intended to join or cross the Carolina Central Railway, was thought to give the area the necessary transportation links to desirable markets.[40] In December 1875, Wilmington's *Evening Review* reported on the latest developments with the Raleigh & Augusta. Noting that it was an extension of the Raleigh and Gaston Railroad, it would connect Raleigh, via Sanford and the coalfields with the Carolina Central Railway at Sand Hill, approximately four miles east of Rockingham. Construction of the Raleigh & Augusta also would con-

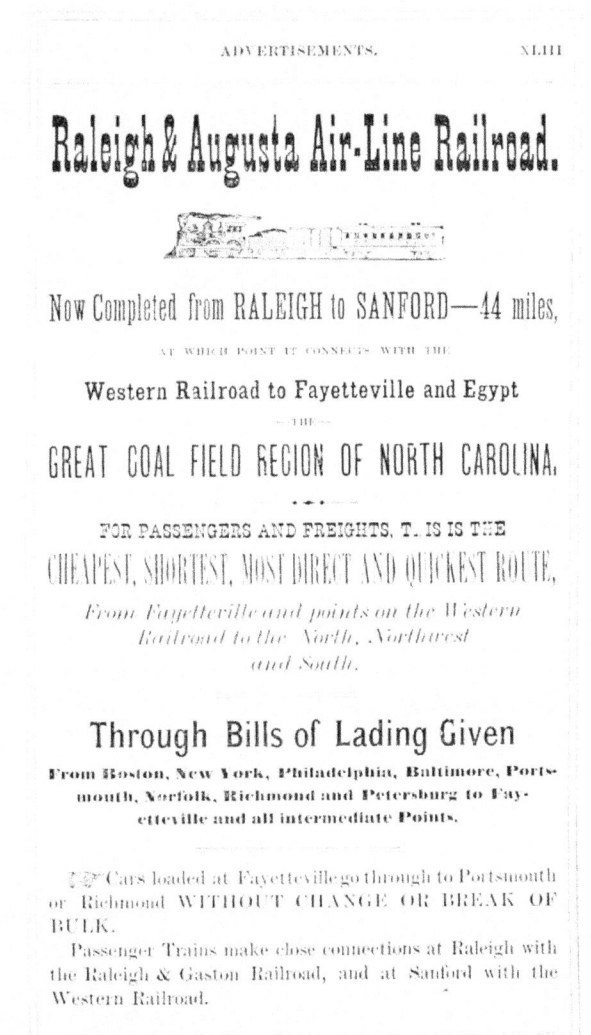

After the war, railroads became a means for those with leisure time to travel for enjoyment and health. Once seen as a major transportation mode for moving goods and commodities, railroads helped create the new industry of tourism. This advertisement for the Raleigh & Augusta Air-Line Railroad appeared in *Branson's North Carolina Business Directory for 1872*. Note the prominent reference to the state's Great Coal Field Region.

nect Wilmington with Fayetteville, the coalfields, and Raleigh all by rail. Pointing out that the area between the Carolina Central and Raleigh was rich in timber and turpentine, the hope was to provide markets for these products as well as rejuvenate interest in mining at Egypt.[41] Still, in 1876, the *Raleigh News*, observing that the Western North Carolina Railroad ran from Fayetteville to Egypt, took issue with the meager attempts at developing the area for transporting coal to markets and opening up the region for enterprise. With the mines in operation, a cheaper fuel source would go far in helping to spur the area's economic growth and encourage new investment in land and industry.[42] With greater access to previously isolated areas, the railroads played a large role in making the vision of growth and prosperity a reality, but the efforts behind this goal had to have real commitment and backing. It was now incumbent for all those involved—railroads, developers, entrepreneurs, and governments—to prioritize their efforts to make North Carolina a contender in a rapidly competitive world.

The anonymous writer of an article appearing in the February 27, 1879, issue of the *Chatham Record* spoke of a visit to the Gulf, particularly the extension of the railroad from Egypt to the Gulf. Noting that the railroad had been completed to Egypt in 1862 and was abruptly halted because of insufficient funds to build a bridge across the river, the writer pointed to the soon-to-be completion of the bridge and the nearly completed laying of track as signs that the railroad was finally coming to the Gulf. Already, construction of stores and buildings had been planned while in some instances, as with enterprising residents such as John M. McIver, storehouses and shops were beginning to see their completion. A post office and warehouse were soon in the works as well.[43] The writer continued that the extension of the railroad would benefit Fayetteville and "carry their large trade from Chatham which now goes to Raleigh and Greensboro."[44] In years prior to the completion of the railroad, trade only went to Fayetteville without a return to Chatham. Now, expectations focused on a wider network of connections to markets outside the area.

In April 1881, the *Raleigh Evening Visitor* reported a story originally appearing in the *Fayetteville Examiner*, noting that Lawrence J. Haughton, who owned valuable mining property at Gulf, in Chatham County, had been obtaining coal from outcroppings on his estate for which he had had demand. To meet orders, he had sunk a shaft near the outcrop and believed that the stratum of coal could reach five or six hundred feet and with sufficient power would raise the needed amount of coal needed to make a profit. He had ordered a steam engine for hoisting purposes and expected to extract twenty tons per day. Haughton claimed that he would be able to furnish coal at $3.50 per ton on the cars and that the cost of transportation, depending on distance, would find it reasonable. If it could be delivered at $4.00 per ton it would be cheaper than wood at one dollar and a half per cord.[45]

The *Chatham Record* reported in September 1881 that Haughton had sold his coal property to a Northern concern for $30,000, and that the new owners proposed to go to work at once. The purchasers, F.P. Dewees, of Pottsville, Pennsylvania, and associates, reportedly spent $10,000 for mining machinery as well as the employment of laborers. The purchase included not only the coal property at Gulf but all of Haughton's property, consisting of 2,700 acres of land. One newspaper touted the recent developments at Gulf, including the newly constructed terminus of the Cape Fear and Yadkin Valley Railroad, as a source of prosperity for the region and state particularly with a knowledgeable individual like Dewees at the helm whose experience in one of the largest regions of coal and iron production in the country made him an ideal operator.[46]

New Geological Surveys

The next generation of geologists interested in analyzing the mineral content and coal deposits at Deep River appeared shortly after the Civil War. The work of eminent scientists such as W.C. Kerr, F.A. Genth, H.M. Chance, and E.G. Tuttle all contributed toward greater knowledge of the state's geology including statements attesting to the wealth of its resources, in many cases expanding on the work of their predecessors. After the destruction of industry and infrastructure wrought by war, there was a movement to identify areas best suited for capital investments and subsequent growth. Speculators throughout the company, especially well-backed Northern interests, were encouraged to relocate their business in the Southern states, especially North Carolina, to take advantage of cheap land along with their mineral content.[47] During this time of post-war economic hardship on the population, it was important for states to realize a steady flow of capital to help rebuild communities immediately after the war in order to be competitive in a new era of imminent prosperity.

Appointed state geologist in 1864 by Governor Zebulon B. Vance, Washington Caruthers Kerr was a native North Carolinian, born in eastern Guilford County to Scotch-Irish parents. Between 1857 and 1862 he taught upper-level courses in chemistry, mineralogy, and geology at Davidson College, in Mecklenburg County. During the Civil War, from 1862 to 1864, he served the Confederacy as a chemist and superintendent of the Mecklenburg Salt Company at Mt. Pleasant, South Carolina, near Charleston. Completing a term of two years as state geologist his position was renewed in 1866 by Governor Jonathan Worth. While at Raleigh, Kerr took interest in the perambulations of earlier geologists, including Denison Olmstead, Elisha Mitchell, and Ebenezer Emmons, whom he believed had not adequately considered the economic value of the resources in the western quarter of the state, where he would spend a year devoted to the study of its agriculture and geology.[48] Indeed, in reporting the progress of the state geological survey resumed by Kerr, in July 1867, the *Raleigh Sentinel*, noted, with unqualified enthusiasm, his attention to the economic geology of western North Carolina. Echoing Emmons and others who preceded Kerr, the paper stated that the key to accessing the "indefinite wealth and prosperity" represented by the state's mineral resources called for the implementation of internal improvements, an old chestnut brought out from time to time as a precursor to mining and transportation efforts.[49]

Later in the year, in an open letter to *The Wilmington Daily Journal*, Kerr responded to critics who voiced their displeasure that his studies in western North Carolina were made at the expense and interests of those residing in the east. Reminding his detractors that Ebenezer Emmons had given most of his attention to the eastern and midland counties, Kerr took the opportunity to point out that he had spent the previous year devoted to the western area of the state which he saw as undervalued and underdeveloped.[50] Though Kerr never took on the role as an apologist for the western counties he did see the potential for enterprise and prosperity in an area to which he believed his predecessors and contemporaries had given only cursory attention. However, Kerr's survey and recommendations lagged behind those made by his more famous predecessors but his continued observations and commentary on western North Carolina did have their merits.

Advertisement for Egypt coal. Entrepreneur J.A. Porter, who dabbled in real estate and livestock transactions, was the agent in charge for sales of Egypt coal in the Raleigh area. Some of his business dealings may have involved the sale of mules to the owners of the Egypt coal mine (*Raleigh News*, October 17, 1872).

By 1870, Kerr's state geological survey had been completed but an unresponsive General Assembly refused to allocate funding for its publication until 1875.[51] Kerr's consideration and assessment of the coal lands of Chatham and Moore counties were based on an area Emmons had given as 300 square miles. Kerr believed that the coal at Deep River could be mined for gas-making and other by-products. Drawing on a testimonial from C.S. Allman, president of the Norfolk Gas Works, Kerr reported that tests using Chatham coal were very favorable in regard to the quantity and quality of gas produced which a given amount of the coal yielded. Kerr noted the lignite bed on the Tar River, in Granville County, which was a continuation of the Chatham coal formation in the opposite direction. Bituminous shales, Kerr pointed out, were plentiful and with the lignite deposits, represented a fuel supply over and above that of coal already exploited.[52] Four years later in 1879, Kerr authored a report based on a convention held in Charlotte in which northern capitalists, giving their personal experiences during their residence in North Carolina, attested to the rich agricultural, mineral, and industrial resources of the state.[53]

Born in Hesse, Germany, in 1820, Frederick Augustus Genth received his education at the University of Heidelberg, completing his Ph.D. at the University of Marburg. He studied chemistry under the eminent chemist Robert Bunsen and later became his assistant before immigrating to the United States in 1848. In a relatively

6. Post–Civil War Considerations

short period of time Genth became an esteemed chemist and mineralogist in his own right who early in his career spent a year as superintendent of the Washington (Silver Hill) Mine in Davidson County, North Carolina. He later taught at the University of Pennsylvania and directed the Second Pennsylvania Geological Survey while there.[54] Genth was a prolific contributor of scientific articles on subjects of chemistry and mineralogy and was the first to discover and characterize many new minerals. In *The Minerals and Mineral Localities of North Carolina*, co-authored with W.C. Kerr, published in 1885 as Chapter I of the second volume of the *Geology of North Carolina*, Genth's aim was not only to categorize the minerals county by county but to provide analyses of minerals of the state in scientific publications as well as studies conducted in his laboratory at the University of Pennsylvania.[55] A "mineralogical map," included in the study, was a useful resource for locating minerals and their distribution throughout North Carolina.

Though Genth's report was not as detailed as those supplied earlier by scientists such as Mitchell and Emmons, it was, nevertheless, a useful tool for mineral identification throughout the state. In his analysis of mineral coal, Genth reported the occurrence of anthracite, or hard, coal in the vein rock at the Clegg Copper Mine in Chatham County. The bituminous coal beds of the Deep and Dan rivers in some areas near trap dikes were deprived of hydrocarbons, "often approaching true anthracite."[56] He confirmed that the greater portion of coal in the Deep River beds was bituminous while the Dan River coal in Stokes County was semi-bituminous. Lignite frequently met with the marl beds of the eastern counties, in Granville County, on the Tar River, and on Brown's Creek in Anson County. In his report, Genth was primarily interested in cataloging and analyzing minerals in the state without as much attention to sampling and market analyses.

One individual, Peter Mallett Hale (1829–1887), older brother of Edward Joseph

Washington Caruthers Kerr (1827–1885). Born in what is now Guilford County, North Carolina, he graduated from the University of North Carolina in 1850. In 1864 he was appointed state geologist by Gov. Zebulon Vance. He worked on the state's geological survey from 1866 to 1882, leading to greater understanding of the geology and climatology of the region (*Journal of the Elisha Mitchell Scientific Society*, Vol. IV, July-Sept 1887).

Hale and editor of the *Fayetteville Observer*, contributed uniquely to the work of geological surveys in the state with his 1883 publication of *In the Coal and Iron Counties of North Carolina*, representing a compilation of the geological reports submitted by Emmons and Kerr as well as Laidley and Wilkes's reports to the war and navy departments, respectively. Additional chapters contained census data and sketches of each county's mineral resources submitted by qualified individuals on their subject. Hale intended to make the state "thoroughly known to its own people" based on the scientific analyses of eminent scientists and the data collected on the state's counties.[57] His role as a promoter of internal improvements and the state's natural resources combined with his approach as a "lay geologist" to offer investors a vast expanse of natural resources—in this case coal and iron—which could be parlayed into successful manufacturing ventures ultimately leading to "large towns, large markets, and great and prosperous states."[58] Although Hale provided the people of North Carolina with useful statistical information, the thrust of his work was to attract investment in the state, representing but one in a number of enterprising individuals whose enthusiastic approach to the industry and commerce of his own state appealed to capitalists and speculators alike.

At its regular quarterly session in Raleigh in April 1884, the State Board of Agriculture took up the cause of the coalfields at Deep River for the purpose of surveying and assessing the value of the natural resources in the region. With the completion of the Cape Fear and Yadkin Valley Railway's connection from Deep River to Greensboro, the board recognized the importance of development even more than previous attempts had undertaken.[59] Employing the services of Dr. Henry Martyn Chance, a geologist of considerable expertise from Pennsylvania, the board, with the sanction of the governor, had chosen one of the most experienced and knowledgeable men in his field.[60] *The Raleigh Register* minced no words when it called for prompt action to investigate and report "the extent, value and quality of the coal deposits" and present to the board a study to determine "the extent, value and quality of the coal deposits of the state, and what would be the probable cost of such a determination."[61] The board voted an appropriation to pay for a survey, undertaken by Dr. Chance, to determine once and for all the need to pursue further inquiries on the subject of the Deep River coalfield. Looking to private investment and capital to develop the area, the paper concluded that until the name of a party "that carries authority" could be ascertained, it would be impossible to direct resources to the project.[62] Acknowledging the potential for mining iron ore especially with its close proximity to the coal deposits, the paper referred to it as the "bed-fellow" of the coal, and that it made financial sense to extract both at the same time.

In his report submitted to Montford McGehee, commissioner of the board of agriculture, Chance summarized his approach to investigating the Deep River coalfield at eight locations: Farmville, Egypt shaft, Taylor Place, Gulf, Evans' Place, Gardner Place, Murchison Place, and the Fooshee Place.[63] In many instances, coal workings had fallen into disrepair, and standing water often meant expensive pumping efforts, resulting in the need to excavate new pits. In the Farmville region, Chance sank a

6. Post–Civil War Considerations

total of twenty-two shafts in the upper and lower beds for the purpose of analyzing the coal deposits. Firing the coal to test its heating ability took place at the chemical laboratory of the North Carolina Agricultural Experiment Station under the direction of Dr. Charles W. Dabney, Jr.[64] Established by the North Carolina Department of Agriculture in 1877, the Experiment Station was first located in a chemistry laboratory at the University of North Carolina. On a visit to the Egypt shaft, a frustrated Chance described the colliery as a total wreck. The shaft had caved in, the head frame had rotted and fallen into the workings, boilers and engines were rusted, and masonry had disintegrated.[65] Given the obstacles, Chance made new openings, which were done by hand on account of not wanting to incur a large expense using horse- and steam-power. Equipped with windlasses and buckets, drills, picks, shovels, saws, axes, a churn drill, and small implements, Chance and his team found the labor-intensive work slow with attention to drilling bore holes and pits for extraction of samples of "good, firm, black coal."[66]

Frederick Augustus Genth (1820–1893). Born in Hesse, Germany, Genth received his Ph.D. at the University of Marburg, where he studied under Robert Bunsen. His cataloging and analysis of minerals provided greater understanding of their locations and composition (George F. Barker, "Memoir of Frederick Augustus Genth, 1820–1893," *Biographical Memoirs of the National Academy of Sciences*, 1902).

Evaluating the Deep River coalfield in relation to other competitive fields, Chance focused on four areas of comparison: commercial value of coal, quality of coal, workable area, and obstacles to successful mining. In determining commercial value, Chance considered the cost of mining and transportation rates to market. Transportation rates, according to Chance, needed to be sufficiently low in order to defray the costs associated with mining the coal, such as labor, machinery, wages, and property rental or ownership. Proximity to the interior markets of North and South Carolina bode well for Deep River coal reaching and supplying these locations but coastal markets on the Atlantic were better serviced by the Richmond, Virginia, field, he concluded. In terms of quality, Chance favorably compared specimens from the upper coal bed at Gulf, Taylor place, Egypt, and Farmville to those mined in the northern and western states. He reasoned that these areas were of a workable thickness and quality. Outside the area he believed that any other workable coal beds were essentially non-existent despite the enthusiasm shown when coal mines were first established, exhibiting what Chance considered areas of inflated values. In the vicinity from Farmville to the Gulf, Chance discovered that the coal was of sufficient quality to command a high price for blacksmith and foundry use.[67]

Chance believed that if the coal were carefully mined it could compete with the other coalfields and find a market as a coal suited for a variety of purposes. Some areas on Deep River, where the lower bed was accessible, such as at Farmville and Gulf, the coal was considered of the strong steam variety as well as for open grates and cylindrical stoves. However, the upper coal bed held the most promise as a coking coal most suitable for the blacksmith's forge but not as a gas coal because of its sulfur content. The task of mining suitable coal at Deep River, Chance noted, did not seem to have many obstacles other than how the coal seams were situated for easier extraction.[68] Chance estimated the length of the outcrop between four and five miles, comprising an area of 2,160 acres. He calculated that the approximate amount of coal to be mined was 8,640,000 tons per acre for the upper bed and 5,400,000 tons per acre for the lower bed, amounts without hindrances taken into account.[69] However, factoring into the areas ruined by trap dikes, faults, and impure coal, Chance believed that the area had been reduced to 1,100 workable acres with the lower bed consisting of 2,000 tons per acre and the upper bed 3,500 tons per acre.[70] His assessment took into consideration that if his calculations were too large at even 3,000,000 tons, this amount could sustain a daily output of 500 tons for twenty years, and at an average price of $3.50 per ton realize over $10,000,000. Chance believed this figure represented the money that could be kept within the state and used as capital instead of being paid to mining concerns outside the area.[71] He recognized savings to the state as part of his agenda for mining coal at Deep River.

Chance identified six obstacles that would need to be overcome for successful mining of coal at Deep River. These included: variations in thickness and quality, faults, trap dikes, explosive gasses, water, spontaneous combustion, and absence of coal from some areas.[72] Variations in thickness and quality of the coal seams were often abrupt and if not overcome could ruin a mining company in this aspect alone. For instance, a coal seam could suddenly shrink to a third or a half its normal size or impurities in the bed could exist in greater proportions than the measure of mineable coal.[73] Chance noted that provisions for recognizing and avoiding these variations needed to be taken into account in advance of actual mining efforts. Failure to do so could cause mining operations to bog down, resulting in significant costs to the company with few advantageous results to show for it. In that regard, it was incumbent upon mining companies to hire knowledgeable experts to assess and sample the areas to be mined.

For the purpose of coal mining, faults, or breaks in the continuity of the coal seam caused by tensional or compressed force in the earth's crust, were identified in the workings at Egypt and Farmville.[74] In April 1884, the North Carolina Board of Agriculture at its quarterly session published its own surveys of the Deep and Dan River coalfields. In mining the coal at Farmville for the 1884 Raleigh State Exposition, Chance noted two local rolls, which signified inequalities in the roof or floor of a mine.[75] However, he believed that rolls and faults known to exist there, other than those caused by trap dikes, would not hinder successful mining efforts. As with considerations for variations in thickness and quality of the coal bed, serious faulting could jeopardize mining efforts because of the expense involved in overcoming them.[76] However, Chance gambled that samples of his coal would test satisfactorily at the exposition which would prove to be the most convincing test of Deep River coal and perhaps the most universally acclaimed among those in attendance.

Chance saved his biggest concern for the trap-dike issue that intersected the coal beds at the Farmville, Egypt, and Gulf workings. As with faults, he recommended a determination of the extent of the trap dikes before any serious mining operations commenced. The best method for accomplishing that, according to Chance, was by a complete system of bore holes made over the area to be worked, as was being done by the owners of the Gulf property. Bore holes provided the most comprehensively accurate means of assessing quality and depth of coal in a particular area. Additionally, bore-hole exploration was becoming standard practice for proving the positions of old workings and faults, and in some instances letting off accumulations of gas or of water both of which were dangerous and expensive to circumvent.[77]

Many coal mines, in addition to those at Deep River, were hazardous on account of the presence of firedamp, a highly combustible, invisible gas formed by the natural decomposition of coal which when accumulated in large quantities could ignite and explode when mixed with the atmospheric air. Chance had already noted a number of explosions of firedamp occurring at the Egypt shaft. Mining larger seams such as at Egypt, where the bed was three feet in thickness, necessitated costly maintenance of large airways to help dissipate and remove the gasses from the workings.[78] Chance's observations into firedamp provided some of the earliest considerations as to how a mine should be vented in order for workers to perform their tasks safely. The fledgling U.S. Bureau of Mines, which was created in 1910, would conduct important investigations into the nature of mine gases at the behest of pioneering scientists such as Chance.

In the event of standing water inside the mine, mechanical pumping of some means was required to discharge the water from the workings. While Chance found that the rocks did not carry sufficient water to be of concern, miners working a room of coal could accidentally break through the wall of an abandoned mine filled with water. Depending on the velocity of the water, men, machinery, and animals could be swept away, resulting in death and destruction of work areas.[79] Chance's assessment that heavy pumping machinery was not necessarily needed could, if proved wrong, mean disaster to a mining company's output and safety of its workers. In some instances, mine owners and operators relied on maps of the workings to avoid known abandoned workings but maps, if they did exist, were sometimes out of date and of no use for current mining efforts. Mine maps also could not account for rotted roof timbering, explosive gases, faulty cages and cables, and insufficient ventilation. Standing water could signal the death knell for a mining operation especially when a significant area of the mine was breached by water.

What Chance referred to as spontaneous combustion in a mine, examples of which had been reported in the Richmond, Virginia, coalfield, was combustion of coal without external ignition usually caused when coal was exposed to oxygen without sufficient ventilation for cooling.[80] Chance's concern was that the similarities between the Deep River field and the Richmond field might lead to events of spontaneous combustion. His advice to employ precautionary measures did not spell out any specific tasks except to anticipate the danger as best as possible. While spontaneous combustion could occur at any given time whether the coal was still in the ground or stored above ground in bunkers or cars, some coals were more susceptible, especially those having a higher content of volatile material and where firedamp was appreciably present.

In surveying the Deep River coalfield, Chance relied on visible outcroppings of coal as a guide to determining the extent of the coal beds at Deep River.[81] In some instances, outcroppings deteriorated over the years due to atmospheric conditions and age. Citing a failure to find outcroppings between Egypt and the Taylor place and between the latter place with the Gulf workings, along with the already proven absence of coal at some places in Farmville, Chance believed it was incumbent upon each company mining an area to define the limits of their coal bed. Fearing that some companies would assume that coal might or might not be found in a certain tract, resulting in unnecessary expenditures, Chance affirmed that completing a series of bore holes short distances apart could best indicate the existence of coal in a particular region.[82] His recommendation that the area in question be "completely riddled with bore holes to determine the local condition (thickness, quality, depth) at every point" was critical before attempting actual mining operations.[83] Chance believed that slopes should be used instead of shafts unless the area was severely faulted or contained a significant amount of trap dikes, which meant drilling deeper and more expensive workings. The use of airways held promise to remove dust and dangerous gasses from the workings, thus negating the possibility of explosions. His advice that ventilation fans should be used instead of furnaces to promote mine ventilation reflected a growing trend in mine efficiency and safety.[84]

In an article appearing in the November 10, 1894, issue of the *Engineering and Mining Journal*, Edgar G. Tuttle, of Pennsylvania, noted both the Deep River and Dan River coalfields as contributors to both the state's industry and commerce but that unsatisfactory business conditions in the last few years concerning the Deep River field had impeded development of the area.[85] However, their favorable location gave them an advantage of three hundred to four hundred miles in haulage as compared with the nearest West Virginia and Tennessee fields "on coal delivered to the principal railroad lines, and to Raleigh, Wilmington, and other points."[86] Deep River coal was being used by gas companies, industrial, and other works at Raleigh, Wilmington, and Fayetteville. Tuttle stressed that an existing network of railroads provided an outlet for coal haulage from Egypt "over the Cape Fear & Yadkin Valley Railroad and the Raleigh & Western Railroad, connecting with the Seaboard Air Line at Sanford

6. Post–Civil War Considerations

and Colon respectively." Further, the Cape Fear Railroad also connected the coalfield with Fayetteville and Wilmington and other important points in the state. These readily available outlets meant that transportation costs could be held down for the coal company without having to transport shipments long distances and compete with other coalfields for favorable rates.[87]

Tuttle reported that production at Egypt had reached approximately 100 tons daily and with the planned improvements to the land and plant completed, production was to reach 500 tons daily. Tuttle also favorably noted the new machinery in place, all at a considerable cost to the owners of the Egypt Coal Company, such as a newly erected headframe, a platform with revolving tipples for dumping mine cars, a Litchfield fast motion engine with a Beach's patent balanced slide valve, and machinery for pumping water out of the workings. Two small underground engines raised the coal from the workings by two slopes to the gangway level leading onto the cages which could accommodate 150 cars daily. An airshaft measuring 8 ft. × 12 ft. had been sunk at a depth of 158 feet for ventilating the mine. Openings in the coal seam at Egypt pointed favorably to "its extent for considerable distance in different directions" while openings in the four-foot seam were made at Farmersville, some two or three miles northeast of Egypt. To the south approximately two miles an opening in eighteen inches of coal was believed to overlay the four-foot seam at a considerable distance. To the southwest, approximately one and one-half to three miles, openings in the four-foot seam were made at Gulf, where a small amount of coal was being used for blacksmithing and other purposes.[88]

REPORT
ON
AN EXPLORATION
OF THE
COALFIELDS OF NORTH CAROLINA,
MADE FOR THE
STATE BOARD OF AGRICULTURE.

BY H. M. CHANCE.

WITH THREE PLATES, AND THIRTEEN CUTS IN THE TEXT

PUBLISHED BY ORDER OF THE BOARD.

RALEIGH:
P. M. HALE, STATE PRINTER AND BINDER.
1885.

Title page of Henry Martyn Chance's study of the North Carolina coalfields. He was involved in producing the Second Geological Survey of Pennsylvania (1884) before coming to the state to investigate the Deep River field. He believed that if Deep River coal were carefully mined, it could compete with better known varieties.

7

The 1880s Through 1900

A showing of North Carolina coal presented itself at the North Carolina State Exposition held at Raleigh from October 1 to October 28, 1884. A wooden tank located in the machinery hall, consisting of 25,000 gallons of water to supply the boilers for working the moving parts of the equipment, relied on furnaces fueled by North Carolina coal. As this was an exhibition to showcase achievements in the state, efforts were made to use only North Carolina coal for the task. Further, on display in the main building, a large pyramid of North Carolina coal from Farmville, Chatham County, spoke directly about the extent of North Carolina's coal resources and provided interested investors the opportunity to observe firsthand the quality of Deep River coal. It was H.M. Chance's predilection for the quality of North Carolina coal which provided such an important venue.[1]

By 1887, plans for the resumption of coal mining at Deep River seemed to be real without the caveats and delays that had defeated previous enthusiasm, at least seeming to offer more promise. This time, however, the focus of attention was Farmville, a small community on Deep River near Egypt.[2] Noting that the Farmville Mine was beginning to be worked on a more extensive scale than since the end of the war, the article also claimed that the quality of the coal at Farmville "is of a better quality than any ever seen in Chatham" and it was this coal that had made such an impression at the 1884 State Exposition in Raleigh.[3] The superintendent, W. H. Segroves, with an address at Egypt, hired a considerable number of miners and workers to extract this first-rate quality of bituminous coal. Soon, advertisements in local newspapers began to appear offering for sale Farmville-mined coal.

In 1888, a quarterly publication entitled *The South, an Immigration Journal, Devoted to the Interests of the Southern States, and North Carolina Particularly*, published by J.A. Harrell in Weldon in Halifax County, did its part to entice investors and speculators to take advantage of the abundance of opportunities to tap into the natural resources of the state. Appearing on the title page of the publication was an extract of a speech New York City Mayor Abram Hewitt delivered before the Southern Society of New York City, enthusiastically predicting that the "Southern States of the Union will far outstrip Pennsylvania and the other manufacturing States of the North."[4] If old war wounds had not healed sufficiently for some, they did at least for enterprising individuals who could purchase and develop land in the "banner State of this Union."[5] Harrell, the owner of a job press in Halifax County, who happened

7. The 1880s Through 1900

to be a real estate and immigration agent, published editorials and advertisements on a variety of subjects soliciting individuals to take advantage of cheap land prices and the fairer climate in the South. A self-promoter in more ways than one, Harrell advertised his experience as a qualified land agent who took the reader on a tour of every major town and city in North Carolina, while noting their industries and advantages. Raising the importance of coal mining in the South to a new level, he cited the Egypt Coal Company as one of the industries that had taken hold in Chatham County along with other industries in the state such as cotton, brick works, furniture, flour and corn mills, and textiles. Indeed, Harrell's catch phrase "Boom! Boom!" served as a waking call for all manufacturing and industrial enterprises to relocate here in what was certain to be a land of plenty.[6]

COAL FOR SALE.

The undersigned is now working the Farmville coal mine near Egypt, and is now prepared to supply in any quantity a first-rate quality of BITUMINOUS COAL, said to be the Best in this State. Send orders to

W. H. SEGROVES,

Oct. 13, 1887.   EGYPT, N. C.

Advertisement in *Chatham Record* from October 1887 announcing sale of coal mined at Farmville near Egypt. The Wilmington *Daily Review* reported that mine superintendent W.H. Segroves had employed a considerable number of miners and was getting out coal of first-rate quality in large quantities.

In March 1888, the *Raleigh News and Observer* announced that the Egypt Coal Company had been incorporated by Charles D. Upchurch, Clerk of Wake Superior Court, for the purpose of mining coal for a term of sixty years. Operations would continue at Egypt while meetings of stockholders and directors would be in Raleigh, hammering out the details. The amount of capital stock, $250,000, was to be divided into 10,000 shares of the par value of $25. The incorporators, James A. Hennessy, A.H. Leftwich, and W.E. Anderson, planned to begin operations on an extensive scale immediately, using the existing shaft along with the gangways. Local newspapers were soon reporting to their readers that they could look forward to having coal mined locally on a large scale again.[7] Flooding of the passageways and tunnels together with neglected, damaged mining machinery from earlier years had taken a huge toll on any attempts to start up mining operations. An appeal to investors for the needed capital and labor force required to put the mine in working order, at least to be able to reach a consistently workable seam within four months, hoped to entice speculators who saw the potential for mining coal to meet local fuel demand as well as industry and beyond.[8]

In July 1888, actual work toward pumping out the water in the mine and shaft had started with thirty laborers on site to complete this phase of work preparatory to resuming mining operations.[9] The *Chatham Record* reported that water had been steadily pumped out at a rate of 600 gallons a minute, and that a railroad track had been constructed from the depot to the mine.[10] In 1889, the Cape Fear and Yadkin

Valley Railway issued a "hand-book" oriented toward speculators and interested parties, pointing out that 300 tons of coal per day were being mined at Egypt.[11] With the anticipated connection of the Cape Fear and Yadkin Valley Railway with the Norfolk and Western Railway and its large freight from the coalfields of southwest Virginia and southeast Kentucky, a coaling station at Southport, North Carolina, near the mouth of the Cape Fear River, would be a boon for ocean-going vessels as the distance for taking on coal there would be much shorter than stops at Newport News.[12] After its discussion of Sanford, which was the intersection of the Cape Fear and Yadkin Valley and Raleigh and Augusta Air Line railroads with Charlotte, Wilmington, and Raleigh, the authors of the handbook next considered the Egypt Coal Company, which had purchased the Egypt mine estate, comprising 1,200 acres of coal land, 1,200 acres of timber land,

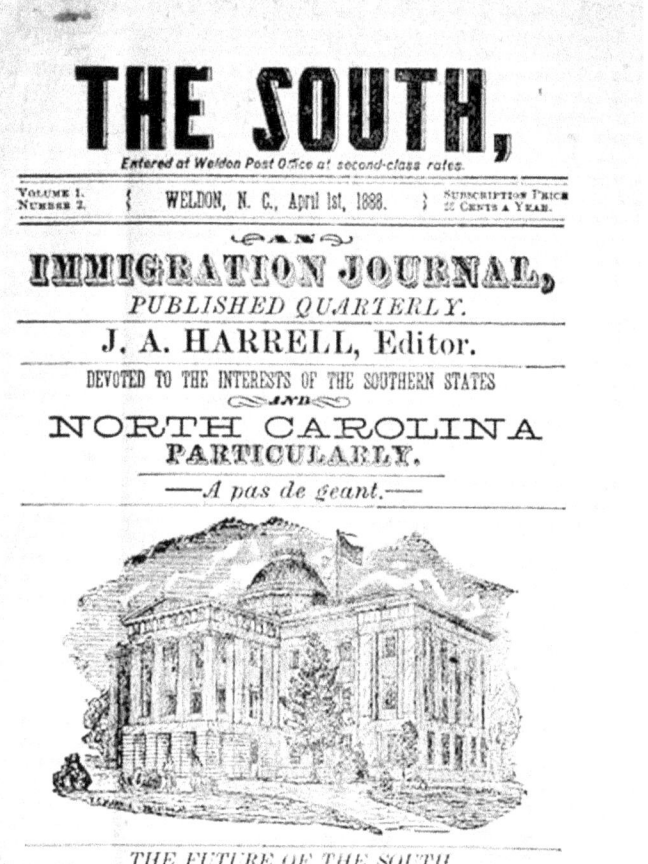

J.A. Harrell's periodical *The South* relied on favorable endorsements from Northerners to attract investment in the southern states, especially North Carolina.

and 300 acres of well-cultivated farm land.[13] With the mining machinery nearly repaired and de-watering of the mine almost complete, together with a shaft of 463 feet in depth, and 1,200 feet of gangways, chambers, and slopes, the commitment to coal mining was as great as it ever had been.

Progress in the repairs at the Egypt shaft allowed mining operations to resume on a small scale in 1889, but not without further delays, as the *Chatham Record*

7. The 1880s Through 1900

reported in May that the machinery for hoisting the water out of the mine broke after half the water had been removed. The damage to the equipment was substantial, necessitating an expenditure of $7,000 to purchase new machinery from the Mecklenburg Iron Works at Charlotte.[14] Improvement of Egypt's infrastructure became a priority in order to have a sufficient labor pool and housing for workers employed in its mines. In January 1890, an article appearing in the *Raleigh News and Observer* reported that the Egypt Coal Company had made available free lots to any manufacturing concerns, schools, or churches who wished to build there. The company had distributed circulars with information about the mines and property and that lots had been laid off for the building of a town.[15] The *Hickory Press and Carolinian* noted that the company had also purchased the electric-light company and gas works with the intention of enlarging both for its community.[16] The Egypt Improvement Company had plans to build a large hotel nearby to accommodate the visitors on business and leisure.[17] The company no doubt had ambitions to establish a community for prospective residents and businesses having the conveniences for its employees of an established town. The creation of a town infrastructure was paramount for the success of the Egypt Coal Company and its long-term prospects of attracting investors.

Coal Production Statistics for Deep River (1889–1899)

Year	Number of mines	Loaded at mines for shipment	Sold to local trade & used by employees	Used at mines for steam & heat	Total product*	Total value (dollars)	Average price per ton (dollars)	Average number of days active	Average number of employees
1889	1	—	—	—	350	—	—	—	—
1890	1	—	—	—	10,262	17,864	1.74	—	—
1891	1	18,780	600	975	20,335	39,635	1.93	254	80
1892	1	6,679	—	—	6,679	9,599	1.44	160	90
1893	1	15,000	—	2,000	17,000	25,500	1.50	80	70
1894	1	13,500	1,000	2,400	16,900	29,675	1.76	145	95
1895	3**	23,400	600	900	24,900	41,350	1.66	226	61
1896	1	5,356	295	2,162	7,813	11,720	1.50	220	18
1897	1	21,280	—	—	21,280	27,000	1.34	215	51
1898	1	9,852	304	1,339	11,495	14,368	1.25	—	—
1899	1	24,126	486	2,284	26,896	34,965	1.30	210	70

Data from Twentieth Annual Report (1898–1899), Twenty-First Annual Report (1900–1901) of the United States Geological Survey, Pt. VI, 58. *In addition to the Egypt Mine, the two other mines for which statistics were reported most likely were the Glendon & Gulf Mining and Mfg. Co., and the Kohinoor Coal Co., both located at Gulf (*Branson's North Carolina Business Directory, 1896*). Yearly totals are not always consistent with reportage from other sources.*

*Long tons (2,240 lbs.) No other statistical information available. All other weights in short tons (2,000 lbs.). Output reduced in 1896 due to mine explosion, 1895; mine fire, 1898.

In March of that year *The Wilmington Messenger* reported that H.B. Peters, general sales agent for the Egypt Coal Company, was in town to announce and promote uses for coal from his mine at Egypt. Tests made for steam purposes by the Cape Fear & Yadkin Valley Railway and the steam tugs *Marie* and *Lawrence*, along with the railroad's ferry boat, were in progress at the time of his visit. Peters stated that sixty- to seventy workers had been employed to work in the mines and that they were hoisting from sixty- to seventy tons of coal a day. With new levels of coal being opened, he estimated output would result in or exceed 200 tons per day. The operations consisted of the shaft sunk 463 feet previously, and from which two levels had been run out from the coal vein, one a distance of 460 feet and the other 1,280 feet. The newer levels had seen an improvement in the quality of the coal with less sulfur and a harder, flint-like quality. Egypt coal was already being used to great advantage in both municipal and industrial applications as seen in Greensboro at the ice factory, gas works, electric light plant, iron works, and in industrial applications where steam power was required. Peters claimed that the Cape Fear & Yadkin Valley Railroad's guaranteed low haulage rate from mine to market would ensure that cheap coal was within easy reach of Wilmington, Fayetteville, and Greensboro and would result in more industries locating in the region than any other factor.[18]

On June 14, 1890, the North Carolina General Assembly chartered the Egypt Railway Company, which opened its route October 15, 1891.[19] The road began at Egypt, a point on the Cape Fear and Yadkin Valley Railroad, and extended eight miles to Colon, in current Lee County, a point on the Raleigh and Augusta Air-Line Railroad. In September 1893, the road was leased to the Raleigh and Western Railway, which was another short line running from Egypt to Harpers Crossroads in Chatham County. Inventory consisted of one locomotive, one combination passenger car, two box cars, twenty-three coal cars, one hand car, one caboose car, and one road car. Capital stock was set at $120,000 and the cost of the road $226,000.[20] Railroads into the coalfields were now becoming established with several carriers available to transport coal to market, indicating a significant improvement not lost on investors.

On February 25, 1891, the General Assembly chartered the Glendon and Gulf Mining and Manufacturing Company for the purpose of determining and developing mineral properties "to mine, smelt and work all minerals, ores, metals and earth ... to erect works of any kind for mining purposes, to manufacture metal, earthenware or brick of any kind." The charter also authorized the company to build, maintain, and operate a railroad from a point near Fair Haven, Moore County, to a point at or near Gulf, in Chatham County. If required, the company could extend the road to a point on the Raleigh and Augusta Air-Line Railroad to be determined by the directors of the company. Along with new developments in railroad service to the coal regions, competition among the roads became more evident which damaged the shared good will among the owners as seen with the Glendon and Gulf Railroad and the Raleigh and Western.[21]

In April 1891, *The Jonesboro Leader* reported that several new houses had been built at Egypt with thirty- to forty more planned to be used as residences of the miners and workers.[22] Three weeks later, the *Chatham Record* updated its readers on the activity at Egypt by noting that seventy men were employed in the mine with the number estimated to increase significantly. Three car loads of coal were shipped every day with the expectation that the volume would increase. As in most fledgling coal-

7. The 1880s Through 1900

GLENDON AND GULF RAILROAD COMPANY.			
John B. Leming, Pres., Bridesbury, Pa.			
Station.	Dist.	Station.	Dist.
Gulf	0	Haw Branch	7
Palmer	3	Riverside	8
Carbonton	5	Glendon	10

EGYPT RAILWAY COMPANY.			
Samuel A. Henszey, Pres., Cumnock, N. C.			
Station.	Dist.	Station.	Dist.
Egypt Junction	0	Endor	5.5
Lobdell	1.3	Oakdale	6.0
Millport	2.2	River Point	6.5
Clarendon	3.0	Egypt, Fourth St.	7.0
Boudinot	4.5	Egypt, Myrtle St.	8.0

Time schedules for the Glendon and Gulf Railroad Co. and the Egypt Railway Co. The two companies had regular stops to the mining towns of Egypt and Gulf (*Branson's North Carolina Business Directory*, 1896).

mining operations, means had to be provided for transporting the coal from the mine to the awaiting empty coal cars. In the case of the Egypt mine, as much as 1,300 pounds of coal could be hoisted to the surface in a small car or bucket, rolled onto an elevated platform, and dumped into the cars standing below on the railroad track to be transported to its destination. With the completion in October of the Egypt Railway Company's connection from the coal mines to a link on the Raleigh and Augusta Air Line Railroad halfway between Sanford and Osgood, a distance of nearly eight miles, a ready, reliable supply line made coal available to areas previously underserved. The *Chatham Record* praised the "intelligently directed energy" of Samuel A. Henszey, president of the Egypt Railway and the Egypt Coal companies, whose vision to complete the rail line was anticipated to serve and open up the area for further development.[23]

In August, the recently incorporated Greensboro Coal and Mining Company employed Rody Maher, of Bedford County, Pennsylvania, to examine their property located in the Dan River coalfield near Walnut Cove in Stokes County.[24] A former member of the State Board of Mine Boss Examiners of the Sixth Bituminous District of Pennsylvania, Maher came to the job with impeccable mining credentials and experience to determine the extent and value of the coal deposits as well as assume management of mining operations at the site. Maher's cursory report to the company showed promise as to continuing mining operations in the area. He found the coal at one site, where the company had already commenced work, to be of a very good semi-bituminous type and that it promised to have excellent coking capability. However, one caveat Maher provided the owners was that his assessment of an unworked vein took into consideration only the outcroppings at or near the surface and not coal found at greater depths where overall quality tended to be better.[25] The company continued to make exploratory drillings and indicated that it would construct a plant with a daily output of 150 tons.[26] In a follow-up correspondence to the *Coal Trade Journal*, Maher added that the two veins being worked simultaneously and the close proximity to two railroads enabled the company to meet demand and that if the coal coked well it could supply Greensboro's steel furnaces.[27]

The death of state geologist W.C. Kerr in 1885 represented the demise of the state geological survey, which had its beginning in 1823 when the General Assembly approved a study of the state's mineralogical and geological resources. However, eventually realizing a need to examine the extent of its mineral and timber resources for the purpose of developing new industry and spurring economic growth, the North Carolina General Assembly in 1891 re-purposed the geological survey into the North Carolina Geological Survey for that reason.[28] Recognizing that the state's slow economic recovery from the end of the Civil War to the present was holding back capital investment in the state, the General Assembly embarked on a comprehensive program to encourage economic growth through development of the state's natural resources.[29] The survey's mission to determine the nature and extent of the state's natural resources, particularly mineralogy, forestry, and water management, was commensurate with promoting economic recovery and growth. Its first director, Joseph A. Holmes, who was professor of geology and natural history at the University of North Carolina, had garnered national acclaim for his leadership of the earlier geological survey. In addition to promoting conservation methods, especially through educational programs, Holmes saw the need to attract investment in the state and with the promotion of solid conservation practices, investors and speculators could be enticed to engage themselves in new business opportunities. Taking the initiative to issue biennial reports to the governor and General Assembly, the Geological Survey also issued press bulletins and economic papers on the resources of the state and the need for expedient conservation practices. Articles appearing in newspapers and magazines throughout the state helped bolster the survey's mission and gain a consensus for its causes.[30] With an approach that conservation and development were not mutually exclusive, the state began to attract rather than deter new growth among its cities and towns.

Joseph A. Holmes (1859–1915). His contributions to the state's geological surveys and pioneering work in mine safety paved the way for his appointment as the first director of the U.S. Bureau of Mines in 1910 (*American Forestry, the Magazine of the American Forestry Association*, January 1916).

The year 1894 saw a resurgence of mining activity in general in the Deep River coal district.[31] The Egypt Coal Company, under the direction of Samuel A. Henszey, was producing 100 tons of coal an hour and that with the newly installed equipment and machinery, as reported by the Raleigh *Evening Visitor*, a greatly increased amount would soon be realized. Negotiations were

7. The 1880s Through 1900

being made with the Seaboard Air Line Railroad for supplying it with 300 tons of coal per day for use in its locomotives. The increased output of coal meant additional jobs and with the recent grading of the Raleigh & Western Railroad to Harpers Crossroads completed—a distance of fifteen miles—optimistic projections of increased railroad traffic running to that point bode well for the company. When completed, the road would run from Egypt to Asheboro, taking on new passenger service in the process in addition to its freight haulage.[32] Also in 1894, the Kohinoor Coal and Iron Company opened a mine near Carbonton, in Chatham County, and the Gulf and Glendon Mining and Manufacturing Company opened a mine in Moore County.[33]

In December 1894, the *Wilmington Star* carried a story from special correspondent D.A. Tompkins, with the *Charlotte Observer*, reporting from New York that the Egypt mine and approximately 4,300 acres of coal land at Deep River would be worked by S.B. Langdon, president of the United Mines Company, of Pennsylvania, and that the name of the firm would be the Langdon-Heneszey Coal Company, with a capital stock of $1,000,000, and offices in New York and Pennsylvania. In the proceedings, the name of the coal was changed from Egypt to Chatham coal and would reach tidewater at Wilmington on the Cape Fear and Yadkin Valley Railroad. It was hoped that the Seaboard Air Line Railroad would put in a spur to the mines. The name of the station at the mines was to be called Langdon.[34] In January 1895, the Wilmington *Weekly*

Gulf Depot, ca. 1974. The community of Gulf was serviced by the Norfolk Southern Railway line. The depot building no longer exists (photographic postcard from author's collection).

Messenger reported that the Cape Fear and Yadkin Valley Railway was constructing coal chutes at their Point Peter terminal.[35] A dredge was employed for cutting docks into the rice fields in order to accommodate vessels coming into the port. Provisions were made to cut the docks sufficiently long and deep so that vessels of the deepest draft could take on coal from the Chatham coalfield. Improvements, such as docks and chutes, were put in place in anticipation of handling large shipments of coal and Wilmington becoming a coal-exporting port.

Experienced mine workers from Pennsylvania and Tennessee were hired to operate in the old shaft.[36] Houses and other buildings, including a hotel, storehouses, together with equipment- and work-shops, were being built near the mines on a large scale. Langdon-Heneszey also added new equipment and facilities to their operations. The arrival of one-hundred rail cars made in Pennsylvania especially for the coal industry anticipated business on a grand scale. The plans for becoming a major coal center included exporting large quantities of coal to South America and the West Indies. S.B. Langdon, a co-director of the firm, had experience shipping coal to the former, and Wilmington's closer proximity to the southern continent made for easier and more profitable passage. Another consideration in shipping coal from Wilmington was that vessels would avoid the dangers associated with Hatteras or any other navigational challenge north of Wilmington. The newspaper exhorted its citizens and businesses to immediately take advantage of introducing manufacturing facilities in order to induce capital into the region.[37] For North Carolina to compete in the global market, a viable port was essential.

On March 13, 1895, five business executives met at the Horton Hotel in Wilmington to discuss their proposition to capitalize on the Southern coal business, including exporting coal from the ports at Wilmington. In attendance were J.W. Fry, of Greensboro, general manager of the Cape Fear and Yadkin Valley Railway; W.R. Kyle, of Fayetteville, general freight and passenger agent of the Cape Fear and Yadkin Valley Railway; Samuel P. Langdon, of Philadelphia, president of the United Collieries Company, president of the Altoona and Phillipsburg Connecting Railway, and president of the Langdon-Henszey Coal Company, of Cumnock, Chatham County; Samuel A. Henszey, of Philadelphia, vice-president of the Langdon-Henszey Coal Company, president of the Raleigh and Western Railway, of Chatham County; and, Col. Charles B. Evans, of New York, general agent, United Collieries Company, export agent for the Langdon-Henszey Coal Mining Company.[38] An updated report on the construction of the coal chutes showed that from January to March seventy-five carpenters and laborers had completed five chutes at a great expense to the railway in anticipation of handling large shipments of coal from the mines at Cumnock.[39] Col. Evans informed the newspaper in an interview that Wilmington was to be the company's shipping point and that vessels would be loaded with coal bound for Cuba, the West Indies, Mexico, and Southern ports. He emphasized that Wilmington's harbor was sufficiently deep to accommodate large ocean-going vessels and that vessels previously going to Norfolk and Newport News would have a safer route by avoiding the reefs and storms

7. The 1880s Through 1900

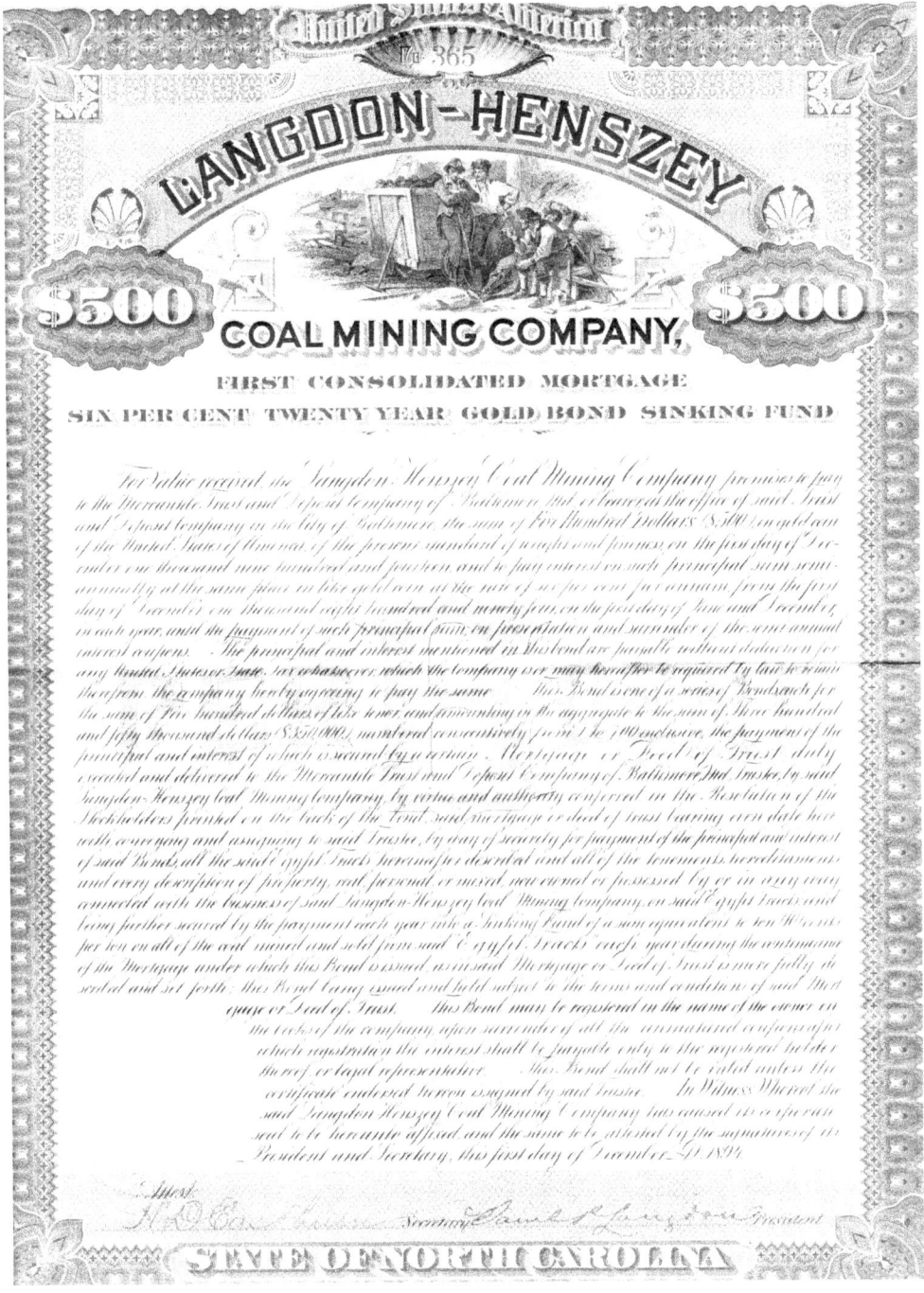

Stock certificate issued by the Langdon-Henszey Coal Mining Co., December 1, 1894 (author's collection).

associated with Cape Hatteras. In other words, it was safer and cheaper for ships to dock at Wilmington and take on coal there.[40]

All signs thus far were encouraging for a vigorous coal trade. The close proximity of the mines at Cumnock to Wilmington meant that coal could fuel local industry or be exported to markets. Early completion of the chutes meant immediate benefits to Wilmington as an important center of exportation, with the goal of attracting investment in the area. In his report to the Commissioner of Labor in 1895, C.H. Beckham, representing the Brotherhood of Locomotive Engineers, Division 339, had recommended to railroad officials in 1894 that Egypt or North Carolina coal be used as it would save the company a considerable expenditure. Addressing criticism that North Carolina coal was unfit for locomotive use, Beckham stated that it was now being used successfully.[41] The seemingly unlimited amount of superior coal to be extracted at the Langdon-Henszey mine, together with a transportation network in place to ship the coal to market, meant that commerce could be conducted on a large scale, hitherto unseen in eastern North Carolina or the entire state for that matter. Investors and speculators were the key remaining components needed to bring the overall goal of prosperity to fruition.[42]

In the several months ahead, all attention focused on Langdon-Henszey and its efforts to mine and transport its coal to Wilmington and beyond. In its April 26, 1895, edition, Wilmington's *Weekly Star* noted that the coal chutes at Point Peter had been completed and that three carloads of Pocahontas coal had arrived and placed on the chute for delivery to the British steamer *Corinthia*, which had arrived to take on a cargo of cotton. A newer, more powerful locomotive was at the port to position the coal in the chute, which was thirty-two feet above the level of the track.[43] There was no mention of coal mined in North Carolina being delivered at the time but the Pocahontas field's output was considerably greater than the combined totals of the North Carolina mines. Newspapers in April and May reported promising numbers in coal production with fifty men employed at the Egypt Mine and a daily output of 150 tons. A month later daily output doubled, reaching 300 tons.

The *Seventeenth Annual Report of the United States Geological Survey* for 1895–96, reported total coal production for the state amounting to 24,900 short tons with a value of $41,350, which exceeded by over fifty-percent the previous year's totals. The report also noted that principal efforts went toward much needed development to the area instead of increasing production. This observation, no doubt, had everything to do with establishing new infrastructure in the coal-mining regions, where none previously existed or was beyond repair due to flooding, inactivity, and loss of manpower since earlier mining efforts. A contributing writer to the *Manufacturers Record of Baltimore* projected that the Deep River region had a capacity of 35,000,000 tons of mineable coal.[44] With an expansive transportation system in place, coal could be shipped to many locations in the state or to markets beyond.

The entire production of the Egypt field was estimated at 80,000 tons from its opening in 1850 to the resumption of mining. The present output was approximately

7. The 1880s Through 1900

Steamship *Martin Mullen* of the Pioneer Steam Ship Co. coaling, Houghton, Michigan, ca. 1900. Coal was loaded into the ship's hold and transported to its destination or used as its own fuel source. On some occasions mined coal was loaded onto larger capacity barges (Library of Congress).

100 tons daily and with improvements to the plant completed the output would be over 500 tons daily. Among the developments were a newly erected headframe, a platform with revolving tipples for dumping mine cars, which were of one-half ton capacity with nine-inch wheels, and tracks of two feet six-inch gauge. A new eighteen-by-thirty-two inch Litchfield Foundry fast motion engine had been erected and fitted with Beach's patented balanced side valve.[45] The overall investment in new machinery for the mine cost over $140,000, a significant sum for that time, while at one point 250 men were employed in the mines.[46] In its August 21, 1895, edition, *The Greensboro Patriot* announced that the Langdon-Henszey Company had added another shaft two miles west of their mines at Cumnock in order to increase production. With the added growth, a large store and four buildings had been recently constructed. A hotel was promised to be opened soon along with the installation of an electric light system.[47]

At its December 1895 meeting, representatives from the State Board of Agriculture agreed to revise the *North Carolina Hand Book* in an edition that more accurately reflected the enormous growth in all branches of industry within the state. The purpose of the new publication, entitled *North Carolina and Its Resources*, was to "note the advance made all along the lines of enterprise since 1893."[48] In a tone that practically chastised earlier entrepreneurs for succumbing to indifference and defeat, the latest efforts at Deep River to mine coal were poised for enterprise and success. One of the features of the book was to present to the reader an overall consideration of the state's natural resources and encourage development in areas such as agriculture, horticulture, fishing, geology, and mineralogy. Industry and manufacturing were examined in regard to their current state and their potential for future development. An assessment of each county provided useful information such as historical data, agricultural accomplishments, transportation availability, population statistics, area and acreage, and taxation usage. The book also contained a great deal of updated information previously unseen in earlier editions. The underlying theme of progress reverberated throughout the book with the intention of attracting new investors and speculators as well as new residents.

Citing the Cumnock mines in Chatham County as the only operating collieries in the state, the writers gave credit to Samuel A. Henszey for installing an efficient and modern plant as well as practicing sound business methods, which in previous efforts at mining had been lacking.[49] Under Henszey's direction, machinery for hoisting, pumping, and ventilation were in working order and every safeguard for the protection of life and property had been considered. The colliery contained two perpendicular shafts, one for ventilation only, measuring eight by ten feet, intersecting the coal seam at a depth of 226 feet. The second shaft, which was the main working shaft, measured eight by twelve feet and intersected the seam at 464 feet. Total production capacity was established at 1,000 tons per day, considerably more than was previously reported in various projections. The mines were directly connected with the Cape Fear and Yadkin Valley Railway at Cumnock, and the Seaboard Air Line Railway at Colon via the Raleigh and Western Railway. Progress was being made with the construction of an extension with the Southern Railway near Randleman in Randolph County.

The writers emphasized the excellent qualities of the coal and its abundance near iron ore deposits, which had created additional interest in the area. The workable veins reached a thickness of six feet, consisting of two benches of four- and six feet respectively, separated by two feet of black band iron ore. Experts estimated that 47,300,000 tons of coal existed in the land owned by Henszey.[50] Testimonials as to the quality of the coal came from the business and industry sectors alike, such as the

Opposite: **The top photo (A) shows a steam locomotive of the Raleigh and Western Railway with hopper cars at the loading platform taking onboard recently mined coal. The bottom photo shows the steam plant (B), which contains the hoisting engine that raises and lowers the cage located inside the mine shaft. The headframe apparatus (C) sits stationary over or near the shaft which contains the sheave wheels (D) that wind the steel cable attached to the cage where the men are lowered and raised into and from the shaft. In addition to coal, men, supplies and equipment could also be lowered and raised with equal facility (*North-Carolina and Its Resources, Illustrated*, 1896).**

7. The 1880s Through 1900

CUMNOCK COAL MINES

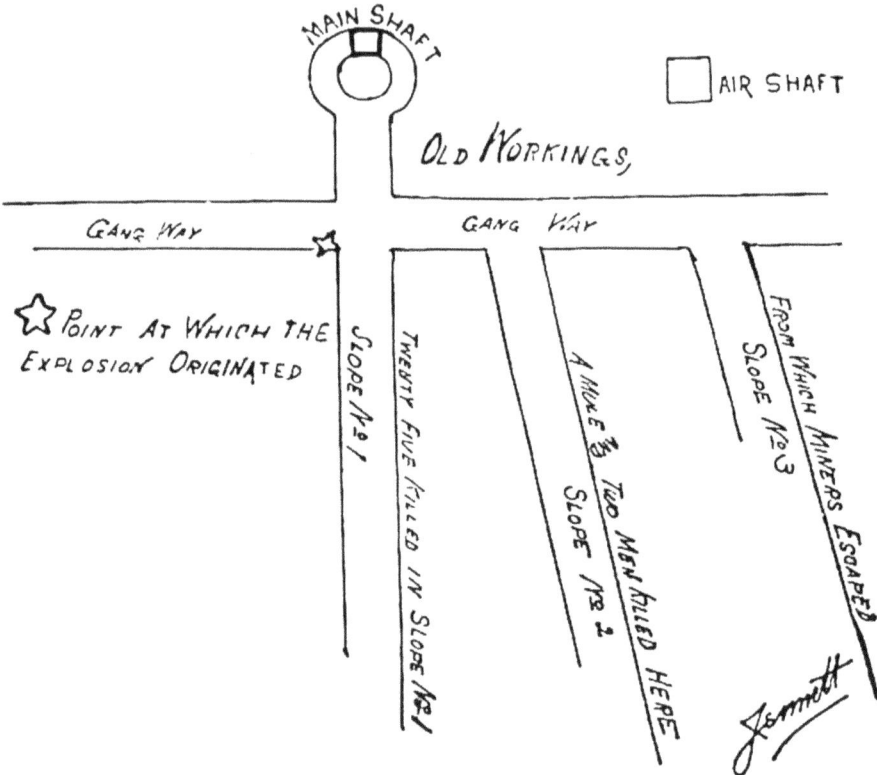

Diagram of workings sketched by mine employee showing location of explosion near slope 1, where the majority of the miners were killed. Note that slope 3 is where most miners escaped.

Greensboro Gas Company, claiming that Deep River coal made 9,700 cubic feet of gas, eighteen and one-half candle power, and forty-nine bushels of "good, clean, hard coke," which could be used in the iron and steel industries. The superintendent of the Central Division of the Seaboard Air Line Railway attested that Cumnock coal surpassed Pocahontas coal in steam tests, and that as a blacksmithing coal it was shipped to local points on the Norfolk and Western Railway. For domestic use, it burned cleanly without a trace of soot and smoke.[51]

On December 14, 1895, Samuel P. Langdon, president of the Langdon-Henszey Coal Company, and George T. Cant, superintendent, arrived in Wilmington to make arrangements for operating the coal chutes, which were built in the summer for the purpose of coaling ships and loading vessels for export. F.S. Fouse was appointed on-site tide-water agent, with fifteen carloads of coal expected to arrive during the week.[52] Five days later, on the morning of December 19, an event that no one anticipated or expected devastated the Cumnock community, when an explosion in the mine took the lives of thirty-nine workers.[53] Noting that the mine owners utilized

7. The 1880s Through 1900

the most "prudent precautions," newspaper accounts made real the personal tragedy and destruction that caused it to be the second worst mining disaster recorded in the United States for the year.[54]

The mine had four shafts—main, air (ventilation), hoisting, and pumping—and as was generally the case with mine explosions the gangway—the main passage into the mine workings—was often made impassable by fallen rocks, timbers and dangerous gases.

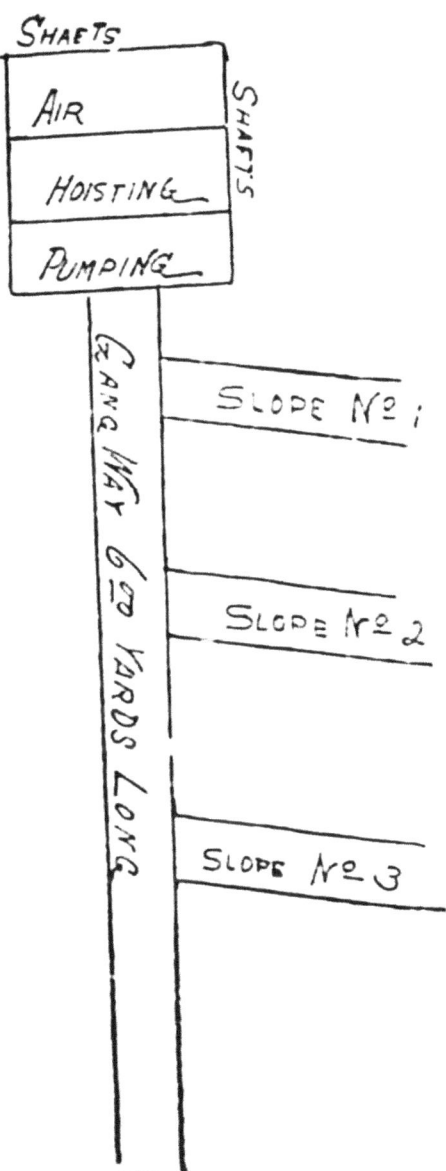

8

The 1895 Mine Explosion at Cumnock

On December 19, 1895, *The Raleigh Press-Visitor* was the first newspaper to report the news of the morning's deadly explosion in the Langdon-Henszey Company's coal mine at Cumnock. The newspaper reported that the explosion occurred between 8 o'clock and 9 o'clock in the morning with forty known workers in the mine. H.A. London, editor of the *Chatham Record*, was a passenger on the train that stopped briefly at Moncure, where news of the explosion was first announced. Among the inconsistencies in the first news reports were the exact number of workers killed and injured and the cause of the explosion, although preliminary reports stated that only six or eight men perished. Noting that the mine consisted of four tunnels, reaching a depth of 450 feet, the newspaper announced that one of its staff members, Thomas J. Pence, was already en route to the mine.[1]

Soon, additional coverage of events started to arrive at newspaper offices in an outbreak of frantic activity. The *Raleigh News and Observer* reported at least forty-three killed and six injured of a total of sixty-seven men inside the mine.[2] Cumnock, six miles west of Sanford, became the focal point for rescue and recovery efforts, as first responders pumped fresh air into the mine and the first of several rescue parties were organized to venture into the mine and ascertain the whereabouts of any survivors.[3] The first group of rescuers who entered the mine brought out twenty-five men from slopes two and three, five or six of whom were badly injured. Two men and a mule were killed in slope number two. After considerable delay, according to the paper, a search party entered slope one, where they encountered dead men, most of whom were mutilated and partially covered by coal, timber, and other debris. The search party returned from the scene and reported their findings to the authorities and mine officials. At approximately 4 o'clock, a rescue party consisting of ten- to twelve men descended into the mine for the purpose of recovering the dead. Reports indicated that among the victims were "negroes and foreigners and the rest natives of North Carolina."[4] Eight of the victims were from Pennsylvania, who had been recruited by mine officials for their experience. A "quantity" of dynamite in the mine was determined to have exploded, wrecking coal cars and splitting timbers into small pieces. On a final note to the day's coverage of events, the paper exclaimed, "Excitement runs high."

8. The 1895 Mine Explosion at Cumnock

Meanwhile, the *Charlotte Observer* reported thirty-eight deaths and included in its coverage comments made by H.A. London, who stated that he heard the noise of the explosion at Pittsboro and that medical assistance was being rushed to Cumnock.[5] Picking up the story as covered by the Southern Associated Press, the *Observer* stated that forty-three workers "were caught in the explosion" and that rescue efforts had been abandoned on account of the presence of firedamp, which nearly overwhelmed several of the rescuers.[6] Later reports arrived claiming that thirty-eight were killed and twenty-five had escaped. The Cumnock mine had been tapped by the Seaboard Line Railway to furnish its coal and earlier in the week the company began to transport surplus coal to ships at Wilmington. Medical assistance was being rushed to the area and special trains were dispatched to the mines bringing aid. A temporary morgue was set up in the mine's engine room where the bodies were laid out for identification.

The Raleigh Press-Visitor, which first reported the news of the tragedy at Cumnock, affirmed in its December 21 issue that all victims had been located—thirty-nine total—and that the cause of the explosion was due to a detonation of firedamp. Officials announced a preliminary cause that a worker touched off a gas pile with an open flame, which resulted in the explosion and that the dynamite exploded afterwards. As it turned out, the approximately 250 pounds of dynamite inside the shaft was found intact. The difficult task of identifying the dead often required the assistance of family members, friends, or co-workers before returning the bodies to their respective homes and families for burial. The victims were all shrouded from public view and the coal company announced that it would pay all funeral expenses, as many families, no doubt, found it beyond their means. Sadly, seven of the miners killed in the explosion had planned to leave for their homes in Pennsylvania that day.[7] The *Raleigh North Carolinian*

Thirty-Eight Men Reported Killed in a Disaster in North Carolina.

Rescuers Could Not Make Any Headway Against the Overpowering Firedamp.

RALEIGH, N. C., Dec. 19.—News has been received here of a bad accident in the Cumnock coal mine, located about forty miles from here. Shortly after the day shift began work a violent explosion of firedamp took place.

Advices at 6 o'clock this evening from Colon regarding the coal mine disaster at Cumnock were to the effect that forty-three persons were caught by the explosion and that all hope of rescuing the men was abandoned, owing to firedamp, which at several times nearly overcame those who attempted to rescue the imprisoned men.

Only two of the rescuers were in sight of some of the victims, but had to retire in the face of the overpowering firedamp.

A telegram was received here at 7 o'clock saying there are thirty-eight dead and that twenty-five escaped. Telegraphic communication with the scene of the disaster is difficult.

News of the explosion spread quickly as far as the West Coast as seen in this article printed in a California newspaper (*San Francisco Call*, December 20, 1895).

included in its coverage two maps of the interior furnished by Superintendent Cant of the Langdon-Henszey Coal Company which showed the mine as it appeared when the explosion occurred along with a depiction of the gangway. The newspaper stated that the mine foreman and fire boss made their rounds about 7:45 a.m. before the workers arrived and reported everything was safe and in good working order, including the air shafts and ventilation. Approximately twenty minutes later, the explosion occurred.[8] In continuing coverage of the explosion, the consensus seemed to be that the cause of the explosion was firedamp. The coroner's inquest issued a verdict of gas explosion of unknown origin.

Relief work became the norm after the cleanup was completed and the bodies of the coal company employees returned to their families for burial. Appeals to the public, benefits, and church and civic events all sought to raise needed funds for the wives and children whose loved ones were killed in the mine. In early January 1896, Raleigh Mayor William M. Russ arrived at Cumnock to offer his and his city's support to the grief-stricken families. Raleigh's *News and Observer* was at the forefront of fund-raising efforts, though local committees, comprising both white and black communities, organized themselves to receive and distribute contributions to those affected by the tragedy. Some committee members provided transportation for those wanting to move in with family or friends.[9] One relief committee, whose efforts on behalf of the survivors was reported in the local newspapers, expressed a certain amount of frustration at the absence of mine co-owner Langdon and superintendent Cant at meetings but greater frustration still when severely injured survivors seemed to be ignored by the company.[10] One miner, named Joe Mills, had been terribly burned in the explosion and received only a small amount of linseed oil from the company to help soothe his burns. A woman, whose husband was killed, received a trifling amount of food staples from the company but her continued residence in her own home was uncertain. All of the residents interviewed by the mayor seemed to blame the coal company for not providing sufficient assistance during their time of need.

In the final analysis of the tragedy at Cumnock, the outlook for the Langdon-Henszey Coal Co. did not appear encouraging. Within only a few months of the disaster, damage suits were filed in courts against the company on behalf of the widowed and orphaned for compensation. One such case was that of one of the victims, Charles Poe. T.B. Fowler, attorney, filing suit in Chatham County Superior Court on behalf of Poe, alleged that Langdon-Henszey Coal Co. knowingly and recklessly endangered the life of his client in the December 19 explosion. Poe was employed by the company as a coal digger and common laborer. The complaint strongly condemned Langdon-Henszey for neglecting safety measures and procedures leading to the loss of life and damage to the mine. The document provided insight into the struggle of the miner as he, or in this case, his representative, confronted his employer, the coal company, with charges of neglect and criminal actions. Unfortunately at the time of the 1895 explosion, the state of North Carolina did not have laws in effect for

8. The 1895 Mine Explosion at Cumnock

the inspection of mines which might have made the mine safer for the miners and workers.[11] Adding to the company's woes, soon thereafter the company went into receivership. In a separate, much publicized case, mine co-owner Samuel P. Langdon was arrested for the murder of his mistress in Philadelphia. Also, the two partners, Langdon and Henszey, were entangled in protracted litigation in which Langdon accused Henszey of giving Henszey's private secretary $600,000 face value of stock without any consideration.[12] Lastly, the Langdon-Henszey Coal Company was sold in 1896 to the New York Gas Coal Company of New York.[13] The on-and-off again exploits of the owners continued to play out in the public eye without regard for the deceased miners' families.

Though it was reported in newspapers that the mine itself did not suffer extensive physical damage, efforts at restoring production to previous levels were slow and wrought with setbacks. Henszey, it was noted, claimed that he would pay off the fifty miners now employed at his mine and put a large force to work at bringing production to 300 tons daily.[14] In its annual statistical analysis of coal mining, the United States Geological Survey, Department of the Interior, reported total coal production for North Carolina in 1896 at 7,813 short tons, down more than two-thirds from the overall total of the previous year.[15] The *North Carolinian* announced in its June 10, 1897, issue the adjudication on June 9 of claims made by representatives of the victims against Langdon-Henszey Coal Mining Company. The eighteen actions for the recovery of damages resulted in the company's liability of $8,000 plus costs of the actions, $3,000 of which was to be paid by the receiver, Samuel A. Henszey, within thirty days of the agreement and the remaining $5,000 to be paid as soon as funds were available.[16] The decree of foreclosure on their property, which was issued by Thomas Richard Purnell, judge of the United States District Court for the Eastern District of North Carolina, called for the appointment of three commissioners—A.H. John, W. Hinsdale, and Thomas B. Womack—to sell the mining plant and property at Cumnock on or about September 1st for the purpose of settling the company's indebtedness of $350,000.[17] Some good news, however, emerged when it was reported that the Cumnock coal mines during July sold more than 2,000 tons of coal, most of it to the Seaboard Air Line Railway.[18]

In September, announcement was made of the public auction on October 8, 1897, at the company's store house with "all the property real, personal or mixed, and premises, rights, immunities and franchises lately belonging to and owned by the Langdon-Henszey Coal Mining Company and the Egypt Coal Company."[19] On October 9, *The Press-Visitor* announced to its anxiously awaiting readers that the Cumnock mines were sold to George Pendleton of Baltimore for $40,000, though it was noted that Langdon did not submit a bid and that the Henszey interests still controlled the mines. The commissioners named to oversee the proceedings made the recommendation that the court approve the sale.[20] On November 11, it was reported that Judge Purnell had confirmed the sale of the coal mine at Cumnock but not the 125 acres constituting the town site of Cumnock.[21]

Mine Safety Legislation

Coal mining has always been an extremely hazardous occupation. As early as 1865, a motion was introduced in the U.S. Congress to create a Federal Mining Bureau, but little interest was shown and the plan eventually was scrapped. It would take until 1891 that Congress passed the first substantial federal statute governing mine safety, thus marking the beginning of a long series of federal and state legislation regulating mining activity below and above ground. Though unenforceable by the government, the legislation did set standards in some areas, such as requiring underground ventilation and prohibiting coal operators from employing children under the age of twelve years.[22]

Beset by the large numbers of new immigrants desiring to work in an industry that provided relatively steady employment and good wages compared to other industries, coal mining, despite its risks, appealed to men recently arrived from the old country because they could start at the lowest level, with little or no experience, and work their way up the ladder in hope of attaining the coveted job of miner.[23] Experienced

Mine ventilation fan, ca. 1910. These huge apparatuses directed a steady flow of fresh air into the mine workings to help dilute harmful mine gases, render harmless and carry away dangerous accumulations of explosive and toxic gases as well as coal dust. The fan was installed on the surface in an incombustible housing (A) and connected to the mine opening (B) for continuous air circulation via the fan's blades (C) (from the 1922 Jeffrey Mining and Manufacturing Company catalog, courtesy Dresser Industries).

8. The 1895 Mine Explosion at Cumnock

miners from coal-producing countries such as England, Wales, and Scotland could find work in the mines almost immediately as their skills were invaluable to the company. Saddled with the burgeoning ranks of men and contending with a language and cultural barrier, operators dealt with mine safety on an individual basis, sometimes investing a great deal of money in new equipment and safety programs.[24] While many mine operators and owners agreed with the passage of mine safety laws, an attitude prevailed among them that because miners were inherently careless legislation "cannot remedy the evils which result from the perversity of human nature."[25] In the early days of mining legislation, it would be left to the employer and employee to determine the extent of compliance even when such laws were mandatory.

At the time of the Cumnock disaster in 1895 no mining laws or inspectors' reports existed in the state of North Carolina. In fact, there were no mine inspectors on the state payroll. Included in the minutes of the coroner's inquest was a plea to the legislature for passage of a law to protect the miner. No doubt spurred by the events at Cumnock, the North Carolina General Assembly, at its 1897 legislative session, passed an act "to provide for the inspection and regulation of mines" within the Bureau of Labor Statistics. In this much-needed legislation the Commissioner of Labor Statistics became the de facto mine inspector for the state of North Carolina regardless of his qualifications to carry out his functions.[26] The mine inspector was charged with examining all mines in the state as often as possible to verify that all due diligence was met and carried out in accordance with the law. He was required to keep a record of examinations made and report the condition in which the mines were found, note the extent to which all mining laws were observed or violated, give the status of progress made in the security of life and health of workers, record the number of accidents and deaths in or about the mines, ascertain the number of mines in the state and the number of workers employed in and about the mine, together with providing "all such other facts and information of public interest, concerning the condition of mines, development and progress of mining in the state as he may think useful and proper, which record shall be filed in the office of the inspector."[27] Additionally, the information collected by the mine inspector was to be published in an annual report under the auspices of the Bureau of Labor Statistics.

The mining law gave the inspector of mines a certain amount of latitude previously unknown by having the right to enter any mine "at all reasonable times" for the purpose of a personal inspection."[28] The mine inspector was required to keep in his office all records, maps, surveys, reports, and papers required by law to be filed by him which then were consolidated and made available to the governor of the state of North Carolina as an annual report. In conjunction with the mine inspector's duties, the mine operators and owners were responsible for providing the mine inspector on or before November 30 a report of data and figures of his mine, including accidents, as stipulated in the mining law. The law also mandated owners and operators to note any areas of improvement "for the better preservation of the life and health of those engaged in such an industry." Non-compliance on the part of the mine owner or operator often resulted in being found guilty of a misdemeanor, but, again, no enforcement was in place to impose the law.[29]

With the 1869 disaster at the Avondale coal mine in Luzerne County, PA, still fresh in peoples' memory, the new North Carolina legislation called for two outlets in a shaft operation for every seam of coal mined.[30] This provision helped to ensure that two openings existed in the mine in the event of an emergency and could be accessed for escape or entry. In addition to the miners having a way out of the mine, rescue work was contingent on having at least one opening for easier access to attend to those injured inside the mine workings. Cages were to be available at all times, especially where a mine had only one opening, and kept in good working order. In no instance was an air shaft with a ventilating furnace at the bottom of the mine to be construed as an escape shaft. In the event that a mine did not have two openings, the outlets were to be provided within twelve months "after shipments of coal have commenced from such mine."[31]

The new legislation required all mines to maintain an amount of ventilation "of not less than 100 cubic feet per minute per person employed in such mine as to dilute, render harmless, and expel the poisonous and noxious gases from each and every working place in the mine."[32] Methane, or firedamp, was an especially noxious mine gas because it was colorless, odorless, and tasteless. Highly flammable and explosive, firedamp could exist in large amounts inside the mine and sometimes outside at the mine's entrance. The mining legislation stipulated that in mines known to contain firedamp, the fire boss, or an equally qualified mine official, must check for the gas in every working place before any of the workers were allowed to enter the mine. The mine foreman was responsible for measuring the ventilation in the mine at least once a week at both inlets and outlets and at or near the face of all entrances. The readings were to be recorded on blank forms and forwarded to the mine inspector.[33]

Designated engineers were responsible for lowering and hoisting the cages in the shaft, and no more than two men were allowed to ride on any cage or car at one time. Some mines had lamp houses where safety lamps and other types of lighting devices were stored and checked out to employees working in the mine. In some mines, a lamp-check system, using metal checks stamped with an identifying number which the miner received at the beginning of his shift, kept track of lamps as well as miners. At the beginning of his shift, the miner presented the check to the designated person at the lamp house in exchange for his lamp. The check was then placed on a board with other checks and returned to the miner when he turned in his lamp at the end of his shift.

The lamp man was responsible for examining the lamp before being taken into the working area for any damage, notably cracked glass. This system also could serve as a "roll call" after a disaster struck to account for a miner's whereabouts. Lamps were outfitted with a key- or magnetic-locking device to deter miners or workers from opening the lamp inside the mine, where an exposed flame could ignite flammable gases. Only the lamp man, mine foreman, or his designated proxy, was allowed

8. The 1895 Mine Explosion at Cumnock

to possess a magnet or key on his person for unlocking the lamp. Implements called strikers, which were used in the mines to light safety lamps, were also kept at the lamp house and carefully monitored.[34]

Mine operators were responsible for reporting fatal accidents to the mine inspector as well as the coroner of the county in which the accident occurred. Any type of accident or explosion was to be reported to the mine inspector within twenty-four hours of the occurrence, including the nature and cause of the accident and the name or names of individuals killed or injured, with a description of the extent of the injuries. Upon receiving notification of a death, the mine inspector, or his representative, was instructed to go to the mine in which such death occurred, and make a written report, which was filed by the inspector in his office as a matter of record and future reference.[35]

Another critical area of the new legislation was that the owner was required to report work on any new opening in the mine, abandoned mines, reopening of mines

Miners lining up at lamp house to receive a lamp for their shift (ca. 1935). The lamp room supervisor was responsible for the upkeep and inventory of all lamps taken into the mine (National Archives and Records Administration).

	Killed.		Killed.
NUMBER KILLED UNDERGROUND.		**NUMBER KILLED UNDERGROUND**—contd.	
1. Falls of roof (coal, rock, etc.):		9. Animals.......................	
(a) At working face................		10. Mining machines (other than 8c)....	
(b) In room or chamber............		11. Mine fires (burned, suffocated, etc.)..	
(c) On road, entry, or gangway......		12. Other causes:	
(d) On slope......................		(a) Fall of person...............	
2. Falls of face or pillar coal:		(b) Machinery (other than 10)...	
(a) At working face................		(c) Rush of coal or gob..........	
(b) On road, entry, or gangway.....		(d) Falling timber..............	
3. Mine cars and locomotives:		(e) Suffocation in chutes........	
(a) Switching and spragging......		(f) Miscellaneous...............	
(b) Coupling cars................			
(c) Falling from trips............		Total number killed inside of mines...........	
(d) Run over by car or motor......			
(e) Caught between car and rib....		**NUMBER KILLED IN SHAFT.**	
(f) Caught between car and roof while riding...............		13. Falling down shafts or slopes	
(g) Runaway car or trip...........		14. Objects falling down shafts or slopes..	
(h) Miscellaneous................		15. Cages or skips.................	
4. Gas explosions and burning gas:		16. Other causes:	
(a) Due to open light..............		(a) Overwinding...............	
(b) Due to defective safety lamps ...		(b) Breaking of cables..........	
(c) Due to electric arc.............		(c) Miscellaneous..............	
(d) Due to shot..................			
(e) Due to explosions of powder....		Total number killed by shaft accidents...........	
(f) Miscellaneous................			
5. Coal-dust explosions (including gas and dust combined):			Yards, shops, etc.	Strip pits.
(a) Due to open light..............				
(b) Due to defective safety lamps ...				
(c) Due to electric arc.............		**NUMBER KILLED ON SURFACE.**		
(d) Due to shot..................		17. Mine cars and mine locomotives....		
(e) Due to explosions of powder....		18. Electricity....................		
(f) Miscellaneous................		19. Machinery....................		
6. Explosives:		20. Boiler explosions or bursting steam pipes............		
(a) Transportation...............		21. Railway cars and locomotives.....		
(b) Charging....................		22. Other causes:		
(c) Suffocation..................		(a) Explosives.................		
(d) Drilling into old holes.........		(b) Fall of person..............		
(e) Striking in loose rock or coal....		(c) Falling objects (derricks, booms, etc.)..		
(f) Thawing.....................		(d) Suffocation in chute, bin, or culm..		
(g) Caps, detonators, etc..........		(e) Falls or slides of rock or coal....		
(h) Unguarded shots..............		(f) Steam shovels..............		
(i) Returned too soon.............		(g) Hand tools................		
(j) Premature shot...............		(h) Miscellaneous.............		
(k) Sparks from match, lamp, or candle...............				
(l) Delayed blast.................		Total number killed on surface or in strip pits..........		
(m) Shot breaking through rib or pillar................		Grand total.....................		
(n) Miscellaneous................				
7. Suffocation from mine gases.........				
8. Electricity:				
(a) Direct contact with trolley wire.................				
(b) Bar or tool striking trolley wire.................				
(c) Contact with mining machine				
(d) Contact with machine feed wire.................				
(e) Contact with haulage motor....				
(f) Miscellaneous................				

Sample death report form used by most states for reporting accidents and deaths. Many states were required by law to submit such a form on a regular basis (U.S. Bureau of Mines).

after abandonment, or when a squeeze or crush made a mine unsafe for workers.[36] Boys under the age of twelve years were prohibited from working in the mines and inspectors were given the authority to ascertain the age of any minor in question seeking employment in the mine. At any time an inspector could halt work in a mine for noncompliance of the law. Although the mining act went farther than previous attempts to provide mine safety, its provisions did not apply or affect any mine where less than ten men were employed at the same time.[37] However, inspectors were granted full authority to inspect a mine and enforce any regulations pertaining to safety in accordance with the mining law, which provided for fines and imprisonment in cases of non-compliance of the law, unreported accidents and fatalities, sabotage to workings and equipment, refusal of the coroner to hold inquests, or any willful act in which the health and welfare of the mine workers were endangered.[38]

In 1898, the state legislature chartered the North Carolina Coal and Coke Company, of Glendon, Chatham County which sought to add coke ovens for making coke used in the manufacture of pig iron. The stockholder was listed as Samuel P. Langdon and others with $100,000 capital. In its February 13, 1898, issue, the *Coal Trade Journal* announced that the name of the Piedmont-Cumnock Coal Company, recently organized by the Norfolk Southern Railway, had been changed to the Cumnock Coal Mining Company with capital stock increased from $500,000 to $1,000,000.[39] The Norfolk Southern Railway soon became a steady client of Deep River coal for haulage and fueling its own locomotives. In early February, Raleigh's *Morning Post* reported that Samuel A. Henszey, manager of the Cumnock Coal Mines, was in town to provide an update on the new ownership and reorganization of the Cumnock mines. According to Henszey, the January output of the mines was the largest in their history, reaching 400 tons daily with the company furnishing the Seaboard Air Line Railway 100 tons per day as part of their contractual obligation. The mines employed 200 men, up from 163 the previous month, with twenty-three expert miners from Pennsylvania having recently arrived. Henszey hoped to add a new shaft with an output of 1,000 tons daily. To accommodate the increasing labor force, Henszey stated that a new boarding house, in addition to the existing ones, was being built.[40]

However, several weeks later, on Sunday, February 27, the company was beset by a fire that damaged the outbuildings and equipment near the mouth of the mine. Fortunately, no workers were in the mine at the time as work ceased on Sundays. One observer noted that had the mine been operating at full capacity, the number of deaths would have exceeded the thirty-nine men killed in the December 19, 1895, explosion.[41] Not to be deterred, the owners of the Cumnock Mine soon invested $30,000, a considerable amount of money at the time, to replace the burned, unusable machinery and predicted that work at the mine would resume May 1st.[42] As it turned out, mining operations would not commence until late June, when the *Asheville Citizen* observed that the testing of the machinery supplying air to the lowest workings of the mine was successful and that hoisting of 300 tons daily would begin June 27.[43]

Again, the resumption of mining operations planned for late June did not materialize. It was reported that on August 1 the output of the Cumnock mine would be 300 tons daily and that all coal mined at Cumnock was to be shipped to North Carolina clients only. Customers such as the Seaboard Air Line Railway, the Raleigh Gas Company, and an undisclosed Raleigh cotton mill would be the recipients of most of the company's coal supply. Henszey was determined to make good his contractual obligations with local companies which would put him in a more favorable position to expand operations outside the state later. Henszey not only owned the product but his control of the Raleigh and Western Railway, of which he was president, gave him an upper hand over Langdon.[44] However, Langdon was not about to be deterred when he organized the North Carolina Coal and Coke Company, situated on the Taylor place near Gulf, to compete with Henszey.[45]

In 1898, the North Carolina Bureau of Labor Statistics reported that there were eight coal mines in the state with five in Chatham County, and only one working, the Carolina Coal & Coke Company. The mines included, in addition to the Carolina Coal and Coke Company, the following: Taylor Place Coal Association, Chatham Co.; Cumnock Mines, Cumnock; Kohinoor Coal Co., Gulf; Glendon & Gulf Mining & Manufacturing Co., Gulf; John Dye, Sanford; M.C. Starbuck, Sanford; and, C. Haninton, Walnut Cove (Stokes Co.). Production for August, September, and October reached 2,528 tons.[46] In January 1899, it was announced that Samuel A. Henszey, owner of the Cumnock mines and president of the Raleigh and Western Railroad, had moved his headquarters from Cumnock to New York City, and that John Connolly had been appointed superintendent of the mines. Edward H. Barnes became general superintendent of the Raleigh and Western Railway and general agent of the Cumnock mines.[47] In terms of overall quality, the Bureau of Labor Statistics in 1898 reported that 40,000 pounds of Cumnock coal did as much work as 52,000 pounds of Pocahontas coal.[48] According to the Annual Report of the United States Geological Survey for 1899, North Carolina accounted for 28,853 short tons of coal.[49]

Coal mining often attracted individuals who had on-the-job experience, and there was typically enough work available to easily replace or add men in and around the mines. Pay was usually better than what could be earned in most non-skilled professions though the hazards associated with mining were greater than in most other kinds of work. Those from the larger coal-mining states and countries were especially valued for their work experience. The 1900 U.S. Federal Census for Oakland Township, Chatham County, near the location of the Cumnock mine, showed that thirty-six employees of the mine, of American, German, and Austrian descent, were living within the township. Of the Americans, two were from Pennsylvania and one from New Jersey. Another resident, a physician, who lived in a boarding house with six mine employees, may have been the coal company's doctor. One miner named William Hill, who was born in England, included in his household a boarder, also a miner born in England. Seven miners were residing in Enumeration District 0008, Gulf Township, and one in Enumeration District 0011, which comprised Matthews and Siler City townships.[50] There were probably additional miners or mine laborers in other locales of Chatham County who worked on a seasonal basis in other occupations such as farming where there was a hiatus between the planting and harvest seasons.[51]

Following the May 22, 1900 explosion at Cumnock, the General Assembly in 1905 passed additional legislation to amend the existing mining law of 1897 which represented the state's first attempt to establish legislation to protect miners.[52] For example, in the 1897 legislation, boys under the age of twelve were not allowed to work in mines, but with the new 1905 legislation the inspector had the authority to examine under oath such person and his parents or other witnesses as to his age.[53] Notices of accidents no longer pertained to only loss of life but to personal injuries as well.[54] No less than twenty amendments were made to the 1897 mining law, some dealing with specific safety issues, while others were of an administrative nature. Still, the legislation failed to go far enough to protect miners or provide accountability for mine accidents. Nothing was in place to ensure that accidents could be avoided or contained; the Commissioner of Labor and Printing, the state's mine inspector, was not required to have any mining experience, and, as in the past, no enforcement existed behind the legislation. Not a great deal was improved upon the 1897 law in the 1905 legislation and it would be years later before stricter laws were passed and put into place to ensure mine safety.

U.S. Bureau of Mines

Legislation, particularly of the safety type, can be one of the more anticipated results following a tragic event. From roof cave-ins to explosions to flooding to fires, coal mines were susceptible to a myriad of conditions and events, making it one of the most dangerous, if not unpredictable, occupations in the industrialized world. Miners "won" the coal by hand using a variety of methods, notably pick, shovel, and blasting. Beginning in the early- to mid–1800s, mining typically paid more than other unskilled labor-intensive jobs and throughout the nation's history mining has attracted men of diverse nationalities who set out to make their fortune in the gold fields or silver mines or simply to have the security and stability of readily obtainable employment. Most miners were aware of the risks associated with their occupation which could sometimes lead to complacency and accidents. The introduction of machinery in the mines in the late- nineteenth and early-twentieth centuries resulted in improved efficiency and dramatically increased coal output where it was used; however, this new technology had a darker side as human carelessness or faulty equipment could often lead to electrocutions, haulage accidents, and entanglements in machinery.[55] Mine owners, intent on time efficiency and cost cutting, often disregarded the most basic safety measures while maintaining the prevailing attitude that a miner's death was part of the cost of doing business. A man could be replaced more readily and cheaply than an expensive piece of equipment.

As stated previously, mining laws were created and maintained by the individual states, usually with no enforcement or funding to back them. Such was the case with North Carolina's mining law of 1897 following the December 19, 1895, mine explosion at Cumnock. Moreover, there was no central federal agency that was responsible for regulating the industry much less concerning itself with safety measures. That was about to change when on December 6, 1907, an explosion at the Monongah 6 and 8 mines at Monongah, West Virginia, resulted in the deaths of 361 miners, representing the nation's worst coal-mining disaster.[56] Accidents like those at Monongah and Cherry, Illinois, in which 259 men and boys died in 1909, along with increased demand for coal to fuel industries and generate power for municipalities, were the stimuli for mining reform in the early twentieth century.[57]

Government investigations into the prevention of mine explosions began in the Technologic Branch of the United States Geological Survey, of the Department of the Interior, in 1908, as the result of a succession of disastrous mine explosions that resulted in the deaths of 1,148 miners in 1907. In 1910, the work of the Technologic Branch was transferred to the newly created federal agency, the United States Bureau of Mines, with Joseph A. Holmes, former head of the Technologic Branch, appointed by President William H. Taft as its first director. Having served as chief of North Carolina's Geological Survey, Holmes's experience in mining applications, as well as having a reputation as an organizer and tireless crusader for safety, his appointment was a culmination of years of research, experimentation, and application. Under Holmes's leadership at the bureau, the first experimental coal mine, complete with observatory, was opened at Bruceton, Pennsylvania, near Pittsburgh, for the purpose of conducting large-scale tests in an actual mine to determine the causes of and prevention of mine explosions. Soon thereafter, the bureau took up development of rescue methods at disasters and education of operators and miners in safety techniques.[58] Holmes was

U.S. Bureau of Mines building at Pittsburgh, PA. Built in 1915, the building is now part of the Carnegie Mellon University campus (courtesy National Register Information System, National Register of Historic Places).

8. The 1895 Mine Explosion at Cumnock

personally responsible for introducing the "oxygen breathing apparatus" for mine rescue work which was also adopted by manufacturing plants and fire departments. Because of Holmes's tireless efforts, the Bureau of Mines could quickly respond to emergencies on the East Coast, having in its charge six mine-rescue stations, eight mine-rescue railroad cars, and one rescue motor truck among its arsenal of safety apparatus.[59] Also, of equal importance was his investigation into mine lighting, which resulted in a list of "permissible" portable electric lamps for use in gaseous mines.[60]

In addition to dispatching mine-rescue cars where needed, the Bureau's other function was as an experimental agency, conducting two types of tests. First, it examined samples of coal dust and mine gas received from mining companies and provided analyses of and technical data for company engineers and occasionally recommendations if the dust proved particularly dangerous. For coal mining safety products, the Bureau tested, among other things, explosives, motors, switches, flame safety lamps, gas detectors, and coal-cutting apparatus to determine if the product conformed to Bureau standards. However, the Bureau's most important contribution to mining

U.S. Bureau of Mines safety car (ca. 1920). The Bureau of Mines maintained a fleet of railroad cars specially outfitted for emergencies which could be dispatched to a scene within twelve hours. This photograph shows Bureau of Mines employees disembarking from a car to assist with rescue efforts at an unidentified location (U.S. Bureau of Mines).

safety was through scientific examinations and education, leading to improvements such as rock dusting and investigation into explosions.[61] No longer did safety measures have to rely on actual explosions in mines to be implemented. Using the bureau's experimental mine, these measures could be tested and examined without posing a threat to human life.

9

The 1900 Explosion at Cumnock

On Tuesday afternoon, May 22, 1900, an explosion at the Cumnock mine wrapped yet another shroud of death around this mining community in Chatham County. Newspaper reports of people walking about the area dazed, shocked, and confused were all too reminiscent of the events of the December 19, 1895, explosion. This latest disaster at Cumnock had far-reaching ramifications. The *Raleigh Times* reported that the devastating effects occurred mostly in the east slope, where mine superintendent John Connelly, of Pennsylvania, met his death.[1]

The *Greensboro Telegram*, in its May 24 edition, reported that the sound of a whirlwind came up the shaft followed by fumes of escaping gas. Soon, the survivors began to appear at the mine entrance and messengers were summoned to obtain medical help at Gulf and Sanford. It would be 10 o'clock that night before the first body was brought to the surface. Ironically, it was that of the mine superintendent, John Connolly, whose back was broken in two places, presumably by having been hurled with great force against the wall.[2] Tragic stories soon dominated the scene as families witnessed husbands, brothers, and sons reduced to lifeless forms. Had Cumnock become cursed or jinxed as some thought it to be?[3]

In its annual report for 1900, the North Carolina Bureau of Labor and Printing provided coverage of the events at the mine. It theorized that part of the blame should be shared by the legislative committee that had stricken the provision for an appropriation to carry out the work stipulated in the mining law, resulting in noncompliance by the mine owners. Citing Ohio and Pennsylvania mining law, the commissioner of the bureau, Benjamin Rice Lacy, noted that their respective legislatures had made large appropriations for enforcing the mining laws.[4] Indeed, two Raleigh newspapers reported that upon being approached about regular inspections of the mine, Commissioner of Labor Statistics B.R. Lacy responded that lack of an appropriation to carry out this provision of the 1897 led to no inspection of the mine. Furthermore, Lacy was criticized for not inspecting or receiving regular reports of any mines, particularly from the mine superintendent at Cumnock, which again ran contrary to prevailing mining law.[5]

W.J. Tally, general agent of the Cumnock Mines, in compliance with section six of the 1897 mining legislation, reported to Lacy the details of the May 22 explosion at the Cumnock mine. At approximately 4 p.m., an explosion ripped through the mine, immediately killing twenty people and injuring five additional men. The "gas inspector," or fire boss, had completed his tests for the presence of gas and determined that the mine was safe to enter. The miners' lamps were all examined and cleaned and deemed in good working order. After the explosion, a safety lamp was found with a broken glass, which may have been the cause of the explosion, or its result. The mine employed a state-of-the-art ventilation fan at the top of the second opening as well as a new condenser in the main shaft. As to the mine's ventilation, Tally stated that adequate air was circulated through the mine so as to dilute and expel poisonous and noxious gases in every shaft, and that the equipment was in working order.[6]

TWENTY-THREE MEN KILLED AT CUMNOCK

Heart Rending Scenes at the Mouth of the Shaft

SUPERINTENDENT CONNELLY AND
CONTRACTOR McCARTHY KILLED

Headlines in local newspapers reported the explosion at Cumnock in 1900. Here, one article announces that mine superintendent John Connelly was killed in the explosion. The coal company hired Connelly, a native of Pennsylvania, for his mining experience (*Raleigh Times*, May 23, 1900).

Tally gave the opening testimony at the coroner's inquest, first reporting that one victim injured in the accident had since died. He then provided a list of those killed in the explosion, including the mine boss, John Connolly. J.D. Hart, Sr., who was next to appear before the inquest, was the mine inspector and stated that he was in charge of monitoring the gas in the mine and noted that gas was always present in one form or another. Hart was an experienced mine inspector and familiar with the Cumnock mine. He estimated the location of the explosion occurring in a twenty-inch seam of coal underneath the workings. Hart, who was not in the mine at the time of the explosion, entered the mine and workings to investigate the conditions within an hour and experienced a great deal of heat, which was causing the mine timbers to burn. Connolly was found twenty feet from the compressed-air engine, which was used to pull the coal up to the surface from the dip. Hart attributed the explosion to firedamp and the workers' deaths to afterdamp, which formed after a mine fire or explosion of firedamp and was irrespirable.[7] Coal was extracted using pick and blasted with a battery, or blasting machine. James McCarthy, another victim of the explosion, was a contractor who was in charge of the hired help.

George N. McNath, the lamp man, was responsible for the repair and tracking of the company's lamps taken into the mine. McNath testified that the lamps were in various forms of disrepair but that he did not allow any lamps with cracked glass to enter the mine workings. According to his testimony, lamps in use at the mine were completely safe.[8] After hearing testimony of several others, the jury concluded that the explosion was caused by gas, and that the deaths of the men were attributable to

9. The 1900 Explosion at Cumnock

afterdamp. How the gas was ignited was not determined, though W.E. Faison, Assistant Commissioner of the Bureau, speculated that gas had accumulated in the mine during the day and somehow entered the lamps of the miners and burned until one of the lamps exploded, causing the glass to break. The flame then touched off the gas in the mine resulting in the explosion. In terms of the deaths, afterdamp was deemed the most likely cause.[9]

Judge T.B. Womack, attorney for the mines, when interviewed by the press, stated that the explosion was the result of an unavoidable accident. Citing the evidence demonstrated in the proceedings, Womack added that gas was constantly generated in the mine and that safeguards had been established to render it harmless. The fire boss, who was responsible for checking the workings for gas, testified that occasionally there was an unexpected rush of gas from a hidden reservoir and that the only recourse was to divert it as rapidly as possible into the main air shaft.[10] The more experienced miners concurred that a sudden generation of gas in one of the workings entered the room in which Sim McIntyre was working and that his lamp overheated and cracked a portion of the glass chimney, "enabling the flames to reach the gas."[11]

Womack continued his scenario of the explosion, stating that it occurred in or near this room and "extended a space of three hundred feet." McIntyre's body was the last found and within ten feet of it was his lantern, securely locked, but with a hole in the glass chimney. This scenario indicated that the glass was not

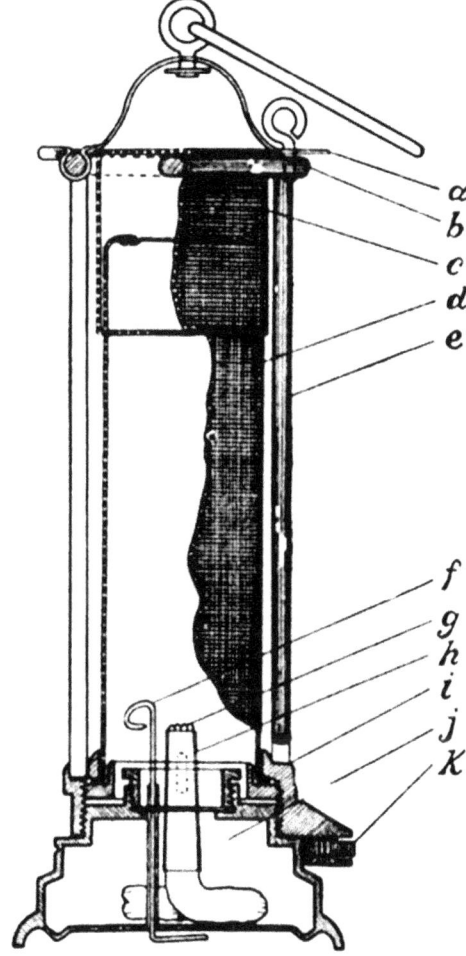

FIGURE 7.—Davy lamp, unbonneted: *a*, Brass hood; *b*, brass ring, top; *c*, gauze cap; *d*, main wire gauze; *e*, iron standards (three); *f*, wick snuffer (pricker); *g*, cotton wick; *h*, wick holder; *i*, brass ring, lower; *j*, brass oil font; *k*, key lock

Cut-away diagram of an early model Davy safety lamp invented in 1815 by Sir Humphrey Davy. Mine inspectors and foremen have used safety-lamp technology in mines to the present to provide illumination as well as test for gas. There have been a number of versions of the Davy lamp through the years but its design has remained basically the same (James W. Paul, *The Use and Care of Miners' Safety Lamps*, Miners' Circular 12, Department of the Interior, Bureau of Mines, 1913).

broken by the explosion but from the interior outward. Those killed came in contact with afterdamp as they were attempting to flee and had they remained in their own rooms, would have been saved, as the air in other portions of the mine was uncontaminated before and after the explosion. One miner who heard the explosion while underground continued to work at his place in an area removed from the blast some half hour or more after the explosion.[12] It was acknowledged that the officers of the mine did all they could to help alleviate the suffering and distress exhibited by the families. Connolly, the mine superintendent, was a graduate of a school of mining and held the highest certification of proficiency in his field granted under the Pennsylvania mining laws. Under his management, coal production at the Cumnock mine had largely increased and but for the explosion would have realized the greatest in the history of the mine. Summing up the proceedings, Womack declared that demand for coal far exceeded output and that the mine would be put operational the following Monday.[13]

The recent explosion occurred when the Cumnock mine was operating at peak production. There is never a convenient time for a mine tragedy, much less any tragedy, especially when lives are at stake. The twenty-three deaths in 1900 and the thirty-nine in 1895 brought the total number of mining-related fatalities in the state to sixty-two in less than five years and disrupted coal production, hampering an industry that was once seen as a boon for the area's economy.[14] For the decade of reporting, from 1891 to 1900, North Carolina ranked fourth in the nation in the number of explosion-related mining deaths, a statistic not too favorable for future investors, not to mention the communities that were devastated.[15] Perhaps even more worrisome was the anticipation mine owners must have felt that another tragedy could be waiting to happen.

By the end of May, prospects for the continuance of the Cumnock mine appeared encouraging. The Wilmington *Semi-Weekly Messenger* announced that a new company from Philadelphia, PA, would take ownership of the mine.[16] On June 6, the same newspaper carried news that no damage suits resulting from the disaster would be filed against the company, thus avoiding costly litigation for owners and shareholders.[17] In July, a deed was registered in the Chatham County courthouse by W.L. Austin, of Philadelphia, to James H.M. Hayes, of Philadelphia, for the sale of the Cumnock mine and associated property with it.[18] A new direction for the company meant that production would resume and hopefully more safeguards put in place to avoid another tragedy that cost dearly in terms of the number of lives lost.

In October 1901, Raleigh's *News and Observer* printed a story announcing that H.K. Meyers, president of the Chatham Coal and Iron Company, had purchased the Cumnock mines and organized the company, having taken possession of the property on September 1st. Myers stated that the completely reorganized company would be adding new Harrison compressed air mining machines along with new hoisting engines and pumps. He anticipated immediate production to reach 600 tons a day and that several more veins would be opened, reaching a capacity of 2,500 tons daily, certainly a very ambitious prediction given the performances of previous years.[19] However, late in the month, Judge T.R. Purnell issued a temporary injunction, restraining the Chatham Coal and Iron Company from selling or using as collateral "any of its recently authorized bonds or disposing of any funds on hand."[20] Even in the midst of these imposed orders, the paper affirmed that mining operations would not be disrupted.

The *News and Observer* reported in late December 1901 that a coal scarcity existed at the Cumnock mine as a result of a shortage of coal cars, crippling miners' strikes elsewhere in the country, and miners quitting work before the holidays. The Seaboard Air Line Railway, which had contracted with the company some years previously to purchase Cumnock coal for its locomotives, decided to undertake the expensive proposition of converting its engines from coal to wood on the Hamlet to Wilmington route. The newspaper further noted that if the coal shortage did not improve, engines designated for the Charlotte route and other points would experience a similar fate. The Seaboard, which had taken de facto charge of the Cumnock mines, was using as much as seventy-five tons of coal daily and brought in forty additional miners to work the seams with the expected result of at least keeping up with demand. The capacity of the mine was 300 tons daily with its current machinery and the Seaboard intended to do all it could to bring the yield up to that figure.[21]

The *Chatham Record* reported in its October 23 issue that S.A. Henszey, president of the Raleigh and Western Railroad, announced that the road would be extended from Cumnock to Greensboro, giving the Seaboard Air Line an entrance into the Gate City.[22] On October 30, the Cumnock mine reported current output of 100 tons daily and a promise of 300 tons soon to follow. Optimism prevailed when some experts predicted a daily output of 1,500 tons for forty years to more than meet demand. During twenty days in September, output at Cumnock reached 2,150 tons.[23] However, the same newspaper reported that coal was no longer supplied to the Air Line Railway, which, during most of the year, had been taking the entire output of the Cumnock mine, and that all production was being diverted to industrial plants at Greensboro, Durham, Winston-Salem, and Raleigh. Still with the loss of railroad business, total output for 1902 reached 23,000 tons, a considerable increase over the 3,723 tons mined in 1901.[24]

On January 1, 1903, the *News and Observer*, touting "Cheap Fuel Here" in its coverage of the Raleigh and Western's imminent buyout of the Cumnock mines, enthusiastically endorsed the sale and turned its attention to the details in an interview with the Raleigh and Western's chief engineer, George C. MacGregor, who explained that the surveyed line from the connection with the Seaboard Air Line at Colon, thirty-nine miles southwest of Raleigh, to Greensboro had been located and adopted and that of this line eight miles between Colon and the Cumnock mines were now in operation with fifteen additional miles toward Greensboro graded. Additionally, two iron bridges of 100 feet in length, one for use at Cumnock and the other at Ramseur, at both of which points the line crossed Deep River, were waiting to be placed and assembled. The goal was to have access to industrial and manufacturing enterprises along the route for development and as carriers of Cumnock coal.[25] When asked whether he believed that Cumnock coal would exceed all other coals entering the region, MacGregor confirmed that it was being shipped to the Winston-Pocahontas territory and that the "home product" would monopolize the coal consumption at all the planned points on the line as no other coals could successfully compete with it in regard to the convenient location of the mines and "independent means of transportation." With businesses adjacent to the line benefitting from cheap coal, MacGregor stated that their inbound and outbound freight could be handled by the Raleigh and Western road. That neither the Southern Railway nor any interests connected with it were keen about building a road, all other enterprises disposed toward and fully able to build a route independent of the Southern Railway expected real profits from coal tonnage and "incidental sources."[26]

With the railroad's acquisition of the Cumnock mines, plans for sinking a shaft of large proportions were expected to realize an output of 1,200 tons daily, with the objective of supplying coal to adjacent markets and converting any surplus to coke for use in metal manufacturing, to be determined after a thorough investigation of the coal requirements and the costs involved in sinking a new shaft. Claiming that a large demand exceeded the supply on hand, officials of the railroad believed that output would be absorbed by industries along the route of the railroad. With all contingencies considered and in place, cost of coal at $1.50 per ton would not only be possible but practically assured. With the support of businesses along the route, cheap coal and competitive transportation were to be realized in the near future, and both bode well for the state.[27]

10

Coal Production in North Carolina, 1900–1925

From 1900 to 1905, the *North Carolina Geological Survey* reported statistics of the mining industry in North Carolina. The purpose of the reports was to describe the condition of the mining industry together with an account of the different types of minerals being mined in the state. For 1900, the Cumnock mines of the Chatham Coal and Coke Company at Cumnock were the only mines operating for the year. Nearly 18,000 tons of coal were mined with a total value of $22,500. However, as a stipulation, due to the number of obstacles in mining coal, it was recommended that any significant operations should be based on a thorough testing by boring. The total tonnage of coal raised for the year fell just behind that of iron, which, at the time, had a greater value than coal.[1] For 1901, coal production had dropped off considerably at the Cumnock mines from the previous year, realizing a total of only 3,723 tons with a value of $5,585.[2] Successful mining ventures in Moore County were anticipated to make up for the shortfalls but these never realized their potential and were eventually abandoned.

The year 1902 saw a considerable increase in the total amount of coal produced for the state at 23,000 tons valued at $24,500, or $1.10 a ton. A local state market existed for all the coal that could be produced.[3] For the year 1903, the total amount of coal produced for the state was 17,309 tons valued at $25,300, or $1.46 a ton. It was noted that due to the high sulfur content of the coal, washing could increase its quality and value, an area of concern that had a long history. The price of coal had seen a gradual rise since 1901, when it reached $1.07 in 1901, $1.10 in 1902, and $1.46 in 1903. The total production of coal in 1903 was down 5,691 tons but saw an increase of $800 in value. With promise of similar coal measures in Stokes County, near Walnut Cove, no efforts had been made to extract it though there were plans to do so in the near future.[4]

Distribution of the coal product from 1901 to 1905 showed that in 1901, 10,000 short tons were loaded at the mines for shipment, while 2,000 tons were used at the mine for steam and heat; in 1902, 20,400 tons were loaded at the mine for shipment, 100 tons were sold to the local trade or used by employees, and 2,500 tons used at the mines for steam and heat; in 1903, 14,429 tons were loaded at the mines for shipment, eighty-seven tons were sold to the local trade or used by employees, and 2,793

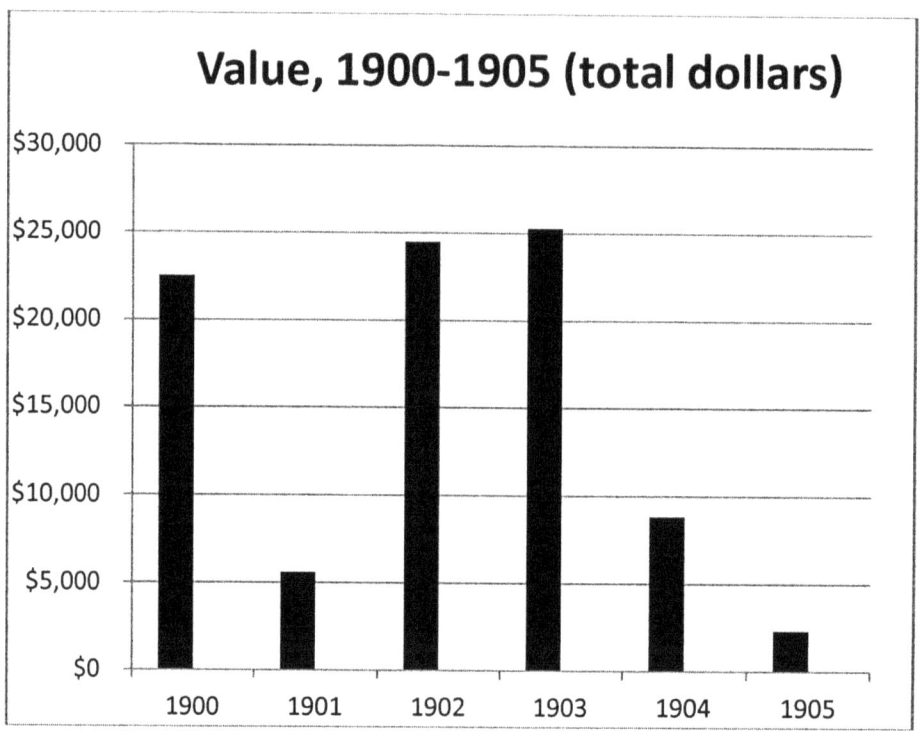

tons used at the mines for steam and heat; in 1904, 4,600 tons were loaded at the mines for shipment, 300 tons were sold to the local trade or used by employees, and 2,100 tons used at the mines for steam and heat; in 1905, 461 tons were loaded at the mines for shipment, 1,096 tons sold to the local trade or used by employees, while no tonnage was reported for use at the mines for steam and heat.[5] Coal measures in Stokes County had been prospected for a while but no real development had taken place to determine the quantity and value of the measures. Exposed coal seams were five-feet wide containing three feet of a hard, compact coal and the other two feet running into fines.[6]

On March 6, 1905, the legislature ratified an act to create the North Carolina Geological and Economic Survey with an annual appropriation of $10,000. The newly appointed state geologist was charged with the task of appointing, with board approval, experts and assistants "to carry out successfully and speedily" the work of the survey. Among the agency's objectives were examinations of the state's natural and material resources, geological formations with regard to their economic potential; road-building materials and the best methods of utilizing them; soils, forests, and physical features in regard to "their bearing upon the occupation of the people"; streams and water power; scientific and economic issues deemed of value to the people of the state; preparation of reports whose results could be put before the state's citizens; and water supplies of the state with respect to sinking deep and artesian wells. The state geologist, again with the approval of the geological board, was to arrange and accept cooperation from federal agencies in completing topographical surveys of the state, as well as carrying out further provisions of the act.[7] With the passage of the act, greater emphasis was placed on the economic value of the survey as opposed to being only a research and statistical agency within state government.

In January 1906, various newspapers reported that Samuel Henszey, president and principal owner of the Raleigh and Western Railway, had plans to build a line of his railroad from Cumnock to High Point. Grading had been completed but construction faltered due to Henszey's health problems.[8] In April, Henszey, owing to his deteriorating health, had been tendered an offer to sell his business interests in North Carolina to an unnamed syndicate.[9] Also during the year, D.P. Bible, of Newport, NC, began to explore the coalfield in Moore Co., to the west of Cumnock, managing to extract 200 tons of coal from the area for sampling though none was sold or shipped.[10] Coal production had decreased to all-time lows, particularly over the previous three years, while nearby states, like Kentucky, were showing significant gains.[11] Indeed, coal mining in North Carolina was performed on an irregular basis and only met a local need. No statistics for coal mining were reported in the state for 1906 and 1907 due to the closure of the Cumnock mine.

However, the year 1906 brought changes in ownership at the mine. The *Chatham Record* reported in August that Edward W. Shedd had purchased the mine along with the Raleigh and Western Railroad. Described as the promoter of the Randolph & Cumberland Railroad Co., a small former logging road, which had an eight-and-a-half-mile link from Carthage to Hallison, North Carolina, Shedd retained Samuel Henszey as chairman of the board of directors of the Raleigh and Western Railway Company.[12] The charter granted by the legislature gave Randolph & Cumberland permission to build a line 117 miles long from Fayetteville to Deep River, in Moore County, then along the river to such a point that the directors may determine the line passing through Cumberland, Moore, Randolph, and Guilford counties. The charter was for ninety years with capital stock of $1,000,000. Edward Shedd; Morrison E. Caldwell, of Aberdeen; and T.J. Edwards of Providence, Rhode Island were listed as the principal shareholders of the company.[13] Perhaps the most encouraging news came from the *Goldsboro Daily Argus*, which reported the recent investigations into the prospects of mining for oil at Deep River.[14]

More speculation centering on the Cumnock mines arose when Raleigh's *News and Observer* in August 1907 carried a story that first appeared in the *Sanford Express* involving Shedd's plan to divide the Lee County property up into farms of twenty- to thirty acres and settle German immigrants upon them. It was thought that the Germans would engage in farming, tending to the uncultivated land that meant prosperity for the area. The Cumnock mine, which had been shut down the previous year, was now being considered for reopening the shaft or digging a new one near the river. Rumors hinted that the Raleigh and Western Railroad would be extended from Cumnock on to Harper's Creek crossroads and Enterprise Factory, where it would connect with the Randolph and Cumberland Railroad, which was being extended from Carthage to High Point.[15]

No coal production was reported from 1908 to 1910 according to the North Carolina Geological and Economic Survey.[16] The next significant event for the Cumnock mine was announced in the *Chatham Record* in November 1909 concerning the sale of the mine at public auction along with 2,734 acres of property. The mine had not been worked in several years and was in a state of disrepair and filled with water, which seemed to be the nemesis of anyone expressing an interest in resuming operations.[17] The same newspaper reported in December that the Chatham Coal and Iron Company would be sold on January 10, 1910.[18] The sale was made under a deed of trust and brought $61,840, being bid on by the bondholders of Philadelphia.[19]

In January 1912, the *Chatham Record* reported that in late December the Cumnock coal mine property in Chatham and Moore counties had been purchased by Major William A. Guthrie, of Durham, as agent or trustee for undisclosed principals, who were supposed to be John Lennig of Philadelphia and associates.[20] In March, the *Charlotte Observer* reported that mining efforts had commenced at Cumnock on the Foushee place between Glendon and Cumnock in Moore County. A.J. Jones, owner of the property, had a number of men there working the outcroppings and descending the shaft, which was near the track of the Raleigh, Charlotte & Southern Railway and was between 300- and 400 feet deep. The mine was worked on a small scale by the Durham & Charlotte Railway Company, constructed by Lennig, who sold it to the Raleigh, Charlotte & Southern Railway. The coal was found in pockets with the hope that with further digging more deposits might be found. The coal was of a good quality and was being used in the engines of the locomotives that belonged to the Raleigh, Charlotte & Southern Railway Company as well as the manufacturing plants of the talc companies near Glendon in the manufacture of talcum powder. A carload of coal was shipped to Sanford for trial use and was found to be very fine for domestic heating purposes.[21]

The year 1912 realized only a small increase in total coal production at 120 long tons. Between 1913 and 1917 no coal production was reported.[22] Much anticipation in 1916 of a U.S. Government contract to manufacture armor plate at Fayetteville for the war effort revived interest in the Cumnock mine as a source of fuel for the enterprise. Despite the unforetold costs associated with reopening the mine, the article optimistically touted the mine's potential as a source of fuel for all local industry which could help the state reawaken from its "Rip Van Winkle" reputation of lack of progress and initiative.[23] The old canard that North Carolina could become a considerable producer of coal had its roots in a long history of attempts to tap the region's only sizeable coalfield and iron ore deposits in Chatham and Moore counties, each one promising success but ultimately failing which the newspaper article attributed to poor management while completely ignoring the actual mining challenges of faulted seams and constant flooding.[24] Although Fayetteville eventually lost out on the government contract to produce armor plating, it provided community leaders and politicians

10. Coal Production in North Carolina, 1900–1925

with an opportunity to take stock of the many resources the area had to offer businesses and promote enterprise.[25]

The Norfolk Southern Railway announced in 1917 its intention to reopen the Cumnock mine for its own use, mainly to fuel its locomotives, and planned to implement operations under the name of the Cumnock Coal Mining Company.[26] The railroad had acquired the Cumnock property from the Egypt Improvement Company, which was chartered in 1912 by W.A. Guthrie, J.J. Tull, and others, including Pennsylvania capitalists, to develop mineral and timber land at Cumnock.[27]

In February, J.H. Young, president of the Norfolk Southern Railway, along with two of his officials, visited Cumnock where the Piedmont-Cumnock Coal Company was expanding efforts to bring the mine into working condition for the purpose of testing the quality of the Cumnock coal to be used in its steam locomotives. A sufficient amount of coal had been extracted to begin engine tests with the result that the Cumnock coal was of good steaming quality. Plans to expand production were made and the work of pumping out the water in the workings had begun. Upon dewatering the mine, repairs were required to the old machinery before bringing in new equipment,

Egypt store, Lee County. Owned and operated by John H. Kennedy from 1913 to 1922, the store stocked provisions for miners and their families. Kennedy also served as the secretary-treasurer of the Egypt Improvement Company, which played a large part in developing mineral and timber lands in Lee County (*Sanford Herald*, April 1, 1937).

along with readying the headings and rooms for extensive mining operations. As in previous attempts with this mine, much hope was placed on finding significant deposits of steam coal for the benefit of the Norfolk Southern and all concerned.[28]

The general manager of the Cumnock Coal Mining Company, W.W. Brewer, stated that it would be some time before actual operations could commence because of the neglected condition of the mine which necessitated considerable preliminary work, especially pumping out the standing water, which all shaft mines were prone to. That the mine was located on the lines of the Norfolk Southern Railroad between Raleigh and Charlotte provided a ready outlet between two large population centers east and west.[29] In August 1918, the *Wilmington Morning Star* reported that the Cumnock Coal Mining Company was mining sixty tons of coal weekly from its shaft, which was at the time 180 feet deep. The continued work of removing rubbish and the remaining standing water, along with re-timbering the mine in the 460-foot shaft, was on schedule to be completed by October, when the company expected to produce between 300 and 400 tons of coal daily or 2,100 to 2,800 tons weekly instead of the current sixty tons. Twenty-six new homes were being constructed to accommodate the miners and their families, with all indications pointing to a community conducive for work.[30]

In July 1918, the *Chatham Record* announced that the Carolina Coal Company in Chatham County, was being sold to a group of Northern capitalists. Workers had previously been engaged in removing water from the workings and the job of timbering up the mine became a main focus. With those tasks accomplished, it was predicted that the mine would be working at full capacity. Already, some coal was being brought up from the main shaft, and as part of its commitment to the community, the company was preparing houses for workers and their families, including water and sewer service.[31] In his 1919–1920 biennial report, state geologist Joseph Hyde Pratt was sufficiently concerned that the coal and wood supply of North Carolina had noticeably diminished over the years, cutting into reserves and fewer props to bolster mine roofs. The Geological Board of the North Carolina Geological and Economic Survey intervened to seek new sources of power to keep up with the demand brought on by the rapid growth and expansion of towns and cities in the state.[32] Although coal production had increased significantly from its 1918 total of 1,420 long tons to 6,989 tons in 1919 and 11,540 tons in 1920, the state also looked to the utilization of its under-utilized water power as a source for the growing manufacturing industries particularly where they were unable to secure power.[33] It was hoped that together with appropriate coal-mining methods, utilization of alternate power sources could help alleviate the acute fuel shortages already being experienced throughout the state.

Meanwhile, the Carolina Coal Company, started to make inroads into its fledgling mine near Farmville, formerly Farmersville, in Chatham County. Operating on land adjoining the Cumnock property, the company had shipped two cars of coal from its Carolina mine to Pinehurst followed by two additional carloads in order to meet fuel demand for the popular resort town. Some of the coal was used during the winter at the Southland Hotel in Southern Pines.[34] The company appeared to be making more significant strides with its coal operations than was previously seen in the efforts of the Piedmont-Cumnock Coal Company. Indeed, the Carolina Coal Company was adding a larger workforce to dig deeper shafts in order to reach the best quality coal but the amount that had already been extracted was of a sufficiently high quality and encouraging to the management of the company. With both companies—the Carolina Coal Company and the Cumnock Coal Mining Company—working adjacent to each other with optimal speed, the prognosis looked excellent for North Carolina to become independent of outside production.[35]

In January 1921, the *Greensboro Daily News* reported that the Cumnock Coal Mining Company was preparing to mine coal on an even larger scale with the installation of two new Babcock engines with a capacity of 125,000 horsepower each. Forty houses for miners and their families had been built to accommodate the steady stream of workers into the area. Ventilation of the mine was by mechanical fan, which forced fresh air through the shaft and expelled dangerous gases. Between seventy and eighty workers had been employed by the company with a daily output of eighty tons and a monthly payroll of $6,000. Soon, plans were in place for opening a mine on the north side of Deep River in Chatham County.[36] Coal production for the year 1921 reached 23,438 long tons, an increase of 11,898 tons over the previous year.[37] Erskine Ramsay, of Alabama, purchased the Cumnock property in 1922 from the Norfolk Southern and soon after set up his own mining operations.

The Raleigh *News and Observer* announced in November 1921 that the Carolina Coal Company had struck a rich vein of coal in its mine in Chatham County. The vein was found at approximately seven-hundred feet, measuring nearly four-feet thick with an eighteen-inch gap separating it from an adjoining vein two-feet thick.[38] The Carolina Coal Company, of which John R. McQueen was president and Bion H. Butler, vice president, had previously opened the mine near Sanford with favorable results. Stockholders of the company included, besides McQueen and Butler, W.H. McNeill, of Alamance Co.; Howard N. Butler, of Southern Pines; C.M. Reeves, of Sanford; F.A. Lane, of Pennsylvania; W.A. Jones, of Chatham County; the Rev. Angus McQueen, of Dunn; T.J. Purdie, of Fayetteville; Ed Purdie and W.J. Purdie, of Dunn. Lane, who had considerable coal experience and had known Bion Butler from childhood, assumed management of the work until pressing business matters required him to return to Pennsylvania. An operating committee comprised of Reeves, McNeill, Butler, and T.J. Purdie was elected to assist the president in the daily business of the mine. Soon after locating the coal seams, the company began the task of driving air courses along with the nearly 2,000 feet of tunnels and headings. The high-grade bituminous coal that characterized the seams compared favorably with the well-known coals of the Pittsburgh district in Pennsylvania. In order for coal to be loaded at the mine, a mile and a half of railroad was planned for construction.[39]

In an article appearing in the November 23, 1922, issue of *Coal Age*, Butler described the development as a slope driven on a twenty-seven-degree pitch into a seam of excellent, high-volatile coal about four-feet thick. After extended investigation of the Deep River coal bed the company obtained a 1,200 acre tract of land and probed the area with core drills to determine the extent of their deposit and then proceeded to sink their slope and open a mine, realizing approximately 3,000 feet of underground workings. At the Carolina and Cumnock mines, about fifteen- to twenty cars of coal were being shipped weekly, with plans to increase production to 1,000 tons daily when fully developed. The company also owned a large tract of land held in reserve under option or lease or through the ownership of stockholders. A preliminary survey made in November was sufficiently favorable to persuade the federal government to make a more thorough survey of the coal basin in order that the deposits may be brought to the attention of developers and its resources made available for the expanding industries of the state. Noting that the state was becoming a prominent manufacturing center, with substantial industries in cotton, tobacco, and furniture-making, Butler underscored the importance of proximity to fuel supplies and with the Deep River coalfield lying two- or three hundred miles from competing coal the state had the inside track in its market.[40]

One of the most ambitious studies of the Deep River coal region up to that time was conducted in 1922 by geologists Marius R. Campbell and Kent W. Kimball, of the United States Geological Survey, who made a complete geological analysis of the Deep River coalfield.[41] In describing the coalfield, Campbell and Kimball referred to an upper bench and a lower bench. It was in the upper bench, located in the southern half of the basin, where the seam varied in thickness from three feet to four feet, which was considered as commercial coal. The geologists put the estimated recoverable coal at 68,000,000 tons and stated that mining could be carried on profitably to a depth of 2,000 feet. One suggestion in the report, based on laboratory analysis, was to coke the coal as tests showed that the coal would make a very good quality coke, comparable to Pennsylvania cokes. Uses of coke included domestic heating, agricultural applications, and generating and transmitting electrical power.[42]

In determining the quality of Deep River coal, the authors collected samples from both the Carolina and Cumnock mines to send to the Pittsburgh laboratory of the Bureau of Mines for analysis and comparison to coals from surrounding regions.[43] When all testing, representing various categories, had been completed Campbell and Kimball found that the heating value of Cumnock coal was somewhat inferior to Pocahontas and New River coals but about the same as the coals mined at Dante, Toms Creek, and Big Stone Gap, in Virginia, and was considerably better than some of the poorer coals mined in the area. At the request of Joseph Hyde Pratt, state geologist, the U.S. Bureau of Mines made tests of the coking quality of Deep River coal. A sample from the top bench in the Cumnock mine was tested at the experiment station of the bureau for its coking properties and yield of by-products. The result showed that the by-product yield was satisfactory, comparing favorably with that of Pennsylvania. However, the coking quality of the Cumnock coal was called into question because of its high sulfur content. The recommendation for producing a quality metallurgical

10. Coal Production in North Carolina, 1900–1925

coke required adding a mixture of coal containing less sulfur, while noting that the best market for Deep River coke was for domestic use.[44]

Campbell and Kimball next shifted their attention to the prospects of extracting petroleum or gas from the surrounding sandstone in the Deep River coalfield. Based on visiting a number of localities projected to show signs of oil, the authors found none. The black shale and coal of the Cumnock formation, which indicated the possibility of material for the formation of oil, in reality produced only a negligible amount in the past when prospects looked favorable. The authors opined that drilling of boreholes in the area to reveal oil reserves was an expensive proposition with little financial return.[45] In their concluding remarks about their study of the Deep River coalfield, the two geologists offered no encouraging words for future development of the area for oil and coal production. Part of their time was spent trying to find a location to drill a test well but concluded, based on their analyses, that one location was not more favorable than another and that no one, including the most experienced geologist, could definitively state that oil could be extracted from the rocks. Given the conditions and limitations, many wells would most likely need to be sunk with the high probability that efforts would be futile. In their concluding remarks, the authors recommended that the state appropriate funds to improve roads or improve the soil or some endeavor that would benefit the entire community.[46]

Erskine Ramsay (1864–1953). His purchase of the Cumnock property in 1922 to mine coal was beset with problems at the outset and sold within a few years. Even his considerable experience in the Alabama coalfields and substantial resources couldn't compete with the challenges of Cumnock (Daniel D. Moore, *Men of the South: A Work for the Newspaper Reference Library*, New Orleans: Southern Biographical Assoc., 1922).

If ever there were a more experienced coal man to enter the scene for the purpose of mining coal at Deep River, one need look no farther than Erskine Ramsay, whose coal-mining acumen received high acclaim while Vice President and Chief Mining Engineer of the Pratt Consolidated Coal Company of Birmingham, Alabama. Born near Pittsburgh in September 1864 to Scottish parents, Ramsay entered mining at an early age, becoming at age eighteen the youngest superintendent of a mine in the entire Connellsville (PA) coke region when in 1882 he took charge of the Monastery Mine, one of Andrew Carnegie's operations.[47] He could boast among his mentors the likes of steel moguls Carnegie and Henry Clay Frick.[48] Ramsay not only had proven himself a capable administrator but his inventions, including a rotary dumping mechanism for coal cars and a coal and mineral washer, proved to be practical applications developed by a practical man with vision. Ramsay's experience as a mining engineer and coal entrepreneur made him uniquely qualified to solve the inherent difficulties of mining Deep River coal.

Setting to work almost immediately, as was his custom, Ramsay, along with other investors, in September 1922 purchased from the Norfolk Southern Railway the 4,000-acre Cumnock coal property for mining with Ramsay acting as president and his brother Charles serving as first vice president and general manager. C.M. Reeves, vice-president of the company, and John R. McQueen, its director, also served in the same capacity at the Carolina Coal Company.[49] The Cumnock property had a long history of ownership as well as problems from devastating explosions in 1895 and 1900 to periodic flooding resulting in mine closures with little or no production. Bolstered by favorable reports from his own engineers, Ramsay felt confident that he would be a producer on a large scale with this latest enterprise, the Erskine Ramsay Coal Company.

In October 1922, the *News and Observer* enthusiastically recognized Erskine Ramsay as a proven captain of industry and philanthropist whose purchase of the Cumnock property held promise for putting the region on the map in the years ahead. Meanwhile, the Carolina Coal Company, two miles east of Cumnock, had plans to install a new eleven-ton hoisting engine and build an auxiliary plant near the mine with the addition of a spur to connect with the railroad on the Cumnock side of Deep River.[50] This auspicious beginning for Ramsay's company encouraged interested parties to believe that significant coal production could be achieved on a regular basis. Ramsay's investments in mining coal at Cumnock enabled him to become an active producer, yielding 78,500 tons for the year 1922, an increase of 55,062 tons over the previous year's totals while meeting contractual obligations for the railroad.[51]

At a meeting of the stockholders of the Carolina Coal Company in May 1923 at Sanford, the Vass (Moore County) *Pilot* reported a proposition put forward by the Sandhills Power Company for a site on the Carolina Coal Company's railroad near the coal mines and a supply of coal to the power company for "a big steam power plant auxiliary to the water power of the Sandhills company." The steam plant would generate 1,200 horse power at the unit under consideration and provide power at times when high water or scarcity of water interfered with required water power. John R. McQueen, director of the coal company, had already intended to try out the use of steam power with the possible result of "using much of the coal from the mine to make power right there at the mines."[52] It was believed that coal power generated at the mines, taken together with the vast supplies of coal on hand and no cost involved to move the coal to its destination, added incentive for future development of the area.

One concern about transportation was the slow process of moving the coal from

10. Coal Production in North Carolina, 1900–1925

the mine to a truck for delivery to the railroad. However, it was pointed out that grading of the company's railroad from the mines to the junction near Cumnock would be completed soon and that the laying of track had already commenced. Early tests involving a locomotive for use on the new road were encouraging and soon the loading of coal onto awaiting cars at the mine opening appeared to be a reality. The machinery for the new power plant had been purchased, and work toward completing the buildings was well underway, including that of the 150-feet high stack, making it the highest in the area.[53]

However, the prospects for the Erskine Ramsay Coal Company began to decline soon after its purchase of the property. It appeared that Ramsay's engineers who were sent to the area on a fact-finding mission to analyze the coal deposits reported that the coal was of a high quality and ran thirty-five to forty-two inches in the seam with about two and a half inches of fair bone coal at the bottom of the seam. Based on their calculations, the engineers believed that the seam would yield about 3,500 tons of coal per acre which would be extracted at a rate of 400 tons daily. They also surmised that the coal could be sold to a market not affected by competition.[54] Upon further examination of the seam, Erskine and his brother Charles, a mining engineer in his own right and an investor in the property, realized that because of the manner in which the Deep River field was formed, great dikes cut through the coal seam and a substantial amount of coke had naturally formed when boiling lava pushed up the coal vein, making rooms almost inaccessible for mining. Faced with the prospects of removing the coke and locating a mineable seam to exploit, the brothers were confronted with a greater challenge they had not anticipated.[55]

In 1923, after the death of his brother Charles, Erskine began to give large blocks of company stock to his brother's estate, and on December 31 transferred the remainder of his stock in the company and resigned from the presidency and chairmanship of the board of directors.[56] By that time he had already invested $170,000 in the Deep River venture and stood to lose another $50,000 unless the mine began to show a steady profit. A devastating explosion on May 27, 1925, at the Farmville Mine of the Carolina Coal Company at Coal Glen, killing fifty-three men, still did not deter Ramsay's work as his company did revive slightly in 1926, but an explosion at his mine in November, killing two men and injuring two more, were two overwhelming setbacks that caused Ramsay to pull out of the area entirely.[57] Additionally, there was the ongoing difficulty meeting the orders of coal specified in his contract with the Norfolk Southern Railway which Ramsay agreed to supply 120,000 tons of coal a year.[58] A reinvigorated Ramsay invited his brother-in-law Marvin Kelly to come down from Big Stone Gap, Virginia, along with R.T. Bagby, an engineer Ramsay had used earlier to evaluate the Cumnock problem, but financial difficulties brought on by the holders of the Cumnock Coal Company's first mortgage bonds put a hold on further proceedings. Ramsay soon pulled up stakes with his Deep River holdings and left the area permanently after losing another $175,000 in the process.[59]

In 1925, the state geologist again noted that the Deep River coalfield of Moore,

Chatham, and Lee counties was the only one that held any promise of economic value in the state. The greatest hindrances to coal mining were the high freight rates and the inability on the part of the owners to hire skilled labor. Because the North Carolina field was so distant geographically from other coalfields of the eastern United States only the lowest class of miners could be employed and that their unreliability greatly affected production. Dr. Frank C. Vildbrandt, an industrial chemist at the University of North Carolina, found that about 378,000,000 tons of oil shale existed in the Deep River coalfield which translated into 70,000,000,000 gallons of petroleum. With the rapid demand and depletion of oil resources in the United States, the Deep River coalfield was poised to become an important petroleum producer in the region. Shale had added fertilizer value that could boost local crop production since the area adjoining the coalfields already was predominantly farm land.[60]

11

Tragedy Strikes Coal Glen: May 27, 1925

Situated about eight miles northwest of Sanford at Coal Glen in Chatham County, the Carolina Coal Company's Farmville Mine was one of two mines producing coal in the Deep River basin on a regular basis. A spur connected the mine with the Norfolk Southern Railway, with the nearest railroad station at Cumnock, the site of the Erskine Ramsay Coal Company in Lee County about one-and-a-half miles west of Coal Glen.[1] Ramsay's company had started up only several years after the Farmville Mine commenced operations and enjoyed a good working relationship with the company. However, no one was prepared for the events that were about to unfold on Wednesday morning, May 27, 1925, when, at about 9:30, Howard N. Butler, acting superintendent of the Farmville Mine, noticed smoke coming from the direction of the mine's ventilation fan. Together with a mine mechanic named Joe Richardson, Butler ventured down the slope to the second right heading where the door was intact but jammed tightly, indicating that something was amiss as mine doors were intended to act as much as escape hatches during emergencies. After forcing the door open, the two men discovered four workers about eight feet inside the mine along with two others who were all determined to be alive. Butler and Richardson next carried the six men out of the mine into fresh air.[2] What Butler and Richardson did not realize at the time was the extent of the casualties and damage to the workings further hampered by an inadequate supply of first aid equipment and machinery to effectively deal with the mounting crisis at hand.

The International News Service, an organization that was a forerunner of United Press International, soon picked up the story and reported to the eagerly awaiting crowd of officials, reporters, and bystanders that a rescue car from the U.S. Bureau of Mines equipped with first aid supplies had been dispatched from Thomas, West Virginia, along with bureau engineer J.J. Forbes, who was currently in Birmingham, Alabama, taking charge of the rescue and recovery operations upon arrival.[3] Even with preparations being made to dispatch the rescue car with all possible haste, along with orders given to rail traffic to yield right of way, the more critical rescue work would not commence until the following morning, on May 28, when the bureau's car reached Raleigh.[4] Meanwhile, efforts to rush Governor McLean and National Guard troops to the scene amidst the distressed and anxious crowd were hurried.

Closer to home, as the human tragedy unfolded with the attendant shock, confusion, and latent grief, hope that a loved one had survived had all but been extinguished when the grisly task of bringing the bodies out of the mine commenced. Parents, siblings, and children related to the deceased clung tenaciously to every sliver of optimism the event might promise, but no one was prepared for the deaths and destruction on such a large scale that devastated this small community of closely knit families all tied to the coal mine for their livelihood and way of life. Everyone knew the risks of coal mining but never spoke of them as if to do so were to break communion with the mine. In many mining communities, death came as easily as life.

Governor McLean, upon arriving at the scene, dispatched troops stationed at nearby Fort Bragg to keep the crowds under control and authorized rescue equipment to be sent to Sanford by airplane.[5] Adjutant General John V.B Metts and a corps of engineers were rushed to the scene to take charge of all rescue operations. McLean summoned into action every state agency at his disposal, including National Guard troops from Raleigh, Red Springs, Raeford, and Parkton, to provide assistance as needed.[6] Making rescue attempts more difficult were the twenty tons of rock and debris blocking the shaft just beyond where the first bodies were found. A cordoned off area near the mine's entrance held back the 5,000 onlookers anxiously gathered

Onlookers gather at opening of the Farmville Mine awaiting news of the rescue and recovery efforts. It was estimated that 5,000 individuals came to Coal Glen on hearing of the news of the explosion (Ben Dixon MacNeill Photograph Collection #P0078, North Carolina Photographic Archives, Wilson Library, University of North Carolina at Chapel Hill).

11. Tragedy Strikes Coal Glen: May 27, 1925

to watch rescue efforts and listen for news of families and love ones. John R. McQueen, president of the Carolina Coal Company, sent the first exploring party into the mine who reported that the shaft was closed 1,000 feet from the entrance. Poisonous gasses emitted through the ventilation shaft eventually cleared somewhat but the fear persisted that the trapped miners were receiving little or no air.[7]

Raleigh *News and Observer* correspondent and photographer Ben Dixon MacNeill had been early on the scene and it was his reporting that gave the most detailed picture of the aftermath of the explosion. On May 27, MacNeill reported that forty men were trapped eight-hundred feet underground with rescuers hindered by the large amount of debris blocking the entrance to the mine. Farmville Mine superintendent Howard Butler made an attempt to enter the mine after the first explosion, where he saw six men alive but trapped under a mountain of mostly crumbled slate and timber. After a second explosion ten minutes later shook the ground near the opening, Butler had to extricate himself from the scene before a third explosion sealed the shaft, giving the men no chance of escape. Large ventilation fans were set to work through the ventilation shaft with the expectation that any remaining poisonous gasses would be cleared from below the surface. By 8:00 p.m., the first of the six bodies trapped near the entrance of the mine were brought to the surface[8]

The town of Sanford immediately organized relief efforts with the American Legion and American Legion Auxiliary, establishing a first-aid station at the mouth of the shaft. Relief provided by Fort Bragg in Fayetteville included hospital supplies, ambulances, and a detachment of men for patrol duty to help with crowd control and ensure safety. The cordoned off area near the shaft kept the onlookers at bay as they gathered for the latest news of the names of the men remaining below. The number of men working feverishly to remove the debris blocking the mine opening was limited by the sheer amount of rubble. Beyond the entombed men, only conjecture of the causes of their deaths could be considered: burned by fire, crushed by falling rock, or choked by poisonous gas. Any one of the three causes attended most mine explosions and all three were most likely to be present at this particular mine. To assist rescue efforts, Brigadier General A.J. Bowley, at the request of Governor McLean, sent under the command of Major W.A. Burr three hospital relief trucks and one hundred troops, which were placed at the disposal of Sheriff Walter Blair for patrol duty.[9]

At one point, MacNeill noted that ten "twisted and fireblackened" bodies had been retrieved from the mine but that more than two dozen men remained in the mine beyond the 1,700 feet level, where rescuers witnessed the "gruesome heap" of dead miners who were caught in the conflagration before reaching the top of the shaft.[10] Reporting in the May 30th morning issue of the *News and Observer*, MacNeill continued his account of the gruesome task of recovery which had accounted for forty-eight bodies.[11] A May 30 update of events at Coal Glen appearing in the *Albemarle Press* reported that company vice-president Bion Butler had made an announcement that all rescue work had been completed and that fifty-three men had died, nearly half the adult population of the community. The worst mining disaster and industrial accident in the state had made widows of more than forty women and orphans of more than seventy-five children. One African-American miner made the sad journey home alongside three sons and a brother who were all killed in the mine. In another family, two brothers died in the mine and their sister lost a husband of only a few weeks.[12]

In its summing up of events, the *Press* noted that rescue and recovery work was shared by the entire community. The auxiliary of the American Legion post in Sanford was on hand to help with efforts. Women of the community dispensed food and poured coffee for rescuers as they came out of the mine with their heavy burden. A kitchen car manned by soldiers from Fort Bragg had arrived to help relieve those who had succumbed to the exhausting work of removing the bodies of those killed.

On May 31, MacNeill was the first reporter allowed inside the mine, traversing 1,600 feet of scorched solid rock walls where light and shadows played tricks with the men's eyes and the feeble light emitted from their cap lamps provided the only illumination in the mine's dark depths.[13] MacNeill had taken nearly one-hundred photographs of the living and the dead that day. The families representing the living confronted the ghostly retinue of men exiting the mine in cars once filled with coal,

Unidentified mother and children wait to hear news of their loved ones at Coal Glen. African Americans accounted for one half of the men killed in the explosion. Although black and white miners worked alongside each other they lived in segregated communities such as the black community of Frog Town and the white community of Red Town (Ben Dixon Mac-Neill Photograph Collection #P0078, North Carolina Collection Photographic Archives, Wilson Library, University of North Carolina at Chapel Hill).

11. Tragedy Strikes Coal Glen: May 27, 1925

Author and journalist Ben Dixon MacNeill at his desk, ca. 1950. He was assigned by the Raleigh *News and Observer* to cover the tragedy at Coal Glen (Ben Dixon MacNeill Photograph Collection #P0078, North Carolina Collection Photographic Archives, Wilson Library, University of North Carolina at Chapel Hill).

hoping that their family member or friend was one of the lucky ones to have survived. MacNeill's somber but vivid descriptions of the mine showed the dangers rescuers had to confront when negotiating the slippery wet rocks and steeply graded slopes. Though MacNeill was there to report facts, his impressions of the blurry images no doubt stuck in the minds of his readers and especially his own.

The Bureau of Mines report, submitted by J.J. Forbes and C.W. Owings, stated that a request for help was sent to the Bureau of Mines at Washington, D.C.; Pittsburgh, PA; and Birmingham, AL. Car 7 with T.T. Read, E.H. Graf, and T.G. Hunt arrived at the scene at 9:00 a.m., May 28, while D.J. Parker, G.S. McCas, George Groves, and W.H. Forbes arrived at 5:30 p.m. J.J. Forbes and C.E. Saxon arrived at 9:00 p.m. Taking into account the first explosion, there was little velocity and only slight signs of violence resulting in little damage done "in the upper part of the mine." Even some lighting farther down in the mine was not affected and one of the mine officials, H.N. Butler, saw no debris.[14] In the dark and din of a coal mine, lighting could sometimes play tricks with the eyes of the miners who experienced difficulty determining distances based on the type of illumination, both natural and artificial.

According to the Bureau of Mines report, the source of the first explosion was attributed to a blown out shot, as the location of a Hot Shot-brand battery and shooting cable near a group of men inside the mine confirmed.[15] In the mines, an experienced shot-firer, whose specialty was to discharge explosives in a mine in order to bring down the coal, must possess a proper certificate to perform his duties. In well-regulated mines, only the shot-firer was permitted to use explosives in this capacity. Typically, the process involved drilling a bore hole in the coal face. Next, an explosive charge was loaded into the hole followed by a fuse, or squib, to ignite the explosive. A small amount of inert material, such as limestone or clay, was forced into the hole with a tamping rod, a wood or metal bar with a copper or wood head, usually rounded at the head, used to stuff the tamping material into the hole, leaving the fuse or squib slightly exposed to light the charge. The inert material compressed into the bore hole was called *stemming*. After shouting a warning to the men in the mine that a shot was to be fired and to take appropriate precautions, the shot-firer detonated his charge. In

Two miners take a break from their rescue and recovery work. The appearance of the man on the left summed up the tragedy and sorrow that day (Ben Dixon MacNeill Photograph Collection, #P0078, North Carolina Collection Photographic Archives, Wilson Library, University of North Carolina at Chapel Hill).

every case, the shot-firer was responsible for warning others in the mine that a shot was about to be fired.[16] If the tamping material was loose or not tamped properly or the explosive was overcharged, the energy produced by the explosion only forced the material out the drill hole instead of directing its explosive force into the coal.

Forbes and Owings found two drill holes, approximately sixteen inches apart at the face of the heading, vertically drilled into the coal. They concluded that one hole

11. Tragedy Strikes Coal Glen: May 27, 1925

was unmistakably a blown-out shot, as there was no attempt to undercut or sheer the coal before setting off the charge to blow down the coal. The charge in the second hole was originally thought to have failed to detonate but was later proven that it had partially exploded and burned. Mine gases most likely escaped through a crevice in the coal next to where the shot had been tamped. As Forbes and Owings theorized, the blown-out shot from the upper hole emitted a flame into the dust-laden, gaseous atmosphere and ignited the mixture. The lower hole might also have emitted a flame as well, while raising the dust deposited on the seam into a cloud, adding more volatility to the explosive conditions. An undertaker found matches on at least five bodies, and the fire boss, George Anderson, possessed five matches in a box. One match was found upon a miner's person but there was no evidence to substantiate that matches were the cause of the explosion, though it was against regulations to carry them inside the mine.[17]

Forbes and Owings surmised that the detonation of explosives in a car in the same heading most likely caused the second explosion. Either flame or heat from the first explosion entered the heading and ignited the explosives, which burned slowly before detonating. Another theory put forth was that part of the check curtain, used to deflect air into the room, may have ignited and smoldered until fresh air being forced into the entry burned more rapidly thereby igniting and detonating the explosive. The rapid entry of fresh air into the heading would have reacted with the heat or fire leading to faster ignition and detonation. Statements to the effect that a third explosion occurred in the mine were discounted when H.N. Butler, acting superintendent of the mine, reported that while inside the mine he heard only two explosions.[18]

The Farmville Mine was known for its gaseous, explosive conditions since the Carolina Coal Company began operations. Firedamp and dry, fine coal dust posed an extremely hazardous environment for miners and workers. Indeed, on May 26, the day before the explosion, a miner was overcome by a mixture of firedamp and coal dust in the entryway, indicating poor and sluggish ventilation. The fire-boss, hired to examine all working places prior to allowing the men into the mine, made only verbal reports to the mine foreman, which, though not a violation in itself, was considered an unsafe practice.[19] North Carolina's mining legislation of 1897 stated that ventilation must be "circulated and distributed throughout the mine in such a manner to dilute, render harmless and expel the poisonous and noxious gases from each and every working place in the mine."[20] The most critical issue regarding the unsafe conditions was its total disregard.

The state of North Carolina conducted its own investigation into the explosion at Coal Glen separately from the report made by the team sent by the U.S. Bureau of Mines. At the time of the explosion, officials of the Carolina Coal Company informed the Commissioner of Labor and Printing, upon whose office fell the responsibility of mine inspection for the state. Commissioner Frank D. Grist went at once to the scene and collected evidence from witnesses who were near the mine at the time of the disaster and from the individuals who brought the bodies out of the mine. Due to the hazards of entering the mine in its present state, along with the risk of lingering mine gasses, the decision was made to wait for all safety measures to be followed before conducting an extensive investigation. On June 23, W.H. Hill, formerly associated with the Carolina Coal and Coke Company, made a thorough inspection of the mine and presented his findings in a report to the Commissioner.[21]

During his investigation, Hill was accompanied by J.E. Tiffney, Chief Engineer, Explosive Division, U.S. Bureau of Mines; B.M. Rogers, General Manager of the Erskine Ramsay Coal Company, representing the receiver of the Carolina Coal Company; and, Howard Butler, of the Carolina Coal Company, representing the stockholders of the company. After a comprehensive examination of the mine, Hill reported to Grist that in his opinion the explosion resulted from a blown-out shot in the fourth right heading which was consistent with the Bureau of Mines report.[22] Additional circumstances revealed that the explosion was caused by human error coupled with non-compliance of certain regulations specified in the state's 1897 mining law.

The miner working in the heading drilled and loaded two holes, connecting the shot-firing wiring to the holes simultaneously. The lower shot failed to explode and did nothing more than burn in the hole, causing the upper shot to take the path of least resistance, blowing the tamping out with the disastrous results. The two holes were poorly placed and the regulations for firing the shot were ignored.[23] Because a blown-out shot directs its energy outwardly away from the hole, it ignited mine gas and any coal dust stirred up by the explosion. It was this explosion that damaged the ventilation system, making it difficult for dangerous gas to be expelled from the mine and for anyone trapped inside to breathe, leading to unavoidable asphyxiation.[24]

Perhaps one of the most difficult tasks in the aftermath of the explosions was the recovery of the bodies from the mine especially by the family and friends of the deceased workers. As family waited anxiously for news, the recovery job never abated until every last miner was accounted for.[25] In coal mining, as in battle, every effort is made to bring the deceased home for burial regardless of the surrounding conditions. The fog of memory and the dazed condition of all those who were there must be sorted and clarified before anyone can begin to make

Plate Supper and Square Dance
AT LAKEVIEW PARK
Friday, June 12th, 1925

Supper served from 6 to 8 P. M.
Square dance from 8.30 til——
Dance in charge of Mr. Duncan McCrimmon—"nuf sed."
Admission 50c. and this admission price "carries you thru" supper and dance.
Given by the Lakeview Community, and the entire proceeds to go to the benefit of **COAL GLEN RELIEF FUND**.
BE ON HAND AND HELP MAKE THE FUND LARGER

Charitable events such as this plate supper and square dance at Lakeview Park in Moore County helped to raise needed money for the Coal Glen Relief Fund while providing entertainment for the devastated communities.

The difficult task of bringing out the dead often fell on the survivors (Ben Dixon MacNeill Photograph Collection #P0078, North Carolina Photographic Archives, Wilson Library, University of North Carolina at Chapel Hill).

sense of the tragic events.[26] However grim the situation at Coal Glen, the attitude that life goes on never seemed to be a deterrent in the mining community despite the risk of continued death and destruction.[27]

After the death and destruction had abated, the dilemma of whether to reopen or close the workings in the mine became the overriding concern. Families relied on the wages of husbands, sons, and brothers to survive, and when that was taken away, they moved on to an uncertain future. The explosion left nearly three-hundred individuals—mostly widows and orphans—in destitute circumstances. The Lee County chapter of the American Red Cross estimated that a total of $75,000 was needed in

```
                CAROLINA  COAL  COMPANY
                      SANFORD, N. C.

                                     JUNE 8th. 1925.


    My dear Mr. Mason:-

              The Carolina Coal Company- its officials
    and their families, its employees and its friends-
    wishes to thank all who helped them in their late
    trouble.

              Your efforts, your tirelessness, your
    co-operation through the dreadful days and nights
    did much toward forwarding the painful work.

              We extend to you our deepest thanks and
    our most heartfelt appreciation.

              The Directors of the Company have
    authorized the sending of this letter.

              Very Sincerely Yours,

              The Carolina Coal Company,

                   per  C M Rine
                        ──────────
                        Secretary
```

Perhaps one of the most difficult tasks in the aftermath of the explosion was the recovery of the bodies from the mine especially by family and friends. Here, a letter from the Carolina Coal Company expressing its appreciation for the help rendered during that time was sent to W.B. Mason, Sr., one of the employees of the company (courtesy W.B. Mason, Jr., Farmville, NC).

11. Tragedy Strikes Coal Glen: May 27, 1925

aid to relocate individuals and cope with the trauma. Everyone in the community from businessmen to factory workers was asked to contribute to the relief fund.[28] Events such as plate suppers and square dances to raise funds for the families were announced in the local newspapers.[29] The commissioner of the Department of Labor and Printing, under whose jurisdiction mining fell, made an argument for the enactment of a statewide worker's compensation law to provide financial assistance to the employees of the coal company and their families.[30]

12

The Nature and Causes of Mine Gases

As mechanization in the early 1900s allowed miners to dig deeper into the coal seam, they were exposed to volatile gases many of which were often harmful if not fatal. Gases are some of the most dangerous elements in a mine simply because most are invisible to the naked eye while some are odorless, leading to lethal doses when breathed or explosive in the presence of an open flame. Most gas types are extremely hazardous especially when found in high concentrations, as in the case of the disasters at Cumnock (1895, 1900) and Coal Glen (1925). Mine gases typically require an ignition source to cause an explosion which in many cases can be attributable to poor mining practices, such as improper use of black powder, smoking, use of open lights, matches, and failure to remove coal dust or use rock dusting.[1] North Carolina's mining law of 1897 made provisions for eradicating mine gases through ventilation, "which shall be circulated and distributed throughout the mine in such a manner to dilute, render harmless and expel the poisonous and noxious gases from each and every working place in the mine."[2] Ventilation was to be provided by force or suction fans, exhaust steam furnaces, or other means "to produce and maintain an abundant supply of air, and all mines generating firedamp shall be kept free from standing gas, and every working place shall be examined every morning with a safety lamp."[3]

Examples of Mine Gases

Gas	Common Name	Cause	Characteristics	Explosive	Combustible
Carbon Dioxide	Afterdamp/ Blackdamp	Respiration of men/ animals, fires	Colorless, odorless, tasteless in low concentrations	No	No
Carbon Monoxide	Whitedamp	Incomplete combustion or blasting	Colorless, odorless, tasteless, poisonous	Yes	Yes
Hydrogen	Hydrogen	Battery charging, incomplete combustion in mine fires	Colorless, odorless, tasteless, poisonous	Yes	Yes
Hydrogen Sulfide	Stinkdamp	Decomposition of some sulfur ores, stagnant water ponds	Sense of smell deadened, paralyzes respiratory system	Yes	Yes

12. The Nature and Causes of Mine Gases

Gas	Common Name	Cause	Characteristics	Explosive	Combustible
Methane	Firedamp	Occurs naturally as a result of decomposition of organic matter	Colorless, odorless, tasteless	Yes	Yes
Nitrogen	Nitrogen	Normal constituent of oxygen	Colorless, odorless, tasteless	No	No
Nitrogen oxides	Oxides of nitrogen	Blasting fumes	Red-brown in high concentrations, tasteless, poisonous	No	No
Oxygen	Oxygen	Atmosphere	Colorless, tasteless, odorless	No	No

U.S. Bureau of Mines

While oxygen and nitrogen make up ninety-nine per cent of the atmosphere in a mine, when nitrogen displaces oxygen, oxygen deficiency can occur, causing buzzing in the ear, rapid heartbeat, blueness under fingernails and on the lips, as well as a dizzying feeling but is normally harmless.[4] However, nitrogen oxides, which are formed at high temperatures, enter the throat and lungs and form an acid-causing irritation. When it collects in sufficient amounts to displace oxygen, nitrogen oxides make the atmosphere unsafe.[5] Methane, which is naturally released from coal and from rocks containing combustible material, is the key ingredient of natural gas. It is light and rises to the roof area. Methane explosions occur in mines when sufficient amounts of the gas come into contact with a heat source and there is not enough air to dissipate the gas to a level below its explosion point. Like nitrogen, methane became an asphyxiant by diluting the oxygen in the air.[6] The explosions at Cumnock and Coal Glen were caused by the ignition of firedamp. Carbon dioxide, also known as afterdamp, or blackdamp, is a colorless, odorless gas that is heavier than oxygen and nitrogen, concentrating near the mine floor. This gas is formed by mine fires and explosions and is a byproduct of firedamp. Federal law dictates that carbon dioxide should never exceed 0.5 percent of a mine's atmosphere. The men who weren't immediately killed by the explosive impact at the Cumnock and Farmville Mines succumbed to the afterdamp that formed after the explosion of firedamp.[7]

Carbon monoxide, also called whitedamp, is produced by incomplete combustion and has a distinctly acid taste. A small amount of carbon monoxide in the atmosphere can be especially harmful when it exceeds 0.005 percent of the total atmosphere, the limit set by the Federal government.[8] Hydrogen, while it does not occur naturally in a mine, can appear after a fire, explosion, or near a battery-charging station. This gas must exist in sufficient quantities (4 percent) for it to be flammable. Like methane, hydrogen is light and rises to the mine's roof.[9] Hydrogen sulfide, or stinkdamp, is sulfureted hydrogen and has a distinct odor of rotten eggs. This gas mostly takes up with water and is particularly dangerous in mines where it occurs in stagnate pools on the mine floor. Most mines are damp by nature and where there is standing water the presence of stinkdamp is more likely to exist. When sufficient amounts of the gas are present, the sense of smell is no longer a reliable indicator as it affects the respiratory system.[10]

The second explosion, mentioned in both state and federal reports, followed the first at an interval of between twenty- and twenty-five minutes. While investigators on both teams ruled out a third explosion, there was some variance as to the cause of the second explosion. The investigators with the Bureau of Mines concluded that the source of the second explosion occurred when the overcharged upper shot blew out and stirred up the coal dust, which with existing gas, ignited and initiated

Electric cap lamp with battery. In 1915, the Edison Flameless Electric Miners' Cap Lamp was the first of its kind to receive U.S. Bureau of Mines approval. The inventor, Thomas A. Edison, considered it his most humanitarian achievement. The introduction of the electric cap lamp was instrumental in reducing mine explosions by 75 percent over the first twenty-five years of use. The miners employed at the Farmville mine were equipped with this type of lamp (courtesy Edison Storage Battery Co.).

the explosion. The state report suggested three possibilities: (1) an electric arc from a shorted cable ignited the mine gas, (2) a burning piece of clothing or timber ignited the gas, and (3) smoldering coal dust or fire from some source ignited a volume of gas drifting out over the area. Because the Bureau of Mines was at the forefront of mine safety and had considerable experience testing for explosions at its facilities in Pittsburgh, the blown-out shot theory was probably the most likely cause. The state report also suggested that the majority of the fatalities occurred as a result of the first explosion as there was little evidence to suggest that the workers had traveled any distance from their respective places.[11]

Other than noting small amounts of coal dust in the entries, which should be gotten up and removed from the mine, the state investigators found the overall condition of the mine to be very good, including ventilation. When tested for gas, none was found and damaged brattices to help deflect air into the workings were to be rebuilt with tile and concrete. It was noted that water had collected at the bottom of the main slope and ordered to be pumped out, along with restoring ventilation to the bottom of the mine, and watering down dusty areas, but there were no glaring issues to preclude the mine's reopening "in the near future." Additionally, Hill recommended adherence to seven safety rules all relating to prevention of explosions.[12] Again, as had been the case since the creation of the state's 1897 mining law, no enforcement was attached to the investigators' recommendations much less a thorough examination of the mine prior to admittance of miners in the workings.

The Bureau of Mines' authority extended beyond that of the state's in the extent of safety procedures. The bureau especially pushed for the use of permissible explosives and equipment that was required to pass certain tests certified by the federal agency and used in accordance with the conditions prescribed by the bureau, meaning that electrical machinery, such as pumps, hoists, and fans, along with explosives for shot-firing, should be deemed "permissible" and comply with the bureau's directives. Care taken to transport explosives or powder in non-conducting fiber or wooden containers into the mine as opposed to metal ones lessened the risk of sparks, which could touch off gas and cause an explosion.[13] None of the recommendations the bureau made for the use of permissibles were included in Hill's report, and as with the state's regulations, conformity of the bureau's recommendations was unenforceable. Non-compliance of safety laws often meant that a mine owner or superintendent put his workers at risk of injury or death, and with owners focused on their bottomline, often no provisions for implementing safety measures were made because of cost and potentially decreased production.

13

The State of Coal Mining After 1925

In 1926, Frank D. Grist, Commissioner of the Department of Labor and Printing, put it most accurately when he stated that greater knowledge of the condition of the mines by proper periodical inspection made by a competent mine inspector with at least five years' experience, "could have prevented the mining industry in this State with a clearer conscience if everything in its power had been done to safeguard the workmen in the mines."[1] His doleful commentary on Coal Glen gives one pause to consider whether the proper inspection of mines would have caught violations or at least made mine personnel more aware of their potential, but, as previously stated, enforcement on the state and federal levels was the key to good safety practices. No mine is safe to enter until the proper procedures are put in place and followed explicitly.

Old Setbacks and New Directions

After the explosion at Coal Glen, the Carolina Coal Company soon went into receivership, with local railroad executive and businessman John H. Kennedy, of Cumnock, appointed receiver. The action was sought to process claims brought about by the disaster as expeditiously as possible in the courts as well as final disposition of all claims.[2] In an assessment of the Deep River coalfield in Lee and Chatham counties in 1927 and 1928, state geologist Herman J. Bryson acknowledged that production on a large scale had not been realized since the explosion at Coal Glen and that flooding, in this case affecting the Cumnock mine, required extensive pumping. More attention, however, focused on the feasibility of mining the black band and shales above and below the coal seams for the purpose of fertilizer fill. Because of North Carolina's burgeoning agricultural economy, the need for a constant supply of fertilizer was "deemed necessary to enable the State to keep step with its aims of progress."[3] Farmers working their crops in the vicinity of the Deep River field hauled off large amounts of shale from the unattended slag heaps left by coal extraction, and found that, broken up, the shale supplemented their fertilizers as much-needed fill. In Bryson's estimation, the addition of a nitrogen-based supplement such as that found in oil shales would allow the state to utilize local sources and not become reliant on supplies outside the state, and as Frank C. Vilbrandt wrote in his study of 1927 of the oil-bearing shales of Deep River, 27,000,000,000 gallons of shale oil could be produced to help meet the petroleum demand of the burgeoning fuel market.[4]

Convict Leasing

Seizing on an approach to make up for the decrease in coal production from 1926 to 1928, state officials implemented the controversial system of prison labor called "convict leasing" to work the Carolina mine at Coal Glen not for the purpose of mining coal but tapping the large stores of oil-bearing shale already present. Beginning in early 1928, the State Prison Department began sending inmates, on a voluntary basis, from Central Prison in Raleigh to Coal Glen for that purpose. The state constructed a modern onsite prison facility "with all sanitary conveniences" for the inmates.[5] The coal company paid the state a set price per ton for coal mined and loaded by the inmates, with the state providing care for the men from the time they left the camp until they returned from work at the end of the day. As an incentive, if the men mined more coal than enough to pay the state, they could share the difference.[6] Still, coal mining was a dangerous occupation requiring a certain amount of skill to handle explosives and machinery safely which usually only could be attained through training and experience.[7]

Later expansion of the facilities at the Coal Glen camp showed an increase in the onsite prison population from 181 inmates in 1928[8] to 196 in 1929.[9] In July 1929, the *Robesonian* reported that Governor O. Max Gardner, citing that employing prisoners in coal mines was contrary to the policies of the current administration, broke a five-year contract with the Carolina Coal Company and removed prisoners from such work. Seven prisoners had been killed in the fifteen months that they had worked in the mines near Sanford—five in an elevator accident, one as the result of electrocution, and one crushed by a coal car.[10] By 1932, due to the poor economy brought on by the Great Depression, reduced demand for prison labor, and the decline in revenue from sources using prison labor, contributed to the shrinking of per-capita earning ability from a high of $212.20 in 1928 to $65.76 in 1932.[11] The financial hardship of the mine owners to implement safety measures further necessitated the State Prison Department's removal of the inmates from Coal Glen and shutting down the camp. The state's plans for the purchase of the Coal Glen property were dropped on account of the cost and inability to provide year-round employment for the prisoners.[12] For the most part, the arrangement was mutually beneficial for both the state and the coal company as having prison labor to work in the mines helped the company maintain a regular source of manpower when hiring miners locally or outside the area became problematic.

New Uses for Coal

Despite the new infusion of convict labor into the Cumnock mine, from 1929 to 1935 coal mining production at Deep River continued to decline, as illustrated in the following table:

Year	Quantity (tons)	Value	Average per ton
1929	58,180	$177,000	$3.39
1930	28,500	100,000	3.51
1931	2,363	9,000	3.81
1932	1,900	6,000	3.16
1933	2,014	7,000	3.48
1934	3,140	9,000	2.86
1935	2,000	4,500	2.25

(The Mining Industry in North Carolina, from 1929 to 1936)

Mining companies who operated in North Carolina from the period 1929 to 1935 included N.C. Coal Mining Corporation, of Carbonton, Chatham Co.; Ralph Jordan and R.G. John, of Gulf, Chatham Co.; Southern Anthracite Corporation, of Sanford, Lee Co.; and, the Anthracite Mining Company, of Carbonton. On account of the price differential of between two and three dollars per ton in freight rates within a one-hundred-mile radius of the field, interest continued in coal mining at Deep River, although attempts to sustain a working and profitable mine had failed because of the limited capital these companies had for investment.[13] A new approach under consideration for optimizing profits in the Deep River coalfield involved reducing coal to a pulverized state, which was used as a substitute for natural gas in furnaces, steel mills, and other industrial applications. Preliminary tests indicated that Deep River coal ranked high with other coals processed in this form.[14] During the pulverization process, coal was crushed, dried, reduced, and conveyed to the boiler or furnace where it was to be used as a fuel. Pulverized coal used in industrial applications had several drawbacks most notably its high volatility and introduction into the atmosphere when not completely burned off, but it provided an immediate heat source more so than other grades of coal.[15]

14

New Challenges for Coal Mining

The 1930s

August 15, 1930, saw flooding at the Carolina Coal Company mine, resulting in the death of Ernest Chapman, who succumbed when a rock fell and crushed his skull.[1] September proved to be a month of continued setbacks for coal mining at Deep River, when three injuries and three fatalities were reported by the Commissioner of Labor and Printing. Alex Hines, employed to lay and repair track, was injured at the Carolina Coal Mine on September 8, when a rock fell on his back, pinning his right hand on the track and breaking the fingers of his right hand. An investigation showed that Hines was negligent after being warned of a weak roof and falling particles. On September 9, Silas Worthy, a miner employed by the Carolina Coal Company at Coal Glen, died when a runaway coal car coming down the slope hit and dislodged a mine prop, striking him on the head, resulting in instant death. The accident was considered unavoidable and cause could not be assigned to one individual. On November 24, an explosion at the former Erskine Ramsay Coal Company's Cumnock Mine resulted in the asphyxiation of Charles Shirley and the death of Sylvester Murchison by burns and asphyxiation. Two other men, John Henry Bailey and Ed McIver, were burned but survived. An investigation revealed that crossed wires in a signal bell short circuited, giving off sparks that ignited methane gas present in the area. No responsibility was assigned to either equipment failure or human error.[2]

The number of fatalities in coal mines nationwide attributed to non-explosive and explosive events in U.S. mines by five-year periods is shown in the following table below. One can understand the focus of the early efforts of the Bureau of Mines to address the unacceptably high numbers of deaths from explosions. The 1906–10 time span reflects the Monongah, WV, tragedy of 1907 in which 362 miners were killed, the worst loss of life in U.S. mining history. Note that explosions contributed to more deaths than non-explosive accidents. From 1895 to 1930, a total of 122 men were killed in coal-mining accidents in North Carolina which was a staggering loss of life given the small coal-mining operations in the area. Despite injuries and loss of life, speculators began to see the advantages of mining anthracite, a virtually untapped source of coal superior to the bituminous variety.

5-year period	No. of Fatalities	
	Disasters	All explosions
1891–95	527	743
1896–1900	461	730
1901–05	1,212	1,524
1906–10	2,104	2,388
1911–15	1,533	1,722
1916–20	618	1,066
1921–25	1,322	1,690
TOTALS	7,777	9,863

U.S. Bureau of Mines

The North Carolina Sandhill's regional newspaper, the *Pilot*, announced to its readers in September 1932 that a new startup company, the Southeastern Anthracite Company of Asheville, had been chartered by the state as a mining enterprise to do business in Moore County.[3] As previously stated, anthracite, or hard coal as it is sometimes called, contains eighty-five to ninety-five percent carbon as opposed to bituminous, or soft coal, which contains seventy- to eighty-five percent of the amount of anthracite. Anthracite burns more cleanly, leaves less ash, and emits less smoke than bituminous, making it a preferred coal for domestic and industrial use. Northeast Pennsylvania has been the country's leading producer of anthracite coal for two centuries, while much smaller deposits occur in Alaska, Arkansas, Colorado, Massachusetts, Rhode Island, New Mexico, Utah, Virginia, Washington, and West Virginia, in addition to North Carolina.[4] The company claimed that the quality of the anthracite in the upper end of Moore County was equal to that of Pennsylvania's finest anthracite. R.P. Simmons, of Asheville, was made company president along with D.B. Long, of Cincinnati, vice-president and general manager. The location of the property was at the Gardner farm, between Deep River and the Norfolk Southern Railroad approximately thirteen miles northeast of Carthage in Moore County.[5]

Long was a mining engineer of considerable experience in the coalfields of West Virginia, Kentucky, and Tennessee and became resident manager and engineer of the company with an office at Sanford. The property consisted of 600 acres and

> What about the question of your fuel for the Winter?
> The Answer is
>
> **Carolina Anthracite**
>
> Mined in Moore County
>
> CLEAN   SMOKELESS   LONG BURNING
>
> **DOUB SUPPLY COMPANY**
>
> Phone 139   Aberdeen

Advertisement for anthracite coal in the Southern Pines *Pilot*, November 11, 1932.

was well situated for railroad and truck access. The new enterprise was arranged by Howard N. Butler, once acting superintendent of the Farmville Mine of the Carolina Coal Company. He and John McMillan McIver, Jr., of Gulf, acquired the property in an undeveloped state and soon began a systematic exploration of anthracite in an area known mostly for its bituminous coal deposits. The two drove a slope and opened cross headings "showing a good quality of anthracite coal several hundred feet into the workings."[6]

Upon arrival at the site, the new owners began ordering material and equipment to have in place prior to starting up mining operations. The expected daily output was 200 tons. A new tipple forty-feet tall with screens, scales, and loading machinery was expected to be installed within weeks and operations commencing soon afterward. Also, the steam hoisting plant was to be supplemented by an electric hoisting engine. Capacity for loading five or six cars at a time was included as part of the plans D.B. Long devised. Five sizes of coal from buckwheat all the way up to egg or larger were produced as an innovation in coal grading in the North Carolina coalfields. All newly mined coal would be screened, separated from any slate and debris, and sorted to size by screens that would grade the entire output.[7] Because of its hard qualities, unlike its bituminous counterpart, anthracite coal typically required additional machinery to break it down before transporting and shipping to markets. Impurities gleaned during the process were transported to a culm pile or bank, which in some instances caught fire and smoldered for years. Thus, anthracite processing plants came to be known as "breakers" for this type of work. Tipples, on the other hand, which usually processed bituminous coal, did not utilize expensive machinery in processing coal, and in many instances the coal could be dumped directly from the tipple into awaiting trucks or railroad cars. While tipples and breakers shared commonalities, coal sizing was done more extensively at the breaker, but in some instances, "breaker" and "tipple" were used interchangeably.

Distribution of the coal and haulage were intended to be made by truck within a fifty- to seventy-five mile radius, while longer haulage would require transport via the railroad. Arrangements were being made for building a loading tipple at the tracks near the mine which would allow a quicker and more convenient process to meet demand. Mined coal could be collected in large yards for storage and distribution, or shipped directly from the tipple to the customer.

Soon after mining operations commenced a considerable amount of the coal was sold to state institutions and private consumers. Sales of anthracite coal mined in Moore County saw distribution in larger markets such as Aberdeen in Scotland County. In September 1934, the *Pilot* reported that the North Carolina Highway Commission had purchased the Elkins and Thomson property in the vicinity of Horseshoe, Moore County, comprising nearly 1,000 acres, for the purpose of mining coal and cultivation. The deal included two tracts of land, one of which was known as the John Elkins place, which later became the property of the late John McMillan McIver, Sr., of Gulf, whose heirs owned the land at the time of McIver's death in 1923. The other tract was owned by Herbert Thompson, formerly of the area, but then residing in Richmond, Virginia. The Elkin tract, consisting of more than 600 acres, was part of 4,000 acres leased by Julian T. Bishop of New York, as a hunting preserve. There were approximately 200 acres of cleared land that prisoners from the convict camp in Moore County were consigned by the state to develop and cultivate. The Highway Commission planned to work the coal residing under the land, formerly owned by McIver, to supply various state institutions. That work was scheduled to begin the following year.[8]

Activity in the Deep River coalfield between 1936 and 1940 consisted mostly of small-scale operations producing coal for local domestic use. In August 1936, Burlington's *Daily Times-News* announced that the Deep River Coal Company near Gulf had sunk a twenty-eight-foot shaft and was mining ten- to twelve tons of coal a day.[9] Perhaps the most noteworthy new venture to appear on the scene was in 1937, when the Carolina Fuel and Transportation Corporation, of Sanford, under the ownership of Goodson and Emmitt C. Williams, was incorporated with an authorized capital stock of $500,000 and subscribed stock of $2,010. Operating the former Cumnock Mine, the investors initially gave no indication as to when operations would begin, but nevertheless felt confident that their business would be successful, though they would be competing with the likes of Pocahontas coal of Virginia and West Virginia.[10] During 1940, operations were reported by Deep River Coal, Inc. and the Reeves Coal Company.[11]

Fifty-pound bag of Deep River anthracite coal, mined and packed by Deep River Coal, Inc., Gulf. This semi-anthracite, smokeless variety was most extensive near Carbonton (photograph by the author, with permission of J.R. Moore & Son, Inc., Gulf, North Carolina).

The 1940s

On June 27, 1941, the *Pilot* announced a new enterprise organized by H.B. Chatfield and Howard N. Butler as the Coal Products Company, which was to commence coal mining at the Carolina Mine at Coal Glen. As the most complex and costly attempt at coal mining in the region since the Erskine Ramsay Coal Company's ambitious but failed efforts in the 1920s, the investors hired the respected Charleston, WV, firm of Robinson and Robinson, mining engineers, to act as consultants in reopening the mine.[12] Replacing older mining technology with newer equipment, such as the outdated steam hoist technology with an electric counterpart, Robinson and Robinson were given the go-ahead to upgrade all machinery and equipment with the latest technologies. Plans for installing mechanical loading machines, chain belts and conveyers, coal-washing facilities, new pumping engines, and the most up-to-date safety equipment promised that the mine would be one of the most advanced and safest mining operations in the South. Perhaps one of the more prominent features was the capability to clean the coal inside the mine as well as at the tipple, removing more slate and bone than was previously accomplished by hand, before shipment to its final destination. The new operators took the lead from Carol Robinson, a known authority on mine safety, to implement steps to reduce the amount of flammable coal dust and firedamp in the atmosphere which had plagued the mine for decades with its history of death and destruction.[13]

Employing local labor, the expectation for the Carolina Mine was to produce 500 tons of coal a day within a year, exceeding the previous highest amount of 300 tons. By the summer, the company had initiated operations to remove standing ground water preparatory to opening the mine. Seeing an even more important role for the use of local coal leading up to the country's involvement in World War II, coupled with the mine's convenient location to transportation outlets, the mine owners set their sights on supplying coal to local military installations, such as nearby Fort Bragg and the Marine base at Jacksonville, an enterprise that would not only open up a new market but achieve much for industrial progress and defense efforts.[14] A year and a half later, the first carload of coal shipped via the Atlantic and East Carolina Railway, known as the "Mullet Line." One of the objectives was to try out the coal in the fireboxes of the railroad's locomotives, while another plan involved developing the iron ore and coal seams for the purpose of producing sponge iron.[15]

In early 1943, Governor Joseph Melville Broughton called a meeting to inform legislators that the state had sufficient amounts of iron ore and coal to make sponge iron. One of the attendees, Herman A. Brassert, a metallurgist and president of the H.A. Brassert and Co., of New York, spoke to the group concerning his findings in the coal and iron regions of the state for that purpose. Brassert estimated at least 65,000,000 tons of recoverable coal in the state and that gas would be a valuable by-product. The document he issued, called the Brassert Report, was favorable to further exploration and recommended establishing coking plants to utilize the coal.[16] Sponge iron units also could be put into production on a modest scale compared with the higher cost of putting a blast furnace into operation. Regarding the Cumnock mine, Brassert believed that if sufficient amounts of coal and iron were close together, the entire process of sponge iron production could be accomplished in one continuous operation at the mouth of the shaft.[17]

In April 1944, the Bureau of Mines began exploratory drilling in the Deep River area and by October four holes had been sunk. Drilling was renewed in April 1945 and two additional holes were completed by August. The Walter A. Bledsoe Company, of Terre Haute, Indiana, began a series of exploratory drilling operations in late June 1945, and plans were soon made to develop the coalfield.[18] Bledsoe's venture concerning the Carolina Mine represented a new corporate interest after a two-year hiatus since the mine closed in 1942 and its reopening in 1945.[19] Bledsoe's company already was one of the Midwest's largest coal producers and his name added prestige to an endeavor that required a great amount of funding behind it to achieve success. Bledsoe soon took to task drilling ten holes and acquiring 2,000 acres of land containing the most mineable coal, including the Carolina Mine.[20] With a staggering 12,000,000 tons of coal mined yearly from locations in Indiana, Illinois, Kentucky, Tennessee, West Virginia, and now, North Carolina, Bledsoe's company was poised to tap reserves in an area that had a proven demand for coal as well as easy access to shipping lanes on the East Coast.

Walter Bledsoe (1875–1950), president of the Walter Bledsoe Coal Co., an Indiana mining firm of national prominence, which took over the Cumnock workings in the 1940s (*Coal Men of America*, 1918).

The Raleigh Mining Company was formed to carry out exploratory work and analyses of the area, and in fall 1947 mining commenced at the Carolina Mine.[21] When confronted with the daunting task of rebuilding the water-filled, caved-in slope, the company, to provide extra safety for men and equipment, lined the greater part of the 1,500-feet entrance portal, which crossed underneath Deep River, with a semicircular corrugated steel tunnel. New rails for coal cars entering and exiting the mine were installed to facilitate the increased amount of traffic allowing coal cars to be carried to the top of a steep hill and dumped into a large loading bin. In November 1949, the *Pilot* reported that 2,000 feet of the old slope had been reopened with the expectation that when completed the slope would reach 3,000 feet and lead to a deposit of 46,000,000 tons of coal. Additionally, two modern coal-cutting machines were installed, with fifty- to sixty men working at the mine, producing fifty tons a day to be increased to 500 tons daily and then 1,000 tons.[22] For the year 1949, approximately 14,000 tons of coal were mined.[23] Deeply faulted veins caused considerable difficulty and expense extracting the coal, and in 1952 the operation ceased. Much of

14. New Challenges for Coal Mining

the mine's coal output was used by the Carolina Power and Light Company in its steam-generating plants for the production of electricity.[24]

The 1949 Strike

On September 22, 1949, state-wide coal dealers prompted the North Carolina Coal Merchants Association to hold an emergency meeting to address a nationwide work stoppage and dwindling supplies of coal as the result of a strike called by John L. Lewis, president of the United Mine Workers of America.[25] Because North Carolina

Photograph dated 1949 showing the entrance of the semi-circular steel tunnel installed by the Raleigh Mining Co. leading into the Carolina Mine. The tunnel provided a safe passage under Deep River to the thirty-six-inch vein where 14,450,000 tons out of a total of over 100,000,000 tons of coal were already completely drilled and mapped. The single-track cable was planned to bring out 500 tons per twenty-four-hour period, with a planned second track accounting for up to 1,000 tons (N.C. Dept. of Conservation and Development, Division of Mineral Resources).

relied mostly on coal mined outside the state to meet its needs, the situation was sufficiently serious when Governor Kerr Scott contacted President Harry Truman about the dire situation that had considerable ramifications for the residents of North Carolina. The state's one operational coal mine at the time, the Carolina Mine at Farmville, increased production without assistance or encouragement from the state government with most of the mine's daily output of 200 tons going to the Carolina Power & Light Company's electric generating plant near Moncure. Some accommodation was made for schools and hospitals but supplying the entire state with coal from the one mine was deemed impractical.[26] In late December, Burlington's *Daily-Times News* reported that overall available supplies had precipitously dwindled from a low of three days to a high of three weeks.[27] Because of the nature of the strike, involving both coal and steel industries, some municipalities experienced shortages of fabricated steel, putting building projects on hold.

The 1950s and 1960s: New Geological Studies

In the 1950s, two important studies of the Deep River coalfield were published under the auspices of the U.S. Bureau of Mines. In 1952, a joint report entitled *Coal Deposits in the Deep River Field, Chatham, Lee, and Moore Counties, N.C.* by Albert L. Toenges, Louis A. Turnbull, Joseph J. Shields, Wilbur A. Haley, B.C. Parks, and R.F. Abernathy was followed in 1952 (published 1954) by a study, *Geology of the Deep River Coal field North Carolina,* authored by John A. Reinemund. After the well-received report Campbell and Kimball submitted in 1923, the two later surveys were initiated as the result of newer technologies and techniques available to scientists and geologists particularly through the Bureau of Mines. Toenges, et.al., included more extensive data about the chemical analysis of core samples collected from various locations in the coalfield, centering on the thickness and continuity of the coal beds, physical conditions in and around the coal beds, and physical conditions in and around the coal beds that would influence mining, and the petrographic and chemical characteristics of the coal. Although their work borrowed extensively from Campbell and Kimball, the authors expanded their surveys to include data on the forty bore hole samples, of which all previous reports had been lost.

Reinemund, in his more detailed study of the Deep River coalfield to date, used text, photographs, and tables, incorporating information about its geology from previous publications, unpublished sources, climactic data, and recent mining and drilling data.[28] Mostly, he contributed extensive mapping of the Deep River coalfield, a more current estimate of the reserves and the value of the coal in the field than his predecessors, and geological data about the problematic faults and folds in the coal seam, providing scientists and investors much needed current data in order to make informed decisions about mining coal at Deep River. The eight geological maps consisted of an aerial map of the field, structure sections maps, maps of mines, and maps showing the nature of the coal. Although both surveys were to some extent similar, they provided new scientific data previously unavailable to Campbell and Kimball.

With planned nationwide strikes always looming in the background, putting North Carolinians at risk by not having sufficient supplies of coal on hand, the Bureau of Mines in October 1952 emphasized to state officials that local coking coals could be mined in the case of an emergency, thus eliminating the need of transporting coal from outlying areas to meet local need.[29] By 1953, coal was still considered the "nation's dominant fuel" according to R.W. Bidlack, president of the American Retail Coal Dealers, stating that the national group was the "guardian of the industry."[30] Though coal may have been king in some peoples' eyes, production in North Carolina was negligible and any further attempts at mining were nowhere to be seen on the horizon, representing a continued dependence on supplies outside the state to meet demand. Two more deaths beset the Carolina Coal Company mine when on January 22, 1952, Arthur F. Devine, of Goldston, succumbed to mine gas.[31] On March 11, James H. Minter, of Oakland, was electrocuted when he grabbed a charged guy wire outside of the mine.[32] As a result, whatever mining activity remained was shut down by the company brought on by a continued loss of life unacceptable for the time.

Several attempts at coal mining at Deep River from the mid–1950s to 1967 were overall unsuccessful. However, in 1958, Glenn Arter, of Charlotte, attempted stripmining for coal in the northern part of Moore County, near Carbonton, without success.[33] Also that year, Governor Luther Hodges visited the U.S. Bureau of Mines in Washington to seek studies in economic methods with the expectation that renewed coal mining at Deep River would produce sufficient amounts of coal for use in improved furnace equipment.[34] In 1958, a commodity study of the Deep River coalfield was approved by the Board of Conservation and Development which held great interest because of the undeveloped coal deposits in the area estimated at 100,000,000 tons. A geophysical survey of the field was conducted in July 1959 with the assistance of the United States Geological Survey, which furnished electro-resistivity equipment to produce images of sub-surface structures. During the course of two weeks work, it was found that differences in resistivity of the fault planes and the surrounding rock were not great enough to measure and the project was abandoned. For the fiscal year, 1959–60, $2,699 in state funds was disbursed to the Deep River Coal Survey.[35] From 1960 to 1967 no further attempts were made to mine coal at Deep River

In April 1968, Chatham County was the recipient of a planning assistance grant of $9,000 funded by the U.S. Department of Housing and Urban Development to undertake a land potential study of the county's resources to determine areas of projected growth. The authors of the study—released in July 1970—noted the deposits of good quality bituminous coal in the Deep River region as well as the history of mining efforts, including the issues associated with its demise, such as heavily faulted seams, lack of sufficient capital, inconclusive testing results indicating the location of coal, and a history of mining disasters. While acknowledging the potential for continued mining in the region, the report discouraged any further attempts given the current market conditions and that the region could meet demand with readily available coal in other parts of the country.[36] Additionally, alternate fossil fuel sources existed locally for domestic use and were making inroads at replacing coal as the fuel of choice.

The 1970s and Beyond

Coal for commercial and industrial applications still held some interest in the region as Charlotte-based Duke Energy, formerly Duke Power Company, in the late 1970s looked at means to obtain coal other than the Appalachian variety, to fire its hydroelectric generating plants and investigated known coal deposits in the Deep River field. The consensus was that the undertaking would be unprofitable largely due to the complicated logistics involved in extracting the coal from badly faulted seams. Duke's findings were consistent with those of other companies that attempted to mine coal at Deep River and further efforts were called off.[37] One of the last attempts at finding commercial value in the Deep River coalfield occurred in 1991 when a well was drilled for the purpose of testing the gas potential in the area. While the well produced small amounts of natural gas, operations were halted and the well was plugged. Additional consideration was given to oil shale extraction, which had first been explored in 1927.[38] In 2009, the North Carolina Geological Survey compiled data regarding shale gas extraction in the Deep River basin and determined that shale deposits extended across 25,000 acres at depths less than 3,000 feet in the Sanford sub-basin of Lee and Chatham counties. Six of twenty-eight wells that had been drilled in the Cumnock formation indicated the presence of gas and oil, but controversial proposed methods of extraction resulting in potential harm to property and the environment lost favor among local residents.[39]

Conclusion

Based on annual state reports of the mineral industry of North Carolina, the total output of coal produced at Deep River from 1840 to 1950 resulted in 1,023,692 tons. To put production levels into perspective, Pennsylvania coal production peaked in 1917 when 329,000 miners accounted for 278,000,000 tons of coal.[1] Though coal mining in North Carolina never achieved the rich legacy and tremendous yield of a Pennsylvania or West Virginia, Deep River coal proved to be a readily available fuel that stimulated early growth and prosperity before better fuel alternatives appeared on the scene. Proprietors of iron foundries, such as Wilcox and Endor, discovered Deep River coal to be a cheap and convenient fuel source without having to import coal from faraway regions. These foundries fueled by Deep River coal not only provided tools and utensils for local use, but they also forged weapons during the American Revolution, giving the weapons-depleted Continental Army an available source of material for manufacturing armaments. During the Civil War, Deep River coal fueled local industries in support of the Confederacy's war efforts and its use in blockade runners, though at times controversial, provided a ready fuel source when others had ceased. Deep River coal fueled the forges of the Confederate Navy Yard at Charlotte, and its domestic use proved invaluable when supplies from the North were cut off and citizens were left to make do for themselves. For a time, Deep River coal figured in the grand scheme of North Carolina's vision of becoming an industrial player in the early days of the republic. That it did not realize this vision of a coal-centered economy should not be seen as a complete failure.

The forces of nature had much to do with the inherent difficulties mining Deep River coal, causing numerous enterprises to sputter and eventually stop. Least of all, the deaths of fifty-three miners who died in a single tragedy on North Carolina soil in May 1925 should never have taken place but given the history of attempts at mining coal at Deep River it was a tragedy waiting to happen. Regardless of the blame and responsibility over the tragedy, and there is plenty to go around, the devastation wrought on the small community of Coal Glen is incalculable. The loss of fathers, grandfathers, sons, and brothers left gaping holes in families, who were left to pick up the pieces and continue the best they could the difficult task of survival and confronting the future.

We hopefully learn from our mistakes, as was witnessed in the passage of the much-needed Workmen's Compensation Act of 1929, which was created in the aftermath

CONCLUSION

of the tragedy at Coal Glen, to provide financial assistance to workmen and their families in the event of injury or death while on the job.[2] While attention has turned away from coal mining at Deep River, the area has revealed and yielded clay for the manufacture of bricks, making the area an important source on a national scale. New industries have created new demands and the resources of Deep River have continued to figure in those plans, but at a cost no one had imagined.[3]

Appendix A
The Dan River Coalfield

Experts were in agreement that the Dan River coalfield contained coal of little or no commercial value, leading to few serious attempts at mining in the region. However, this field is important from the standpoint of early investigations into the area by noted scientists such as Olmstead, Emmons, McLanahan, Kerr, Hale, Wilkes, Holmes, and Stone. Emmons went so far as to claim that coal deposits found at the Dan River field did not "differ materially from those of Deep River."[1] J.A. Holmes, who at the time was state geologist and involved in the state geological survey, reporting to R.W. Stone's examination much later than Emmons's, revealed that the Dan River field possessed similar characteristics with the sedimentary rocks of the Richmond Basin's coalfield in Virginia, and that the Dan River field's stratigraphy was similar to that of the Deep River field.[2] Also the significance of the Dan River field's location in Stokes, Rockingham, and Madison counties near the Virginia border offered prospects for commerce and trade between the two states as well as providing for an outlet to western markets.

Responding to the state of Virginia's application to the North Carolina legislature for a charter to build a southerly line through the state, the legislature chartered the North Carolina Railroad Company on January 27, 1849, for the purpose of building such a line as the Richmond and Danville Railroad required. Afterward, the Richmond and Danville sought to effect a physical connection with the North Carolina Railroad at Greensboro, and after the Richmond and Danville surveyed a line through Milton, North Carolina, did not respond to the efforts and it was not until after completion of the main line to Danville in 1856, that efforts were made to obtain a North Carolina charter for the extension. Assuming the risk of building a western extension of their line, the Richmond and Danville prompted the Virginia legislature to pass an act on April 7, 1858, to authorize "connections between the Richmond and Danville R.R. and the North Carolina Central Railroads."[3] The act also granted permission to the Richmond and Danville, or any North Carolina corporation "which might be chartered for that purpose to build a railroad from Danville to Greensboro, to connect the Richmond and Danville and the North Carolina Railroad."[4] In response, the North Carolina legislature chartered the Dan River Coalfield Railroad Company for the purpose of constructing a railroad from a place on the Virginia line near Danville to the coalfields

Appendix A

of the Dan River and further stipulating that the road "shall not run within twenty miles of the North-Carolina railroad."[5] However, military priorities of the Confederate government dictated that the Piedmont Railroad was the preferable connection between the Richmond and Danville at Danville and the North Carolina Railroad at Greensboro. The Richmond and Danville consequently dropped the Dan River Coalfield Railroad Company from consideration but it did not spell the demise of interest in the Dan River field.[6]

Although the Dan River Coalfield Railroad never met the objective for which it was chartered, in 1868 a new charter was approved in an ordinance of the State Constitutional Convention with the same provision as the first, namely, "for the purpose of constructing a Rail Road from some point on the Virginia line near the town of Danville, in Virginia, to the Coalfields of Dan River."[7] In 1869, the state legislature passed an act authorizing the Dan River Coalfield Railroad Company to construct and extend their road to Germanton in Stokes County and "thence to some point on the Western North Carolina Rail Road, at or near Statesville."[8] The act also approved construction of branches with the Richmond and Danville Railroad and the North Carolina Rail Road provided that the gauge of the Dan River Coalfield Railroad was the same as that of the North Carolina Railroad.[9]

While studies of the Deep River coalfield far exceed those of the Dan River coalfield, one of the most extensive investigations into the geology of the Dan River field appeared in 1910 when the eminent geologist R.W. Stone completed a report on the coal and lignite deposits for the United States Geological Survey. His examination revealed that the Dan River field possessed similar characteristics with the sedimentary rocks of the Richmond Basin's coalfield in Virginia, and that the Dan River field's stratigraphy was similar to that of the Deep River field, but further interest proved to be minimal.[10]

Appendix B
*Jackson's Survey
of the Deep River Coalfield*

Map on following two pages.

Appendix B

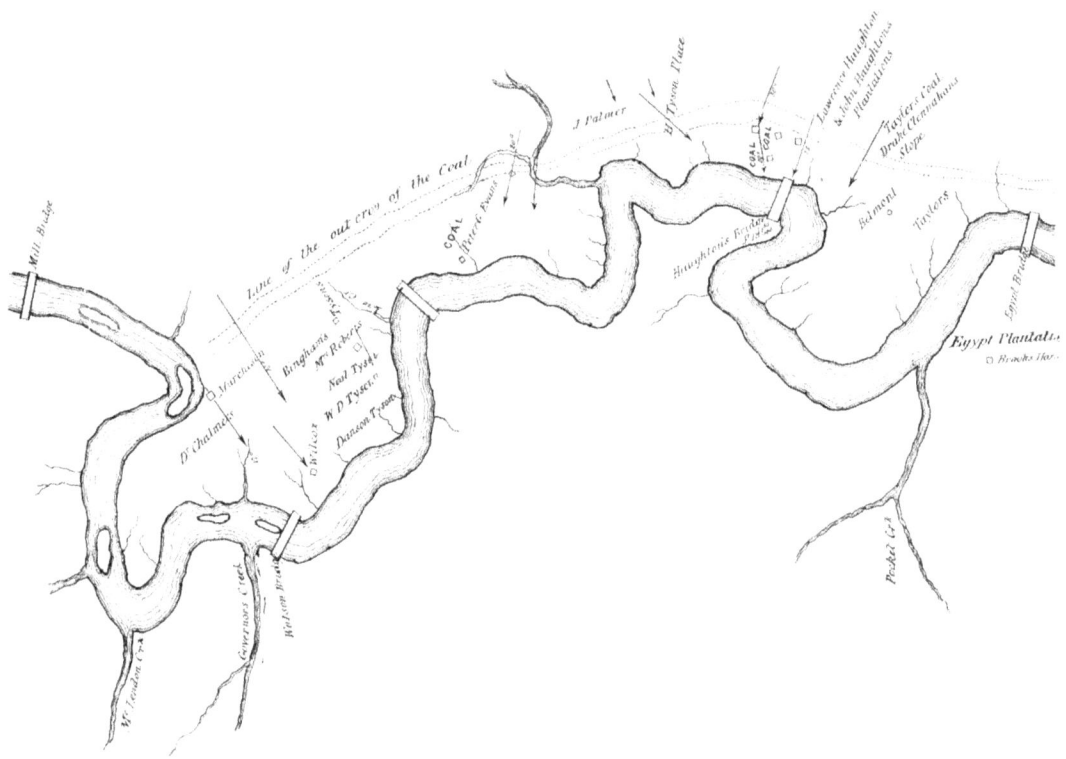

Above and opposite: Geological Map of Deep River Coal field from Recent Surveys, September 1853, by Charles T. Jackson. This map shows a stretch of Deep River which identifies outcroppings of coal together with plantations and areas concerned with mining. The map terminates at Egypt, Chatham County.

Johnson Map of the Deep River Coalfield

Appendix C
*Daddow-Bannon Map
of the Deep River Coalfield*

Map on following page.

Daddow-Bannon Map of the Deep River Coalfield

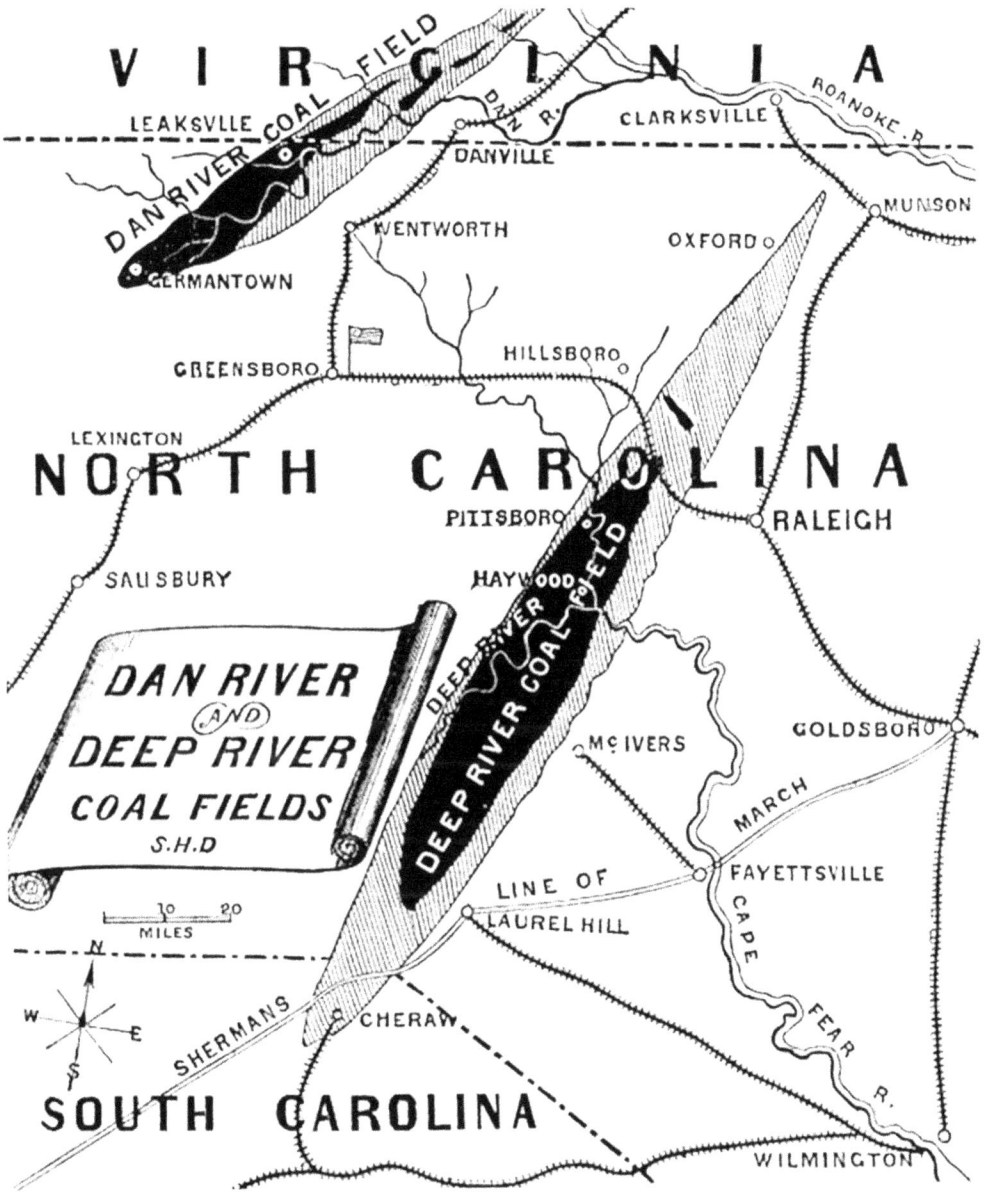

The map showing the Deep River and Dan River coalfields was taken from *Coal, Iron, and Oil; or, the Practical Miner. A Plain and Popular Work on Our Mines and Mineral Resources, and a Text-book or Guide to Their Economical Development. With Numerous Maps and Engravings. By Samuel Harries Daddow, and Benjamin Bannan* (Pottsville, PA: Benjamin Bannan; Philadelphia: J. B. Lippincott & Co.; London: Trubner & Co., 1866). Benjamin Bannan, editor of the *Miner's Journal*, was an influential journalist and political economist on many mining concerns especially in the anthracite coal region of northeast Pennsylvania. His inclusion of the Deep River coal field helped stimulate interest in mining opportunities.

Appendix D
Transcript of Inquest Concerning the December 19, 1895, Disaster at Cumnock

The Chatham Record, December 26, 1895

Thursday, Dec. 19. Explosion at Egypt coal mine, now called Cumnock. Coal explosion which occurs in coal mines "in spite of the most prudent precautions."

Work force divided into two shifts: 7:00 a.m.–7:00 p.m.; 7:00 p.m.–7:00 a.m. Before either shift allowed to enter mine "gas man" or fire boss examines every part of mine to determine if air quality was all right. On day of accident at 7:00 a.m. fire boss gave permission to enter the mine. Sixty-six miners descended into mine and a little over an hour later at 8:20 a.m. forty miners were killed. Outside people heard rumbling noise but did not realize extent of accident. Miners started to ascend and looked dazed. Shaft was 465 ft. deep. The cage or car could transport a dozen miners at a time. After the twenty-six survivors arrived to safety a call was made for volunteers to enter the mine to search for others who had not come out. It was sometime before rescue party could enter mine because of potential danger. The superintendent, George F. Cant, persuaded a few men to accompany him to the mine after fresh air had to be pumped into the mine. All bodies found were dead. Bodies found in different positions and conditions. Some bodies badly burned but others without wounds. Those without wounds had breathed "afterdamp," which is almost always fatal and the result of an explosion. Mine divided into three parts or slopes: No. 1, No. 2, and No. 3. Explosion occurred in No. 1 slope where 37 men were working and all were killed. In No. 2, eleven men worked and two men were killed and four or five wounded. No. 3 no men killed. As soon as men in No. 2 and No. 3 heard explosion, they hurried up shaft to surface as fast as cages could carry them. One of the survivors, Edward Poe, who operated small compress air engine said he heard explosion following by rumbling. He placed himself behind engine and stated he could feel a current of air rush by him and when it reached end of tunnel rebounded and carried him 200 ft. Most of the miners from area, were one half African Americans, and few from Pennsylvania.

Friday, Dec. 20. After bodies retrieved from mine county coroner Dr. H.T. Chapin summoned a jury and held inquest to determine if miners were killed due to negligence on the part of any person or persons.

Transcription of Inquest

E.H. Davis being sworn testifies "I am the underground boss in the Egypt Coal Mine. Have had charge of the mine since June 1st, 1895. I was on top when the explosion occurred. Had I been one minute sooner I should have been right into the explosion. First I thought that dynamite was the cause of the trouble but find that it had not exploded. Mr. Mills, the fire boss, had reported to me that everything was all right. The mine is well ventilated and am willing for anyone to examine the mine. I have been in the mine today and yesterday. I have been in the mining business 21 years. Have worked from the Atlantic to the Pacific. The mine was not in good condition when I took charge, but it is now in good condition, if the weather is heavy there is more danger from gas. The explosion occurred in the west tunnel."

Geo. F. Cant being sworn testifies as follows: "I am general superintendent of the mine. I saw Mr. Davis, the underground boss, at about 7:30 a.m., Dec. 19th, 1895. He told me that Mr. Mills, the fire boss, had just reported that everything was all right. I have had 8 years experience. The mine was properly ventilated. I have had charge of the mine for about 9 days. I was well satisfied with the ventilation and consider the current of air a very strong one. An explosion might occur in the best regulated and ventilated mine. This mine is modeled as much as possible after the Penn mine, and the present company has done as much as possible toward improvement in the time they have had the mine. The cause of the explosion is a mystery. It is customary to use dynamite. The dynamite was not exploded. 37 dead bodies have been taken out, two more bodies are in the mine or not accounted for."

Alfred Cox being sworn says that "I have worked in the Egypt Coal Mine for the past 5 years. The ventilation of the mine is better now or was before the explosion than it ever was or since I have been here. We use dynamite in the mine. It was supposed at first that the trouble was caused by dynamite, but we have found the dynamite and it is not exploded, think that it was a gas explosion. At 7 o'clock a.m. yesterday Mr. Joe Mills reported everything all right. The explosion took place at about 8:20 a.m. Mr. Mills, the gas man, is well up to his business and is as cautious as anyone could be. The management of the mine was in my opinion as good as could be. No man was allowed to go to his place until allowed to go by the gasman."

D.E. Teusch being sworn testifies as follows: "I came to Egypt Jan. 19th, 1895. In 8 or 10 days I went to work in the Egypt Coal Mine, and worked between 6 weeks and two months. I quit work for several reasons, one main reason of my quitting at the time was officiousness of the boss. I have been a miner for 22 or 23 years. Have worked in mines in 13 states and in Canada. Another reason that caused me to quit work in the Egypt mine was on account of the bad ventilation. I warned others to

Appendix D

quit work on account of danger. Mr. Mills, the gasman, is a very careful man, but I do not think he fully understands the ventilation of the mine. The mine is about 465 feet deep. In well ventilated mines an explosion may take place but they always are caused by some one's carelessness or neglect. When an explosion takes place I know that the ventilation is defective. I have not been in the mine since the middle or latter part of last March."

[testimony continues]

"It is customary to use dynamite in coal mines.

"There is no law or laws in this state, so far as I know, protecting the miner or compelling the mine owner to ventilate their mines.

"From the numbers of deaths in this mine in this explosion and the amount of damage done I say that the ventilation was defective.

"I have known miners to come here and go down into the mine and would come right out and would not work on account of danger.

"I enter a plea here to the Legislature to pass a law to protect the miner."

G.B. Hart being sworn testifies that "I know Charlie Poe, that I recognized him as one of the dead persons brought up out of the mine. I work in slope No. 3, the explosion was in No. 1.

Mr. Mills is the man in charge of the ventilation and is a very careful man. I quit working in No. 1 last December on account of the danger but in No. 3 where I now work I think the air and ventilation was all right. I thought the boss in No. 1 did not manage the ventilation as I thought he ought, so I quit and did not work again until a new boss took charge. Considerable damage has been done to the interior of the mine. I understand that the dynamite in the mine was not exploded. The explosion was caused by gas. No man in No. 1 escaped. I work at night and had been out of the mine about 2 hours. I have down in the mine today. Mr. Mills the gas man is burned right badly, had a finger taken off yesterday. I am satisfied that the explosion took place at the bottom of No. 1. There are about 40 men killed in this explosion. All the men in No. 2 except two escaped. All escaped in No. 3."

James Poe being sworn testifies as follows: "I went into the Egypt Coal Mine just before 7 o'clock a.m., Dec. 19th, 1895. I am a miner, worked at the business for considerable time. I worked in slope No. 2. I think that the mine was in good condition. In my section of the mine the air was good and in better condition than usual. The explosion occurred in slope No. 1. I know of no one who was in slope No. 1 that escaped. There were 11 men in my slope. Two died in my slope."

After hearing the evidence of the witnesses, as above printed, the coroner's jury after a short consultation rendered the following verdict: That Charles Poe and other miners came to their death by an explosion in the Cumnock Coal Mine, Dec. 19th, 1895, by explosion of gas, the cause of said explosion being to as unknown, it being shown to us that the mine was in better condition than usual and was well ventilated.

The dead bodies were brought up out of the mine and were decently buried at the expense of the owners of the mine, except the bodies of two of the Pennsylvania miners, which were sent to their families in that state. The following is a list of the killed who were citizens of this county: Charlie Poe, John A. Ganter, Daniel Morris, Henry Morris, George Monroe, John Monroe, Lucian Holland, Wright Tysor, John Obie, Counts Poe, Atlas Andrews and Jesse Lambert. These were white men, nearly all married, and in addition to them there were the following colored men: Walter Haughton, Gaston Lambeth, Jack McIver, Will Jenkins, Jim Rives, Love Schenck, Lewis White, Clay Harris, Arthur White, John Norwood, Joe Thompson, Fisher Rives, Will Baldwin, and Joe White. Among the dead were two young German brothers who had arrived only a few days before, and whose relatives will probably never hear of their untimely deaths.

It is said that over twenty-eight widows and over one hundred fatherless children mourn the victims of this sad catastrophe. And not only were their hearts grief stricken at the loss of their loved ones, but with most of them actual want was staring them in the face. They were entirely dependent on their dead husbands and fathers for their daily bread, and now their only earthly support is gone. Unless timely relief is afforded them, their lot will be hard indeed. In order to illustrate their pitiable condition, one case may be cited. It is that of Mrs. Michael Bentley, an English woman, whose husband and only son were both killed, leaving her with a seven year old daughter alone, far from their native home, without any relative whatever to comfort or to support her. This family had arrived from Pennsylvania only three months ago. It would really seem that the fate of the living is more to be pitied than that of the dead whom they mourn!

Appendix E
Individuals Killed at Farmville Mine, 1925

The names of the fifty-three men killed in the explosion at the Farmville Mine, Carolina Coal Co., Coal Glen, Chatham Co., NC, on May 27, 1925.

Name	Age	Race	Marital Status	Occupation[1]	Place of Death	Place of Burial[2]
Henry Alston aka Lewis H. Alston	32	B	M	Coal miner	Coal Glen, Chatham Co.	Union Grove AME Zion Church Cemetery, Pittsboro, Chatham Co., NC
Johnie aka Johnny Alston	17	B	S	Coal mining	Oakland, Chatham Co.	Piney Grove United Methodist Church Cemetery, Siler City, Chatham Co., NC
Francis Shuble Anderson	37	W	M	Coal miner	Coal Glen, Chatham Co.	Box Cemetery, Ragland, St. Clair Co., AL
George Millard Fillmore Anderson	49	W	M	Coal miner	Coal Glen, Chatham Co.	Box Cemetery, Ragland, St. Clair Co., AL
David Barr	19	B	M	Coal mining	Coal Glen, Chatham Co.	Johnsonville, Florence Co., SC
Thomas Lee Buchanan	Abt. 20	W	M	Coal miner	Coal Glen, Chatham Co.	Cool Springs Baptist Cemetery, Sanford, Lee Co., NC
John Burgess	Abt. 30	B	M	Coal mining	Oakland, Chatham Co.	Bethune, Kershaw Co., SC
W.E. Byerly aka William E. Byerly	Abt. 40	W	M	Coal miner	Coal Glen, Chatham Co.	Farmville Cemetery, Coal Glen, Chatham Co., NC
Ruben Chalmers aka Ruben Chambliss	Abt. 32	W	M	Coal miner	Coal Glen, Chatham Co.	Ragland, St. Clair Co., AL
Wilson Chesney	51	B	M	Coal mining	Oakland, Chatham Co.	New Hope AME Zion Church Cemetery, Sanford, Lee Co., NC

Individuals Killed at Farmville Mine, 1925

Name	Age	Race	Marital Status	Occupation[1]	Place of Death	Place of Burial[2]
June Cotton	20	B	S	Coal miner	Coal Glen, Chatham Co.	Union Grove AME Zion Church Cemetery, Pittsboro, Chatham Co., NC
Thomas N. Cotton	28	W	M	UNKN[3]	Coal Glen, Chatham Co.	Asbury United Methodist Church Cemetery, Sanford, Lee Co., NC
J.B. Curd aka John Benjamin Curd	Abt. 35	W	M	Coal mining	Oakland, Chatham Co.	Garden of Memories Cemetery, Ragland, St. Clair Co., AL
Clifford Davis	Abt. 32	W	M	Coal miner	Coal Glen, Chatham Co.	Ragland, St. Clair Co., AL
Edward Dillingham aka Edward Dellingham[4]	Abt. 30	W	M	Coal miner	Coal Glen, Chatham Co.	High Point, Guilford Co., NC
Walter Dillingham	Abt. 27	W	M	Coal mine	Coal Glen, Chatham Co.	High Point, Guilford Co., NC
Henry G. Hall aka Henry Grady Hall	35	W	M	Coal mining	Oakland, Chatham Co.	Farmville Cemetery, Coal Glen, Chatham Co., NC
Elmer Hayes aka Elmer E. Hayes	26	W	M	Coal mining	Oakland, Chatham Co.	Holly Spring Friends Meeting Cemetery, Ramseur, Randolph Co., NC
Issac Hays	UNKN	B	UNKN	Coal miner	Coal Glen, Chatham Co.	Cumnock, Lee Co., NC
Eliga Hill aka Elijah Hill	Abt. 50	B	M	Coal mining	Oakland, Chatham Co.	Cumnock Community Cemetery, Cumnock, Lee Co., NC
Lee Hodge	Abt. 40	B	M	Coal mining	Oakland, Chatham Co.	Danville, VA
Archie L. Holland aka Archibald L. Holland	Abt. 35	W	M	Hoisting engine operator	Oakland, Chatham Co.	Bethany Cemetery, Gulf, Chatham Co., NC
Albert Holly	UNKN	B	UNKN	Coal miner	Coal Glen, Chatham Co.	Cumnock, Lee Co., NC
Wesley Howard	27	B	M	Coal mining	Oakland, Chatham Co.	Orangeburg, SC
Dan B. Hudson	16	W	M	Coal mining	Oakland, Chatham Co.	Cool Springs Baptist Cemetery, Sanford, Lee Co., NC
Joe A. Hudson	27	W	M	Coal mining	Oakland, Chatham Co.	Cool Springs Baptist Cemetery, Sanford, Lee Co., NC
Will Irick	Abt. 35	B	M	Coal miner	Oakland, Chatham Co.	Fort Motte, Calhoun Co., SC
Claud V. Johnson	48	W	M	Coal mining	Oakland,	Farmville Ceme-

Appendix E

Name	Age	Race	Marital Status	Occupation[1]	Place of Death	Place of Burial[2]
aka Claudius Johnson aka Claude Johnson					Chatham Co.	tery, Coal Glen, Chatham Co., NC
Nathan R. Johnson	26	W	M	Coal mining	Oakland, Chatham Co.	Jones Chapel United Methodist Church Cemetery, Lee Co., NC
Manley Lambert	Abt. 25	B	M	Coal miner	Coal Glen, Chatham Co.	Macedonia AME Zion Church Cemetery, Gulf, Chatham Co., NC
John Edward Laubscher	20	W	S	Coal mining	Oakland, Chatham Co.	Johnson Grove Cemetery, Vass, Moore Co., NC
A.F. Martin aka Albert F. Martin	38	W	M	Coal mining	Sanford, Lee Co.	Oakland Cemetery, Hampton, Hampton City, VA
William Moore	Abt. 45	B	M	Coal miner	Coal Glen, Chatham Co.	Mt. Croghan Cemetery, Mt. Croghan, Chesterfield Co., SC
Jas. Nabors, aka James Nabors	Abt. 65	B	M	Coal miner	Coal Glen, Chatham Co.	Greensboro, Guilford Co., NC
Samuel Napier	27	W	M	Coal miner	Coal Glen, Chatham Co.	Cool Springs Baptist Cemetery, Sanford, Lee Co., NC
Arthur Poe	UNKN	B	UNKN	Coal miner	Coal Glen, Chatham Co.	UNKN
Hollis Richardson aka W. Hollis Richardson	Abt. 20	W	S	Coal miner	Coal Glen, Chatham Co.	Coal Glen, Chatham Co., NC
Zeff Riner aka Zeff E. Riner	23	W	M	Coal mining	Oakland, Chatham Co.	Waxhaw, Union Co., NC
John Shaw	UNKN	B	M	Coal miner	Coal Glen, Chatham Co.	Bishopville, Lee Co., SC
Jas. Spruell, Jr. aka James Spruell, Jr.	Abt. 20	B	S	Coal miner	Coal Glen, Chatham Co.	Union Grove AME Zion Church Cemetery, Pittsboro, Chatham Co., NC
A. L. Stokes	Abt. 50	W	S	Coal mining	Oakland, Chatham Co.	Savage, Howard Co., MD
H. W. Sullivan	55	W	M	Coal mining	Oakland, Chatham Co.	High Point, Guilford Co., NC
Chas. Watson aka Charles Watson	Abt. 38	B	M	Coal miner	Coal Glen, Chatham Co.	Union Grove AME Zion Church Cemetery, Pittsboro, Chatham Co., NC
James Williams	21	B	S	Coal miner	Oakland, Chatham Co.	New Hope AME Zion Church Cemetery, Sanford, Lee Co., NC
Robert Williams	54	B	M	Coal mining	Oakland,	New Hope AME

Individuals Killed at Farmville Mine, 1925

Name	Age	Race	Marital Status	Occupation[1]	Place of Death	Place of Burial[2]
					Chatham Co.	Zion Church Cemetery, Sanford, Lee Co., NC
David Jonathan Wilson	31	W	S	Coal miner	Coal Glen, Chatham Co.	New Center Christian Church Cemetery, Seagrove, Randolph Co., NC
Wade Wilson	UNKN	B	UNKN	Coal miner	Coal Glen, Chatham Co.	Cumnock, Lee Co., NC
Chas. L. Wood aka Charlie Wood	UNKN	W	UNKN	Coal miner	Coal Glen, Chatham Co.	Farmville Cemetery, Coal Glen, Chatham Co., NC
Claud Wood	23	W	S	Coal mining	Oakland, Chatham Co.	Garden of Memories Cemetery, Ragland, St. Clair Co., AL
James Wright	23	B	S	Coal mining	Oakland, Chatham Co.	Rosindale, Bladen Co., NC
Russell Wright	Abt. 16	B	S	Coal miner	Coal Glen, Chatham Co.	Rosindale, Bladen Co., NC
Thedore Wright	Abt. 18	B	S	Coal miner	Coal Glen, Chatham Co.	Rosindale, Bladen Co., NC
Thomas N. Wright aka Thomas Neil	49	B	M	Coal mining	Oakland, Chatham Co.	Rosindale, Bladen Co., NC

Sources: North Carolina State Board of Health, Bureau of Vital Statistics. North Carolina Death Certificates. *Microfilm S.123. Rolls 19–242, 280, 313–682, 1040–1297. North Carolina State Archives, Raleigh, North Carolina.* Find A Grave, *online database, www.findagrave.com.*

[1] The occupation as recorded on the death certificate. The occupation of coal mining may or may not mean a coal miner but someone working in another capacity in the mines, such as a laborer, timberman, fire boss, machinist, or supervisor.

[2] The actual place of burial may in some instances differ from that recorded on the death certificate.

[3] UNKN (unknown).

[4] Two separate death certificates issued apparently for the same person.

Appendix F
Total Tonnage Mined Yearly at Deep River

(based on 1 short ton=2,000 lbs.)

Year	Tonnage	Year	Tonnage
1840*	3	1894	16,900
1850	—	1895	24,900
1851	—	1896	7,813
1852	100	1897	21,280
1853	100	1898	11,495
1854	200	1899	28,853
1855	200	1900	18,000
1856	300	1901	3,723
1857	300	1902	23,000
1858	400	1903	17,309
1859	400	1904	7,000
1860	500	1905	1,557
1861	15,000	1906–11	—
1862	30,000	1912	120
1863	30,000	1913–17	—
1864	25,000	1918	1,420
1865	20,000	1919	6,989
1866	20,000	1920	11,540
1867	20,000	1921	23,438
1868	18,000	1922*****	78,570
1869	16,000	1923	36,019
1870	15,000	1924	57,094
1871	15,000	1925	65,153
1872	12,000	1926	59,936
1873	10,000	1927	53,377
1874–79**	—	1928	60,860
1880***	350	1929	58,180
1881	300	1930	28,500
1882	400	1931	2,363
1883	400	1932	1,900
1884	500	1933	2,014
1885	500	1934	3,140
1886–89****	—	1935	2,000
1890	10,262	1936–48	—
1891	20,355	1949	14,000
1892	6,679	1950–70	—
1893	17,000		

Sources: The Mining Industry in North Carolina, *various years;* United States Geological Survey; The First Century and a Quarter of American Coal Industry *(Pittsburgh, PA: privately published, 1942).*

Total Tonnage Mined Yearly at Deep River

* An early report published by R.C. Taylor in 1848 reported that more anthracite than bituminous coal was produced.

** Any production most likely would have been at the Gulf Mine on a negligible scale.

*** 1880 Gulf Mine only. The 1880 U.S. Census Non-population Schedules shows the following mines named in the Deep River area: (1) Gulf Mine. Operator, E.L. Houghton. Product, 200 tons. Thickness of coal, 6 1/2 feet. Market: Raleigh, Charlotte, Fayetteville, and Lawrenceburg. (2) Egypt Mine. Not operated.

**** Does not include 1889 production from the recently reopened Cumnock Mine.

***** Includes Carolina Coal Company tonnage since operations began at Farmville, in Chatham County, and became an active producer.

Appendix G
Company Records

Mallett-Browne Coal Company Records

The Southern Historical Collection (SHC), located in Wilson Library at the University of North Carolina at Chapel Hill, holds the C.B. Mallett Papers, which contain, among other things, a group of receipts relating to the transactions of the Mallett-Browne Coal Company, in Chatham County, owned jointly by Charles Beatty Mallett and James Browne, of Charleston, South Carolina. The Mallett-Browne Coal Company mined its own coal at Egypt and in the capacity of a wholesaler and retailer sold it to merchant retailers, who set their own prices, or directly to the customer. Only a small number of the recorded transactions kept by Mallett-Browne has survived. In instances where there are discrepancies in figures, I have shown the amount of the written transactions as it appeared on the receipt. Coal prices fluctuated from the beginning of the Civil War, when the average price of one ton of coal was $2.49,[1] increasing significantly to the end of hostilities, and like those of other commodities, were dictated by supply and demand, transportation and labor costs, manpower shortages, accidents, and acts of nature. The SHC also possesses a ledger book once owned and maintained by the Endor Iron Works of Chatham County, containing transactions relating to income and expenditures the company recorded from 1864 to 1868. As with the Mallett-Browne written transactions, it is likely that not all recorded entries of the Endor Iron Works survived. Also, I have included only those transactions that are relevant to the time of the Civil War.

Mallett-Browne Coal Company Sales

Date	Tonnage	Sale Amount*	Customer(s)/Notes
June 13, 1861	Illegible	395.66	Deep River Coal & Iron Company
March 12, 1864	35	Not given	Wilmington & Manchester R.R.
	30	Not given	Wilmington & Weldon R.R.
	19	785.00	Nitre & Mining Bureau
	8	600.00	Central R.R.
	8	600.00	Chattanooga R.R.
	16	1,200.00	South Carolina R.R.
March 16, 1864	19	785.00	Nitre & Mining Bureau
	Illegible	Illegible	Wilmington & Weldon R.R.

Company Records

Date	Tonnage	Sale Amount*	Customer(s)/Notes
	Illegible	Illegible	Central R.R.
	Illegible	Illegible	Chattanooga R.R.
	16	1,200.00	South Carolina R.R.
March 24, 1864	47	1,857.63	War Dept.
April 15, 1864	50	3,810.00	State of South Carolina
April 20, 1864	30	2,250.00	Wilmington & Weldon R.R.
	7	325.00	Charleston (SC) Gas Works
	24	1,800.00	State of South Carolina
May 13, 1864	60	4,500.00	Wilmington & Weldon R.R.
	68	2,069.00	Wilmington & Manchester R.R.
	24	1,800.00	Illegible
July 4, 1864	64	2,250.00; 2,117.50	Illegible
July 11, 1864	65	3,575.00	State of North Carolina
July 14, 1864	35	2,100.00	Illegible
July 25, 1864	85	8,500.00	Blockade runner *Florie*
July 27, 1864	14	1,400.00	East Tennessee & Georgia R.R.
	14	1,400.00	Nashville & Chattanooga R.R.
	7	700.00	Bath Paper Mills (South Carolina)
	19	1,950.00	Georgia R.R.
	65	3,246.25	Illegible
	Not given	32,990.00	Total for month of July
Dec. 17, 1864	Not given	1,970.00	Sale of coal at Egypt**

Source: C.B. Mallett Papers, #3165, Southern Historical Collection, The Wilson Library, University of North Carolina at Chapel Hill.

* Presumed to be in Confederate States of America dollars or state treasury notes. Price fluctuations may be attributed to the value of the Confederate dollar relative to the U.S. dollar on that particular date, or some other unspecified factor.

** May represent sales the company made directly to customers at the company's mine. In these instances, customers usually provided their own means of transportation for haulage.

Endor Iron Works Disbursements for Fuel, 1864–1865

Date	Fuel	Amount*	Cost**	Supplier
April 16, 1864	Coal	15 bush.	33.75	Farmville Coal Co., Deep River, NC
April 25, 1864	Coal	20 bush.	45.00	
April 26, 1864	Coal	35 bush.	78.75	
May 20, 1864	Coal	100 bush.	225.00	
May 22, 1864	Coke	500 bush.	Illegible	
	Coal	700 bush.	Illegible	
May 30, 1864	Coal	625 bush.	1,406.25	
June 6, 1864	Coal	625 bush.	1,366.00	
June 9, 1864	Wood	392 cords	Not given	
July 30, 1864	Coal	Not given	1,150.00	
September 27, 1865	Coal	Not given	Not given	Lockville Mining & Manufacturing Co. Mallett & Browne Coal Co.
Jan. 1—Sept. 27, 1865	Coal	Not given	11,541.00	Not given or illegible

Source: Endor Iron Works Ledger, ##2279-z, Southern Historical Collection, The Wilson Library, University of North Carolina at Chapel Hill.

* One bushel of coal = 80 lbs.

** Presumed to be in Confederate States of America dollars or state treasury notes.

Appendix H
Active Coal Mining Companies, 1867–1896

From Branson's North Carolina Business Directories
(1867–68, 1872, 1887, 1890, 1896)

Year	County	Mine Company	Location	Owner/Contact
1867–68	Chatham	Egypt Mining Co.	Egypt	Robert Payton, engineer
1872	Chatham	Egypt Coal Co.	Egypt	Mr. Atkinson, agent
		Farmville Coal Co.	Egypt	—
		Gulf Coal & Iron	Pittsboro	L.J. Haughton
		Hornville Coal Co.	Pittsboro	J.J. Jackson
		McIver Coal Co.	Egypt	Estate of E. McIver
		Taylor Coal Co.	Gilmer	Caldwell & others
	Harnett	Buckhorn Coal Co.	Chalk Level	Brown & DeRossett
	Moore	—	Deep River	Northern company
		Chambers	Deep River	Northern company
		Murchison	Deep River	Northern company
		Street's (not worked)	Fairhaven	—
		Tyson	Deep River	Frazier & Co. (Charleston)
		Wilcox	Deep River	Alex. McIver
1887	Chatham	—	Egypt	Robt. Peyton, Ag't, NY Co.
		—	Gulf	L.J. Haughton
		Farmville	Egypt	Albany, NY Co.
		—	Egypt	Taylor
	Moore	—	Deep River	Northern Co.
		Murchison	Deep River	Northern Co.
		Chambers	Deep River	Northern Co.
		Tyson	Deep River	Frazier & Co., Charleston, SC
		Fooshee	Deep River & McLindon's Creek	Geo. Fooshee
		Wilcox	Deep River	Alex. McIver
	Rockingham	Leaksville	—	Wade
1890	Chatham	—	Egypt	E. McIver's estate
	Harnet	Caldwell & Co.	Egypt	J.P. Steadman
		Gulf Coal & Iron	Pittsboro	L.J. Haughton
		Buckhorn	Chalk Level	Brown & DeRossett

176

Active Coal Mining Companies, 1867–1896

Year	County	Mine Company	Location	Owner/Contact
	Moore	Chambers	Deep River	Northern company
		—	Sanford	M.C. Stanback
1896	Chatham	—	Sanford	John & heirs of W.W. Dye
		Foushee	Deep River	G.W. Foushee's heirs
		Murchison	Deep River	Northern company
		Street's	Carthage	A.H. McNeill
		Tyson	Deep River	W.G. Tyson
		Wilcox	Deep River	Alex. McIver
		—	Egypt	—
	Harnett	Kohinoor Coal Co.	Gulf	—
		—	Gulf	Glendon & Gulf Mining & Manufacturing Co.
		Buckhorn	Chalk Level	Brown & DeRossett
	Moore	—	Sanford	M.C. Stanback
		—	Sanford	John Dye & heirs of W.W. Dye
		Murchison	Deep River	Northern Co.
		Foushee	Gulf	G.W. Foushee's heirs
		Streets	Carthage	A.H. McNeill
		Tyson	Deep River	G.W. Tyson
		Wilcox	Deep River	Alex McIver
		—	Dan River	—
	Stokes	—	Walnut Cove	C. Hairston

Chapter Notes

Preface

1. See Joe William Trotter, Jr., *Coal, Class, and Color* (Urbana, IL: University of Illinois Press, 1990).

Introduction

1. *News and Observer*, May 28, 1925. The *Durham Sun* in its May 27 edition provided same-day coverage of the event. With unconfirmed information streaming into newsrooms across the state and the nation, newspapers were eager to report information in the public's interest as quickly as possible which was not always commensurate with the facts. For example, newspapers did not accurately report the number of explosions or victims at the outset until mine officials had time to recount events before and during the accident. The official report issued by the U.S. Bureau of Mines stated that two explosions occurred, killing fifty-three miners.

2. Prospectors mining for gold in the California fields happened upon lignite, also called "brown coal," an inferior grade of bituminous coal which met local demand for home and industry use beginning in the 1850s, continuing until the early 1900s. Contra Costa County, near San Francisco, was the site of the Mount Diablo coalfield, the largest of its kind in the state. Similarly, the Deep River and Dan River coalfields of North Carolina would meet a similar local need until more dependable fuel sources became available. For a comprehensive treatment of the Mount Diablo coalfield, see W.A. Goodyear, *The Coal Mines of the Western Coast of the United States* (San Francisco: A.L. Bancroft & Co., 1877), 6–69. North Carolina's less-than-competitive standing as a mining state was more than compensated for by its agriculture, forestry, and animal husbandry industries. In 1920, the total amount of earnings for this category was $397,214 while the category "extraction of minerals" accounted for $1,986. The percentage of earnings for the extraction of minerals compared to that of agriculture, forestry, and animal husbandry was .005. It should be noted, however, that mining was a seasonal occupation whereby activity for some months was negligible or unreported. See Department of Commerce: Bureau of Foreign and Domestic Commerce, "Population—Occupations," *Statistical Abstract of the United States, 1920, Forty-third Number* (Washington: Government Printing Office, 1921), 56.

3. *Chatham Record*, November 24, 1922. In 1922, the name Farmville was changed to Coal Glen to avoid confusion with Farmville in Pitt County. References to the town hereafter will be called Farmville prior to 1922 and Coal Glen after 1922.

4. One such study made by R.W. Stone in 1910 represented the most comprehensive up to that date but Stone concluded that further considerations were a waste of energy and money. See R.W. Stone, "Coal and Lignite. Coal on Dan River, North Carolina," in Marius R. Campbell, *Contributions to Economic Geology (Short Papers and Preliminary Reports) 1910. Part II. Mineral Fuels. Department of the Interior. United States Geological Survey. Bulletin 471* (Washington: Government Printing Office, 1912).

Chapter 1

1. Paul E. Olsen et al. "Rift Basins of Early Mesozoic Age," in *The Geology of the Carolinas: Carolina Geological Society Fiftieth Anniversary Volume*, edited by J. Wright Horton, Jr., and Victor A. Zullo (Knoxville: University of Tennessee Press, 1991), 142. These Triassic basins are comprised of sedimentary rocks formed nearly 190 million to 200 million years ago.

2. Albert L. Toenges, Louis Turnbull, Joseph J. Shields, and Wilbur A. Haley, "Investigation of Field and Estimated Reserves of Coal," in *Coal Deposits in the Deep River Field, Chatham, Lee, and Moore Counties, N.C. Bulletin 515, Bureau of Mines* (Washington: U.S. Government Printing Office, 1952), 3. Much of the Deep River coal seam is not horizontally continuous and is cut up into block-like segments all inclined steeply and lying at different levels, making conditions exceedingly difficult for coal extraction.

3. B.C. Parks, "Petrography of Cumnock Coal," in Toenges, et al., *Coal Deposits in the Deep River Field, Chatham, Lee, and Moore Counties*, 11.

4. Donald L. Miller and Richard E. Sharples, *The Kingdom of Coal: Work, Enterprise, and Ethnic Communities in the Mine Fields* (Philadelphia: University of Pennsylvania Press, 1985), 5. Another geological aspect of the Deep River field was the discovery of iron ore, or black band, deposits, which followed the coal seams and held significant value for metallurgical purposes.

5. Gordon H. Wood, et al., *Coal Resource Classification System of the U.S. Geological Survey*, Geological Survey Circular 891(Washington: U.S. Government Printing Office, 1983), 17. One important factor in the categorization of coal is the presence of volatile mat-

ter, consisting mostly of gases, certain organic compounds, sulfur, and tars, which are driven off when coal is heated at extremely high temperatures in the absence of air. Coals with low volatile matter typically burn more cleanly and consistently than those having more volatile material.

6. *Ibid.*, 136. Pennsylvania also has significant deposits of bituminous coal in the western half of the state and once led the nation in that category.

7. Robert C. Whisonant, *Arming the Confederacy: How Virginia's Minerals Forged the Rebel War Machine* (Switzerland: Springer International Publishing AG, 2015), 135. Some claim has been made that the Virginia semi-anthracite variety fueled the CSS *Virginia* in its fight in 1862 with the USS *Monitor*, although there is no conclusive evidence that it did. Some semi-anthracite coal was mined in several locations in the United States, and Southern blockade runners relied on its ability for faster acceleration to outrun Union blockaders, which used Pennsylvania anthracite.

8. Sir Henry De La Beche and Dr. Lyon Playfair, *Third Report on the Coals Suited to the Steam Navy, Presented to Both Houses of Parliament by Command of Her Majesty* (London: William Clowes and Sons, 1851), 9. The British Admiralty conducted a series of trials in 1848, 1849, and 1851 to determine the coal best suited for the Royal Navy's fleet of steamships, resulting in the selection of Welsh steam coal over the Newcastle, Lancashire, Scotland, and Derbyshire varieties. In a similar study conducted in 1844 for the U.S. Navy Department, Walter R. Johnson, of Philadelphia, completed a series of tests comparing American coals to best determine their application as a fuel in navy vessels. While Johnson ruled in favor of steam coal, the navy eventually selected anthracite as its primary fuel source because of its longer burning capacity, ample supplies, and an extensive transportation network for delivery to markets. However, with the blockade of Southern ports, this became untenable. See Walter R. Johnson, *A Report to the Navy Department of the United States, on American Coals Applicable to Steam Navigation, and Other Purposes* (Washington: Gales and Seaton, 1844).

Chapter 2

1. The Iron Act, passed by Parliament on June 24, 1750, was an attempt by the British government to stifle the manufacture of goods in the American colonies which otherwise could be produced by factories in Great Britain. However, the act openly encouraged the production of pig and bar iron in the colonies which could be imported to Great Britain duty free for use in Britain's iron forges and mills. The gist of the act was to thwart direct competition with the mother country and to suppress any form of self-sufficiency the colonies might be perceived to have. Iron forges working on a smaller scale within the colonies could sometimes circumvent the act to meet their own need, as was the case of the early foundries at Deep River. See *An Act to Encourage the Importation of Pig and Bar Iron from His Majesty's Colonies in America; and to Prevent the Erection of any Mill or other Engine for Slitting or Rolling of Iron; or any Plating Forge to work with a Tilt Hammer; or any Furnace for making Steel in any of the said Colonies* (London: Printed by Thomas Baskett, 1750).

2. *The Journal of the Proceedings of the Provincial Congress of North Carolina, Held at Halifax, on the Fourth Day of April, 1776* (New Bern: Printed by James Davis, 1776; reprint, Raleigh: Lawrence and Lemay, 1831), 30. There existed in the state two additional bloomery forges, one built before the American Revolution in Cleveland County and the other in 1780 in Stokes County, but it is not known whether either or both were producing iron for the American cause. See J.P. Lesley, *The Iron Manufacturer's Guide to the Furnaces, Forges and Rolling Mills of the United States, with Discussions of Iron as a Chemical Element, an American Ore, and a Manufactured Article, in Commerce and in Industry, with Maps and Plates* (New York: John Wiley, 1859), 186, 189. Wilcox was not new to the workings of internal politics and causes as his participation in the Regulation Movement attests.

3. *The Colonial Records of North Carolina: Published Under the Supervision of the Trustees of the Public Libraries, by Order of the General Assembly. Collected and Edited by William L. Saunders, Vol. X— 1775–1776* (Raleigh: Josephus Daniels, 1890), 647–50, 992–96.

4. "Letter from Ambrose Ramsey, Mial Scurlock, and John Birdsong to Richard Caswell," June 6, 1777, Volume 11, 487–88.

5. Colonial and State Records of North Carolina, *Minutes of the North Carolina House of Commons, North Carolina. General Assembly, November 15, 1777–December 24, 1777*, Vol. 12, 438.

6. Robert B. Gordon, *American Iron, 1607–1900* (Baltimore: Johns Hopkins University Press, 1996), 85. Coupled with seasonal fluctuations, the expense of purchasing equipment usually meant that mining was only available to the affluent land owners, who could afford a windlass or drilling tools, primitive as they may be by today's standards. An article appearing in *The Pilot* (Southern Pines, NC) in June 1941 reported that Walter Siler, of Pittsboro, claimed that when Gen. Nathanael Greene marched through the Deep River country with his army in 1780–81, he took, or impressed, a quantity of iron from the Wilcox furnace. In the closing days of the Revolution, Wilcox's furnace and mining machinery were destroyed by the Loyalist David Fanning, whose followers were involved in skirmishes with Whig militia in Orange and Chatham counties. See *The Pilot*, June 27, 1941. No records exist giving any indication as to the extent of the total tonnage or value of coal mined in the area.

7. Alan D. Watson, *Internal Improvements in Antebellum North Carolina* (Raleigh: Office of Archives and History, 2002), 3. In this work, the author provides a comprehensive overview of the state's commitment to and implementation of internal improvements. North Carolina was at the forefront of her sister states when it came to enacting legislation to improve navigation and implement building projects for enhancing access and pilotage of waterways and roads. Some states would not address these issues until after the War of 1812. North Carolina's initial predicament was that for some areas of the state trading with neighboring states was more practical and advantageous than creating markets within their own borders and a reluctance to help shoulder the costs of internal

improvements. In Britain, internal improvements were paid by the merchants and companies who directly benefitted from such initiatives. North Carolina's push to reach its coalfields, however, lagged behind its northern counterparts, notably the three hundred and sixty-three mile Erie Canal, opened in 1825; the Delaware and Hudson Canal, opened in 1828, for the purpose of anthracite coal haulage from northeast Pennsylvania; and the Chesapeake and Ohio Canal, opened in 1831, which ran up the Potomac Valley to the Cumberland Maryland coalfield. See Ronald E. Shaw, *Canals for a Nation: The Canal Era in the United States, 1790–1860* (Lexington: University of Kentucky, 1990), 104. Manville B. Wakefield, *Coal Boats to Tidewater: the Story of the Delaware & Hudson Canal* (Fleischmanns, NY: Purple Mountain Press, 1971). Also, Katherine A. Harvey, *The Best Dressed Miners: Life and Labor in the Maryland Coal Region, 1835–1910* (Ithaca: Cornell University Press, 1969).

8. *The State Records of North Carolina: Published Under the Supervision of the Trustees of the Public Libraries, by Order of the General Assembly. Collected and Edited by Walter Clark. Vol. XIX—1782-'84. With Supplement—1771-'82* (Goldsboro, NC: Nash Brothers, 1901), 605.

9. *Ibid.*, 743. A bill for clearing and opening navigation of the Tar River and Fishing Creek in Pitt, Edgecombe, and Halifax counties passed the House and sent to the Senate.

10. *Laws of North Carolina, at a General Assembly, Begun and Held at Fayetteville, on the First Day of November, in the Year of Our Lord One Thousand Seven Hundred and Ninety, and in the Fifteenth Year of the Independence of the Said State: Being the First Session of the Said Assembly* (New Bern: Arnett & Hodge, 1790), 13.

11. *Laws of North Carolina, at a General Assembly, Begun and Held at New Bern, on the Fifteenth Day of November, in the Year of Our Lord One Thousand Seven Hundred and Ninety-Two, and in the Seventeenth Year of Independence of the Said State: Being the First Session of the Said Assembly* (New Bern: Arnett & Hodge, 1792), 14.

12. *Ibid.*, 15.

13. Watson, *Internal Improvements*, 2. In some cases the state's deficient transportation network linking one section of the state with another for the purpose of commerce created a boon for bordering states like South Carolina and Virginia, which were more accessible, resulting in as much as fifty percent lost revenue. Frustrated legislators from North Carolina believed that commerce should be more of an intra- instead of inter-state proposition.

14. "The Titles of the Private Acts," in *The Public Acts of the General Assembly of North-Carolina. Volume II. Containing the Acts from 1790 to 1803, Revised and Published, Under the Authority of the Legislature,* by Francois-Xavier Martin (New Bern: Martin & Ogden, 1804), 102–03.

15. Tench Coxe, *A View of the United States of America, In a Series of Papers, Written at Various Times, between the Years 1787 and 1794, By Tench Coxe of Philadelphia; Interspersed with Authentic Documents: The Whole Tending to Exhibit the Progress and Present State of Civil and Religious Liberty, Population, Agriculture, Exports, Imports, Fisheries, Navigation, Ship-Building, Manufactures, and General Improvement* (Philadelphia: Printed for William Hall and Wrigley & Berriman, 1794), 180. Coxe first made note of the resources of his native Pennsylvania before turning his attention to the rest of the country. A delegate from Pennsylvania to the Continental Congress in 1788–1789, Coxe, in 1791, had co-authored with Alexander Hamilton the influential tract *Report on Manufactures*. Writing the same time as Coxe, the English-born Benjamin Harvey Latrobe (1764–1820), architect, engineer, and naturalist, also noted during his travels the abundance of coal along the James River in Virginia, eyeing its potential as a fuel source and for trade. Already, from 1758 to 1765, coal exported from James River ports reached 2,748 tons, accounting for both domestic as well as foreign sales to the West Indies and Bermuda. See *Historical Statistics of the United States, Colonial Times to 1970, Part 2* (Washington: U.S. Dept. of Commerce, Bureau of the Census, 1975), 1184. As one of the new republic's earliest trading partners, France permitted coal to be purchased and transported from the United States to its free ports in the West Indies. The Virginia pits sold their coal for 7d. sterling per bushel. In 1797, Edenton merchants named Moody and Avery advertised in the *State Gazette of North-Carolina* the sale of Liverpool-imported coal. Though the coal was sold in small quantities of chaldrons and bushels, it presumably met a local need.

16. *Ibid.*, 181.

17. *North-Carolina Star*, April 13, 1809.

18. *Ibid.*, May 11, 1809.

19. *Ibid.*, July 18, 1811. Charcoal, at this time, was the preferred fuel for forges and industrial applications because of its ability to reach higher temperatures. The *Star*'s editor, Thomas Henderson, Jr., already familiar with the Deep River coalfield, sought to promote the state in a series of articles appearing in his newspaper between 1810 and 1811 on topics such as population statistics and natural resources. Gleaning this information from every county, Henderson found himself in the unique position of conducting his own research while providing the means for disseminating his findings.

20. Watson, *Internal Improvements*, 5. Interestingly, during the formative years of Great Britain's coal industry, capital for coal-mining operations, including internal improvements and transportation, was assumed mostly by land owners, wealthy farmers, and the gentry, who in some instances formed partnerships with miners to share the burden of the cost. See John Hatcher, *The History of the British Coal Industry, Volume I, Before 1700: Towards the Age of Coal* (Oxford: Clarendon Press, 1993), 260.

21. *Ibid.*, 81. Murphey was one of two senators, along with four members of the House of Commons, selected as part of a joint committee appointed by the General Assembly to investigate the feasibility of making internal improvements as a means to establishing a transportation infrastructure within the state to promote industry, business, and resources. In the first decade of the nineteenth century, Pennsylvania had recognized the importance of the nascent coal trade in conjunction with her internal improvement programs. Writing in 1810, William J. Duane, chairman of the state's committee on roads and internal navi-

gation, regarded his state's importation of Virginia coal a costly hindrance to economic development because of high carriage rates, and pushed for internal improvements so that Pennsylvania coal would reach the more populous areas to fuel industry and manufacturing, with the intention of eventually becoming an exporter of coal. See William J. Duane, *The Internal Improvement of the Commonwealth, by Means of Roads and Canals* (Philadelphia: Jane Aitken, 1811), 98. In 1815, an act of the legislature created the Cape Fear Navigation Company, incorporating the powers of the Deep and Haw River Navigation Company, which had been given "exclusive right of all rivers and creeks running into the Cape Fear River to the mouth of said river." A select committee appointed by the legislature in 1830 to inquire into the "situation of the Cape Fear Navigation Company," namely ascertaining the amount of tolls collected by the company, what work had been done toward improving the river, whether the charter of the company had been complied with, and whether the charter was constitutional, reported that efforts to improve navigation had been unsatisfactory and expenditures unacceptable relative to tolls collected. The committee, in turn, authorized the state attorney general to initiate a judicial inquiry into whether the company had violated their charter. This type of audit on the part of the legislature bore out the complaints local governments made about internal improvements, notably, whether their regions and interests were being served adequately. See *Report on the Situation of the Cape Fear Navigation Company* (Raleigh: Lawrence and Lemay, 1832), 6.

22. "Mr. Murphey's Report to the Legislature of North Carolina on Inland Navigation, December, 1816," in *The Papers of Archibald D Murphey, Edited by William Henry Hoyt, A.M., Volume II* (Raleigh: E.M. Uzzell & Co., 1914), 34–47. Hoyt stated that the report was probably first printed in June 1816 but submitted to the Senate on Dec. 9, 1816.

23. Ibid., 35.

24. Ibid., 37.

25. *Laws of North Carolina: Enacted by the General Assembly Begun and Held at Raleigh, on the Eighteenth Day of November, in the Year of Our Lord One Thousand Eight Hundred and Sixteen, and in the Forty-first Year of the Independence of the Said State*, 23. Also, *Raleigh Register*, December 27, 1816.

26. *The Laws of the State of North-Carolina Enacted in the Year 1819* (Raleigh: Thomas Henderson, 1820), 7. According to the provisions of the law, the fund was to be established from the net proceeds of the sale of land acquired by treaty from the Cherokee Nation.

27. *Report of Sundry Suveys Made by Hamilton Fulton, Esq., State Engineer, Agreeably to Certain Instructions, From Judge Murphey, Chairman, and Submitted to the General Assembly, at Their Session, in 1819* (Raleigh: Printed by Tho. Henderson, 1819), 5.

28. Ibid., 6. While living in England, Fulton, a Scottish engineer who had made a name for himself designing canals, met Peter Browne of North Carolina, who happened to be in England at the same time looking for a principal engineer for the state. Fulton reported on the improvements to the Roanoke, Tar, Neuse, Cape Fear and Yadkin rivers, as well as the progress of the Fayetteville canal. He submitted additional reports into the 1820s to the General Assembly advocating internal improvements. See *Annual Report of the Board of Public Improvements of North Carolina, to the General Assembly, December 10, 1822; Together with Mr. Fulton's Reports to the Board, On the Public Works Projected and Carrying on Throughout the State During the Present Year* (Raleigh: J. Gales & Son, 1822). Fulton's reputation earned him future work such as when Georgia Governor Robert Troup in 1825 established a Board of Public Works and hired Fulton as its chief engineer. See Shaw, *Canals for a Nation*, 119.

29. *The Laws of North-Carolina, Enacted in the Year 1822* (Raleigh: Bell & Lawrence, 1823), 25.

30. Fulton, *Report of Sundry Surveys*, 49.

31. "Internal Improvement: Extract of a Letter from a Member of the North-Carolina Catawba Navigation Company, to a Gentleman in Camden, S.C.," *American Farmer: Rural Economy, Internal Improvements, News, Prices Current, Vol. 1, No. 20, August 13, 1819*, 157. In January 1821, Salisbury's *Western Carolinian* weighed in on the topic of navigation, observing that land was cheaper than water carriage and that one of the impediments of the latter was the loss of coal by "frequent removals and the interruptions in the navigation of the river brought down the river," often making it necessary for the boatman to lighten his load. See the *Western Carolinian*, January 30, 1821. In the first of a series of legislative acts to promote inland navigation and internal improvements, the North Carolina General Assembly in 1788 established a company for the opening of the navigation of the Catawba River. Eight years later, however, the legislature repealed the act because the company had shown no sufficient legal evidence of their having complied with the intention of the act, namely clearing obstructions from the river and keeping passageways navigable. For passage of the act, see *The State Records of North Carolina. Published Under the Supervision of the Trustees of the Public Libraries, by Order of the General Assembly. Collected and Edited by Walter Clark, Vol. XXIV. Laws 1777–1788* (Goldsboro: Nash Brothers, 1905), 961. For repeal of the act, see *Laws of North Carolina, at a General Assembly, Begun and Held at the City of Raleigh, on the Twenty-first Day of November, in the Year of our Lord One Thousand Seven Hundred and Ninety-six, in the Twenty-first Year of the Independence of the Said State, Being the First Session of the Said Assembly* (Halifax: Hodge & Wills, c.1796), 32.

32. *Weekly Raleigh Register*, April 18, 1823. Perhaps Raleigh's admonition of a recalcitrant state found the right person in Fisher, a congressman and speaker of the House of Commons who was an ardent proponent of the state's valuable natural resources. The South Carolina state legislature passed a resolve in December 1818 to open inland navigation of more than 1,500 miles with completion projected in the year 1822. See *Plans and Progress of Internal Improvements in South Carolina, with Observations on the Advantages Resulting therefrom to the Agricultural and Commercial Interests of the State* (Columbia, SC: n.p., 1820).

33. Ibid. Raleigh's letter provided an early insight into the importance of merchants requiring open avenues to ship their products to market. He also

pointed to the many manufacturing establishments in Pittsburgh, such as glass factories, breweries, flour mills, potteries, forges, blast-furnaces, rolling mills, and distilleries, which relied on coal as their primary source of fuel.

Chapter 3

1. Michael S. Smith, "The Conflict Between 'Practical Utility' and Geology: Denison Olmstead, Elisha Mitchell and the 1823 to 1828 Geologic Surveys of North Carolina," *Southeastern Geology*, Vol. 38, No. 5 (April 1999), 145. By the late eighteenth century, one approach to evaluating the natural world was through the applied practice of scientific study and methodology. Individuals such as Alexander Von Humboldt, of Prussia, and François André Michaux, of France, explored and wrote about land formations in Europe and North and South America, utilizing scientific instruments for the first time and approaching their subjects from a scientific point of view. Michaux spent time in Burke and Lincoln counties as part of his botanical studies.

2. Charles S. Sydnor, "State Geological Surveys in the Old South," in *American Studies in Honor of William Kenneth Boyd. By Members of the Americana Club of Duke University*, edited by David Kelly Jackson (Durham, NC: Duke University Press, 1940), 86. In 1800, no professorships of geology existed in the universities of America and Britain. In 1802, Yale established a professorship of chemistry and natural science, appointing Benjamin Silliman, a lawyer-turned-scientist, to chair the nascent field of study. Silliman quickly realized that there was virtually no knowledge base or research to draw upon, and pioneered the study of geology which eventually appeared on the course syllabi of American universities. In some instances teachers who taught at the university level could parlay one discipline into two with the assurance that the state would fund field research.

3. In 1809, William McClure, sometimes called the "father of American geology," published a color geological map of the United States with separate notes. In a January 20, 1809 address to the American Philosophical Society, McClure provided an "explanatory" of his geological map, noting geological formations and mineral deposits in the United States. Though McClure's study pre-dated Olmstead's in mentioning coal and iron deposits, there was no mention of coal deposits specifically in North Carolina. However, McClure did refer to a coal formation twelve miles from Richmond which would later have implications for coal mining in North Carolina. See William McClure, "Observations of the Geology of the United States," in *American Philosophical Society, Trans. 1809. Ser. I, v. 6*. Horace H. Hayden's work took a world-view approach with a special interest in the geological features of the United States. See *Geological Essays; or an Inquiry into Some of the Geological Phenomena to Be Found in Various Parts of America, and Elsewhere* (Baltimore: J. Robinson, 1820). The coalfield near Richmond was most likely the Midlothian basin, which had been opened in the early eighteenth century and continued to be the state's main coal-producing region.

4. "William H. Keating," http://en.wikipedia.org/wiki/William_H._Keating. Accessed 26 February 2015.

5. *Consideration upon the Art of Mining to Which Are Added Reflections on Its Actual State in Europe, and the Advantages Which Could Result from an Introduction of this Art into the United States*, by W.H. Keating, A.M. (Philadelphia: M. Carey and Sons, 1821), 3–4. Keating's work was not in the layout of a geological survey or report and did not consider geological formations and abundance of minerals and coal where extraction and transportation costs determined which mines "are to be worked only when and where they are profitable." (73). Keating believed that using England as an example of the correct approach to mining, especially its coal and iron regions, the United States would derive the same advantages and benefits if pursued in a like manner.

6. *Ibid.*, 7.

7. *Ibid.*, 87.

8. "Denison Olmstead," Ncpedia.org/biography/olmstead-denison. Accessed 16 February 2015. Olmstead was the cousin of John Olmstead, father of landscape architect Frederick Law Olmstead.

9. Denison Olmstead, "Red Sand Stone formation of North-Carolina, Extract of Letter from Professor D. Olmstead, of the College at Chapel-Hill, North Carolina, dated Feb. 16, 1820," in *American Journal of Science*, Vol. 2, 1820, 175–176. Olmstead encouraged his students to find specimens of the rock formations they were studying, and it was on one such field trip that a student found coal at Deep River in Chatham County. Portions of the study appeared in the September 19, 1820, edition of Salisbury's *Carolinian*.

10. "An Act Directing a Geological and Mineralogical Survey to be Made of the State of North-Carolina," in *Acts Passed by the General Assembly of the State of North Carolina. At Its Session, Commencing on the 17 of November 1823* (Raleigh: J. Gales & Son, 1824), 17. To address the need for creating and maintaining a transportation infrastructure, involving improved navigation of the state's waterways; creation of roads, turnpikes, and canals; and the construction of a railroad network, the state legislature in 1819 had passed an act "to create a fund for Internal Improvements, and to establish a board for the management thereof." See *The Laws of the State of North-Carolina Enacted in the Year 1819*, 7.

11. Before beginning the study, Olmstead had already published the first account of coal in the Deep River field. Denison Olmstead, "Red Sand Stone formation of North-Carolina, Extract of Letter from Professor D. Olmstead, of the College at Chapel-Hill, North Carolina, dated Feb. 16, 1820," in *American Journal of Science*, Vol. 2, 1820, 175–76.

12. "An Act Directing a Geological and Mineralogical Survey to be Made of the State of North-Carolina," 17.

13. *Ibid.*

14. Sydnor, "State Geological Surveys in the Old South," 89–90. The two studies, which Olmstead considered "statistical memoirs" as opposed to geological surveys, were *Report on the Geology of North Carolina, Conducted under the Direction of the Board of Agriculture. By Denison Olmstead. Part I. November, 1824* and *Report on the Geology of North Carolina. Con-*

ducted under the Direction of the Board of Agriculture. Part II. By Denison Olmstead, November, 1825.

15. "Elisha Mitchell," in *Dictionary of American Biography*, edited by Dumas Malone, Volume 13 (Mills-Oglesby) (New York: Charles Scribner's Sons, c. 1934), 45. Although contemporaries, Mitchell had also studied under Silliman at Yale.

16. *Ibid.* Mitchell, in a letter to his wife, Maria, written from New Bern on December 28, 1827, shared with her his frustration that the "Geological Survey dies a natural death ... and there is no one who takes any interest in the business ... nor, in the present state of the Treasury, did I find there was any the least prospect of succeeding in any application to the legislature and I therefore gave it up at once." Despite his frustrations with the Geological Survey, Mitchell would go on to write influential works such as *Elements of Geology with an Outline of the Geology of North Carolina* (1829), as well as articles for scientific journals. See *Diary of a Geological Tour by Dr. Elisha Mitchell in 1827 and 1828 with Introduction and Notes by Dr. Kemp P. Battle, LL.D. James Sprunt Historical Monograph, No. 6* (Chapel Hill: University of North Carolina, 1905).

17. *Weekly Raleigh Register*, January 29, 1851. Williams liberally referenced findings from R.C. Taylor and W.R. Johnson, two eminent scientists who earlier had reported favorably on the coal deposits at Deep River. Excerpts from Williams's speech before the Senate appeared in this issue. For a fuller treatment of his speech, see J.B.D. De Bow, *The Industrial Resources, Etc., of the Southern and Western States: Embracing a View of Their Commerce, Agriculture, Manufactures, Internal Improvements, Slave and Free Labor, Slavery Institutions, Products, Etc., of the South, with an Appendix, in Three Volumes* (New Orleans: Office of De Bow's Review, 1852), 177–182. De Bow previously served as Director of the United States Census. In 1848, the American geologist Richard Cowling Taylor (1789–1851) published the first comprehensive study "for temporary and specific purposes," showing the geographical distribution of coal measures in the world. His *Statistics of Coal: the Geographical and Geological Distribution of Minerable Combustibles or Fossil Fuel, Including, also, Notices and Localities of the Various Mineral Bituminous Substances, Employed in Arts and Manufactures, from Official Reports of the Great Coal-Producing Countries, the Respective Amounts of their Production, Consumption and Commercial Distribution, in All Parts of the World, Together with Their Prices, Tariffs, Duties and International Regulations. Accompanied by Nearly Four Hundred Statistical Tables, and Eleven Hundred Analyses of Mineral Combustibles, with Incidental Statements of the Statistics of Iron Manufactures, Derived from Authentic Authorities* (Philadelphia: J.W. Moore, 1848) provided statistical data collected from original sources, many of which were unavailable to those associated with the coal trade. Taylor assumed the role of an educator, approaching his subject with the goal of disseminating information to those who might benefit from knowing the specifics of other coal-producing regions. His consideration of North Carolina coal was based on the findings made to Congress in 1840 for inclusion in the Sixth Federal Census of the United States. Although Taylor acknowledged the coal deposits in Pittsboro, as well as small amounts of anthracite coal, he noted that the coal was located in a thin bed and that mining efforts on a marketable scale were non-existent.

18. *Ibid.* Some experts believed two or three veins existed underlying the field. Some geologists, such as Williams's contemporary, Ebenezer Emmons, stated that they could not determine the extent of anthracite deposits, instead analyzing the bituminous seams, which had been exposed to a greater degree. Also, Emmons was not in favor of competing with the anthracite coal producing areas of Northeast Pennsylvania which had a virtual lock on the market.

19. *Ibid.*, In addition to a geological assessment of Deep River coal, Williams's lecture included a cost-analysis of transportation logistics and internal improvements, with particular regard to the northern states. Williams's vision included transporting to the northern states local agricultural products and "fertilizing manures," such as guano, which would help facilitate traffic on Deep River, enabling it to become the "Nile of the South."

20. *Ibid.*

21. Watson, *Internal Improvements*, 69.

22. *Ibid.*

23. Francis S. Drake, "Walter Rogers Johnson," in *Dictionary of American Biography, Including Men of the Time; Containing Nearly Ten Thousand Notices of Persons of Both Sexes, of Native and Foreign Birth, Who Have Been Remarkable, or Prominently Connected with the Arts, Sciences, Literature, Politics, or History, of the American Continent. Giving Also the Pronunciation of Many of the Foreign and Peculiar American Names, a Key to the Assumed Names of Writers, and a Supplement* (Boston: Houghton, Osgood & Company, 1879), 471. Johnson was an early contributor to the literature of coal mining, publishing *Notes on the Use of Anthracite in the Manufacture of Iron. With Some Remarks on Its Evaporating Power* (1841) and *The Coal Trade of British America, with Researches on the Character and Practical Values of American and Foreign Coals* (Washington: Taylor and Maury; Philadelphia: A. Hart, 1850) 161. His work for the U.S. Navy, *A Report to the Navy Department of the United States, on American Coals Applicable to Steam Navigation, and to Other Purposes*, published in 1844 under the authority of Congress, was one of the most important works of its type as it contained detailed results of investigations into the evaporative power of different coals in determining their quality for use in U.S. Navy steamships. For the purposes of steam navigation evaporative power held prime importance, as the length of a voyage relied on the amount of evaporative power of the fuel.

24. Johnson, *The Coal Trade of British* America, 161. The study had a wider scope as Johnson considered coals in foreign countries as well. See especially "Recent Investigations Relative to American and Foreign Coals," 161.

25. *Report of the Coal Lands of the Deep River Mining and Transportation Company, in Chatham and Moore Counties, North Carolina, with Analyses of the Minerals* (Albany, NY: Weed, Parsons, and Company, 1851). As part of the by-laws the mining engineer was to reside at or near the mines and have charge over all property in Chatham County, while the sales agent,

based in New York, would attend to selling coal. Johnson's findings at Deep River were also briefly noted in the February 5, 1851 issue of the *New-York Tribune*.

26. *Ibid.*, 6. Any mining activity that extended below the water level would require drainage by means of mechanical pumping, which was costly and not without its risks to life and property.

27. *Ibid.*, 8. Johnson selected Pit No. 5 for his study on account of the pit being opened more than the others. In measuring the stratum, Johnson included the bottom coal, or fireclay; the intermediate, slate; and the top, coal. A dip is the inclination of a geologic formation, such as bed, seam, vein, or fault, relative to the horizontal plane and is measured downwards at right angles to the strike, or surface. Abrupt changes in a seam dip can create conditions unfavorable to mining.

28. *Ibid.*, 11.

29. *Ibid.*, 13.

30. *Ibid.*, 15.

31. *Ibid.*, 17–18.

32. *Ibid.*, 33. Johnson's acknowledgment of the anthracite deposits portended more expensive costs to access the seams, although the deposits were most likely semi-anthracite.

33. "Act Act to Incorporate the Deep River Mining and Transportation Company," in *Laws of the State of North Carolina, Passed by the General Assembly, at the Session of 1850–51* (Raleigh: Star Office, 1851), 598–599. Lemuel Williams, one of the company's original stockholders, on January 14, 1851, gave a lecture on the Deep River coalfields before the state legislature in Raleigh. A printed version of his lecture appeared ten days later in the *Raleigh Register*. Stressing that the quality, variety, and quantity of Deep River coal is comparable to that of Pennsylvania, Williams proclaimed that production of Deep River coal can match in five years what Pennsylvania has been able to accomplish in twenty. For a printed treatment of Williams' speech, see "Coal Formation of North Carolina—Mr. Williams' Lecture," in the *Raleigh Register*, January 25, 1851.

34. "Report of the Directors of the Deep River Mining and Transportation Company," in the *Fayetteville Observer*, December 30, 1851. Johnson's study was geological in nature and not an economic feasibility study.

35. *Ibid.* The board also recognized the substantial amount of timber on the property which could be purchased adjacent to the borders of Deep River "in view of the expense of building steamers and barges." A healthy climate, springs, cheap provisions and goods, and construction of a depot at Smithville all contributed to the board's recommendation. Interestingly, the board noted a significant amount of anthracite coal on the property which could be mined in competition with the anthracite coalfields of Pennsylvania, but developing this facet of coal mining never materialized. Indeed, the *Weekly Raleigh Register*, in its February 18, 1842, edition, noted that Josiah Tyson had discovered a specimen of anthracite coal on his property at Deep River equal in quality to that of the Pennsylvania variety.

36. "Thomas Lanier Clingman," http://ncpedia.org/biography/clingman-thomas-lanier. Accessed 27 February 2015.

37. Thomas L. Clingman, "North Carolina—Her Wealth, Resources, and History," in *DeBow's Review and Industrial Resources, Statistics, etc. Devoted to Commerce, Agriculture, Manufactures, Internal Improvements, Political Economy, Education, General Literature, etc. Edited by J.D.B. DeBow, Volume XXV* (New Orleans and Washington City: N.p., 1858), 667. Clingman's assessment of domestic iron would culminate in the manufacture of T-rails to supply the burgeoning American railroad industry. Clingman's notoriety extended to an ongoing argument with Professor Elisha Mitchell, over the highest peak in North Carolina's Black Mountain range. Mitchell, in 1844, measured a peak he believed to be the highest in the range, and the highest east of the Rocky Mountains, at 6,708 feet. In 1855, Clingman challenged Mitchell's findings and claimed that his former teacher had measured the wrong peak, while Clingman measured a peak he believed to be the highest at 6,941 feet. Mitchell returned to the area in 1857 and while on a field trip determined to prove Clingman wrong and substantiate his own claim, he met his death at age sixty-three in an accidental fall into a pool of water where he most likely drowned. Nearly twenty-five years later, in 1881–1882, the United States Geological Survey confirmed Mitchell's measurements of the highest peak and named it Mount Mitchell in his honor. The second highest peak in the range was named for Clingman as was the highest peak in the Great Smoky Mountains which came to be known as Clingman's Dome. Continuing his scientific studies, Clingman would go on to research meteors and the height of the atmosphere before his death in 1897 at age eighty-five.

38. In an article originally appearing in the *United States Daily*, state geologist H.J. Bryson noted that iron ore produced in North Carolina was shipped to England as early as 1792. See the *Asheville Citizen-Times*, November 17, 1931. Between 1857 and 1858, North Carolina had only three charcoal and coke works, thirty-six bloomery forges, and one rolling mill, where iron was formed into manufactured shapes. On the other hand, eastern Pennsylvania's and northeastern Maryland's facilities included ninety-eight anthracite furnaces, 103 charcoal furnaces, and 117 forges. The product of the rolling mill for the same time period was 241,484 tons turned out by Pennsylvania compared to North Carolina's 215 tons. North Carolina's total contribution of iron lagged considerably behind the northeastern states and even neighboring Virginia, which had the greatest output among the Southern states at just over 26,000 tons. See Lesley, *The Iron Manufacturer's Guide*, 762.

39. "Ebenezer Emmons," in William S. Powell, ed., *Dictionary of North Carolina Biography* (Chapel Hill: University of North Carolina Press, 1986) 156–157. Emmons's geological surveys of North Carolina produced in 1852 and 1856 would stand out as the most exhaustive and authoritative of any published up to that time.

40. *Weekly Raleigh Register*, October 15, 1851. The specific wording of the legislation called for a geological, mineralogical, botanical, and agricultural survey of the state. Reid's detractors pushed for a resident of North Carolina to conduct the survey, believing that such an individual would be more acquainted with the

state's resources. In the November 12th issue of the paper, one correspondent wrote a justification of Emmons's hiring based on an 1850 visit Emmons made to the state on behalf of New York speculators, who hired him to make an investigation of the land for the purpose of mining gold. During that time Emmons visited eight counties and met with numerous individuals who could assess his qualifications and work, which sometimes included making discoveries that others before him in a similar capacity had missed. The correspondent continued to point out that Emmons's time completing the New York geological and agricultural survey also included attention to botanical production and that this experience made him the right fit for completing North Carolina's survey. Challenging readers to find a more qualified individual within the state to match Emmons's experience and abilities, the correspondent affirmed Governor Reid's decision to hire Emmons and averred that party politics should not be part of efforts with the survey moving forward.

41. Emmons's survey of 1852 was significantly shorter than the other two he produced in 1856 and 1858, respectively. However, local newspapers in the vicinity of where his travels had taken him provided regular updates to their readers, as when Emmons reached Halifax in June 1852 to examine the agriculture and geology of that region. Apparently Emmons was very accommodating and genial in his manner among the residents. See the *Greensborogh Patriot*, June 12, 1852. The September 18, 1852 issue of the *Tarborough Southerner* proclaimed an interest in the second volume of Emmons's *Agriculture of New York*, in regard to how crops exhaust land.

42. *Report of Professor Emmons, on His Geological Survey of North Carolina*. Executive Document, No. 13 (Raleigh: Seaton Gales, 1852), 74. McClanahan had been president of the Cape Fear and Deep River Navigation Company before working as Emmons's assistant. He also served on the board of directors of the Haywood and Pittsboro Plank Road Company, which the general assembly incorporated in 1852. His experience in those two capacities would have been invaluable to Emmons's work. A small excerpt from Emmons's survey of the Deep River coalfield appeared to great acclaim in the March 4, 1852 issue of the *Albany (NY) Evening Journal*.

43. *Ibid*, 173.

44. *Ibid*., 130.

45. Albert H. Fay, *A Glossary of the Mining and Mineral Industry* (Washington: U.S. Government Printing Office, 1947), 602

46. Emmons, *Report*, 130.

47. *Ibid*., 131.

48. *Ibid*., 134. By 1860, railroads were consuming six million cords of wood while steamboats accounted for three million. After the Civil War, fuel wood began to be replaced by coal on a much larger scale, with twenty-five per cent of railroads shifting to coal by 1865. The surge in manufacturing activity brought on more use of coal in iron and steel production, which wood could not equal in terms of heating value and availability. Although railroads, such as the North Carolina Rail Road, continued to rely on wood fuel many eventually switched to coal to the point that by 1900 they were responsible for creating a mass market. See David A. Tillman, *Wood as an Energy Resource* (New York: Academic Press, 1978), 11–12.

49. *Semi-Weekly Raleigh Register*, November 27, 1852.

50. *Ibid*., March 10, 1853. The *Register* pointed out that a more complete analysis of the coal deposits should include findings from bore holes, which Emmons did not carry out. Emmons's report, if anything, the *Register* seemed to be saying, should serve as a rallying call to action. Emmons's decision not to drill into the seam for the purpose of determining the coal's quality and extent would be called into question by Elisha Mitchell later in the year. Mitchell believed that a more accurate and scientific picture of the coal deposits could be obtained by drilling for samples and evaluating their content.

51. Ebenezer Emmons, *Geological Report of the Midland Counties of North Carolina* (New York: George P. Putnam; Raleigh: Henry D. Turner, 1856), xiii. Though Emmons's 1852 survey included data about the coalfields at Deep River, his 1856 survey examined some of the more productive veins "with as much minuteness as the nature of this Report will admit," xvii.

52. The Dan River coal deposits, which cover Rockingham and Stokes counties, was, in Emmons's estimation, inferior and less valuable than the Deep River mostly bituminous deposits which was born out in his report, an assessment in agreement with other geologists.

53. Emmons, *Geological Report of the Midland Counties of North Carolina*, 246. The *Weekly Raleigh Register* had already published in its December 7, 1853 issue a letter submitted by "A Looker On," who sardonically noted that "Dr. Mitchell can no longer shake his incredulous head," given Emmons's and McLean's successful drilling efforts to find coal at Egypt.

54. *Ibid*., 251. Blacksmiths typically favored charcoal for iron work because it generated an intense heat conducive for metallurgical work but because production of charcoal relied on heating wood without the presence of oxygen, which produced a black residue consisting mostly of carbon, an ample supply of timber was required. Concerns of over-timbering the area became a concern and, as a result, a seemingly inexhaustible and readily available supply of coal became the preferred fuel. Walter R. Johnson noted the quality of the anthracite coal at Wilcox's and deemed it on par with the "average of Pennsylvania anthracites." See Johnson, *The Coal Trade of British America*, 166.

55. *Ibid*., 249. Emmons realized that the "black band" ore, which contained a high percentage of iron, ran parallel with the coal seams, and that there were three known seams of it.

56. *Southern Weekly Post*, April 29, 1854, 2. Most of Emmons's comments recommending the area for a national foundry were articulated in *Special Report of E. Emmons, Geologist to the State of North-Carolina Concerning the Advantages of the Valley of the Deep River as a Site for the Establishment of a National Foundry. Made Pursuant to Instructions from Gov. Bragg, 29 December 1857* (Raleigh: Holden & Wilson, 1857). Talk of a national foundry had been bandied about for a number of years, culminating in 1859, with the Hon. John Alexander Gilmore's speech to the U.S. House of Representatives on the location of a national foundry in the Deep River valley. Gilmore's reasoning

was similar to Emmons's in that "coal of the best and most abundant quality" was present in the area along with iron ores that were "equally superior, embracing every variety." Writing in an almost patriotic fervor praising the resources of the South, Gilmer pointed to a favorable climate and proximity to slack water navigation of the Cape Fear and Deep rivers as significant advantages for building a national foundry in the area. See *Speech of Hon. John A. Gilmer, of North Carolina, on the Location of a National Foundry in the Deep River Valley.*

57. *Ibid.* Some early endorsements of coal as a fuel source in industrial applications appeared in local newspapers. In an article in the Raleigh *Weekly North Carolina Standard* for November 19, 1851, the writer called attention to the Novelty Iron Works in Raleigh owned by North Carolina ironmaster Silas Burns (1804–1876), whose most notable work was completing the wrought-iron fence that surrounded the capital building in Raleigh. Here, production at his foundry included locomotive and stationary steam-engines, saw and grist mills, mill gearing, and machinery and castings "of every kind and variety." Burns was already using coal mined at Chatham County in his forges. One blacksmith from England in his employment spoke favorably of the high quality of the coal, including its rich and steady flame. Proximity and economy of Deep River coal were two additional considerations for tapping this readily available fuel source. See the *Weekly North Carolina Standard*, November 19, 1851.

58. *Ibid.*

59. *Greensborough Patriot*, July 2, 1853. Most of the content of the letters had to do with Emmons's statements in his geological survey about the size and wealth of the Deep River coalfields. Among other things, Mitchell claimed that Emmons "will do whatever he can to put me down."

60. *Semi-Weekly Raleigh Register*, October 19, 1853. Emmons's estimate of three miles from the outcropping, Mitchell pointed out, was not based on scientific evidence.

61. Emmons, *Geological Report, 1856*, 246. Emmons is referring to the drilling at Egypt which had uncovered at a depth of slightly over 400 feet a seam of high-grade bituminous coal four-and-a-half feet thick. During the interval between the 1852 and 1856 surveys, Charles T. Jackson, a geologist and chemist who was hired by local landowner Thomas Andrews to conduct an investigation of the area, wrote an article dated September 24 which appeared in the December 24, 1853 issue of the *Greensborough Patriot* based on his examination of the Deep River coalfield. According to Jackson's study, which bolstered Emmons's findings, the outcrops reached a width of between twelve and sixteen miles and that successful boring would determine the depth at which the coal was located. Jackson believed that based on the outcroppings alone an adequate supply of coal was present and that mine owners in the vicinity should be encouraged to keep their mines open. See the *Greensborough Patriot*, December 24, 1853. Jackson's study relied on examination of the coal measures at Peter G. Evans's mine, the Palmer estate, the Farmersville mine, the Belmont mine, and the Egypt mine, which by that time had reached a depth of 285 feet. For a brief overview of Jackson and the Deep River coalfields, see R.C. Lawrence, "The Deep River Coalfields," in *The State Magazine*, May 24, 1941, 5.

62. *Report of the Secretary of the Navy, Communicating the Report of Officers Appointed by Him to Make the Examination of the Iron, Coal, and Timber of the Deep River Country, in the State of North Carolina, Required by a Resolution of the Senate.* 35th Congress, 2d Session, Ex. Doc. No. 26, 1. Capt. Charles Wilkes (1798–1877) had already gained notoriety for leading the United States Exploring Expedition from 1838 to 1842, resulting in the discovery of Antarctica. A controversial individual, whose role in the Trent Affair during the Civil War, led to strained relations between the U.S. and Great Britain, Wilkes retired from the navy in 1866 with the rank of rear admiral. The commission also included Henry Hunt and D.B. Martin, chief engineers, and S.M. Pook, "naval constructor," for the purpose of analyzing the timber land for ship building.

63. *Ibid.*, 18. Wilkes's assessment of Deep River iron ore was most optimistic when he observed, "there is no locality on the eastern side of the Alleganies [sic] where a better article of iron can be produced." Mining considerations for extracting iron ore are similar to those for coal mining namely proximity to market, cost of infrastructure for transporting product to market, and cost of transportation to haul the product. The sandstone formation at Gulf was also of a superior quality.

64. *Ibid.*, 6.

65. *Ibid.*, 11.

66. *Ibid.* Wilkes discovered that many hardwoods had been cleared from the area and spoke favorably of using the middle-size specimens.

67. *Ibid.*, 20. Wilkes did note that Maj. William H. Morrell, of New York, had assumed responsibility for repairs and construction of the locks and dams, and that it came to his attention as he was completing his report that boats loaded with coal, iron ore, cotton, and flour were plying their cargo to market. A select committee appointed by the legislature in 1830 was tasked with assessing the progress of improvements carried out by the Cape Fear Navigation Company, noting that in 1823 the company experienced unacceptable losses in extending navigation to Haywood. See *Report of the Situation of the Cape Fear Navigation Company* (Raleigh: Lawrence & Lemay, 1832), 8.

68. *Ibid.*, 21. Wilkes's associate, S.M. Pook, the naval constructor and shipbuilder, was part of the commission. Ordinarily, the journey took twelve hours.

69. Watson, *Internal Improvements*, 75–76. Indeed, the directors of the Deep River Mining and Transportation Company, in March 1851, prepared a report for its stockholders based on the findings of geologist Walter R. Johnson concerning the coal beds on the company's property at Farmersville, Chatham County.

70. Wilkes, *Report of the Secretary of the Navy*, 21.

71. Watson, *Internal Improvements*, 121. Navigation of the Cape Fear and Deep rivers was to be made possible by slack water pools and locks.

72. Wilkes, *Report of the Secretary of the Navy*, 21. Indeed, Wilkes believed that a railroad to the coalfields would connect with the North Carolina Central Railroad and the Gaston and Raleigh Railroad and

through it the Seaboard and Roanoke, all with the goal of producing a favorable route from mine to market.

73. *Charlotte Democrat*, in its December 28, 1858, issue, reported that the Secretary of the Navy issued an order to the Board of Examiners to report on the findings for a naval foundry in North Carolina, and that the abundant coal and iron deposits would be an inducement to locating there. Unfortunately, Toucey's decision not to build in North Carolina was a blow to the state's efforts to prove itself as an industrial contender in the nation's economy, but with the beginning of hostilities between the North and the South in 1861 the plan for the foundry in North Carolina was scuttled. During the war the Confederate government approved the establishment of a foundry in the area in support of the Confederate States Navy Yard that had relocated from Norfolk to Charlotte for its reputation as a rail center.

Chapter 4

1. *Compendium of the Enumeration of the Inhabitants and Statistics of the United States, as Obtained at the Department of State, from the Returns of the Sixth Census, by Counties and Principal Towns Exhibiting the Population, Wealth, and Resources of the Country*. Prepared at the Department of State (Washington, DC: Thomas Allen, 1841), 179. One bushel of bituminous coal at 80 lbs. × 75 bushels = 6,000 lbs. or 3 tons.

2. Ben Justeen, "Mining Thrived in Days of Yore," *Sanford Herald*, Centennial Edition, 1974. In July 1895, the name of the Egypt mine was changed to Cumnock in recognition of the mining center of the same name in East Ayrshire, Scotland, which was also more "neutral" than the name Egypt where explosions in 1895 and 1900 had occurred.

3. Marius Campbell and Kent W. Kimball, *The Deep River Coal field of North Carolina. North Carolina Economic and Geological Survey. Joseph Hyde Pratt, Director and State Geologist. Bulletin No. 33. Prepared by United States Geological Survey in Cooperation with the North Carolina Geological and Economic Survey* (N.p.: n.p., reprint 1995), 14.

4. *Laws of the State of North Carolina, Passed by the General Assembly at the Session of 1850–51*, 599–600.

5. Ibid., 598.

6. *Laws of the State of North Carolina*, 1854–55, 137, 141, 147, 150–51. The Gulf Coal Mining Company included as one of its directors, John Hooker Haughton (1810–1876), lawyer, businessman, politician, and erstwhile advocate of internal improvements in his state. A native of Chowan County, and a graduate of the University of North Carolina, Haughton practiced law in Chatham County for many years, where he became an active member of the Whig party, representing Chatham in the House of Commons and Senate. An early, outspoken Unionist, he maintained a residence in New Bern where he continued his law practice during winter. The steamship *John H. Haughton* was named for him.

7. In some instances, natural ventilation could substitute for artificial ventilation without either a fan, furnace, or other artificial means. Natural ventilation could pose serious safety risks to miners if airflow were cut off to the mine, leading to explosions and asphyxiation. At this point in the history of coal mining, without legislation backing it, safety concerns could be ignored, often leading to ill-conceived shortcuts with serious repercussions.

8. *Fayetteville Observer*, November 12, 1850.

9. *Tri-Weekly Commercial* (Wilmington, NC), March 6, 1851. These storage facilities, sometimes referred to as "coal pockets," or "retail pockets," were structures that also could accommodate local retail sales.

10. *Greensborough Patriot*, December 24, 1853. Although a practicing physician, Jackson devoted most of his time to non-medical interests, such as chemistry and geology, establishing in 1836 an analytical chemistry laboratory at Boston in 1836. He also served as geological surveyor of Massachusetts and Maine, becoming state geologist of Maine and was later appointed head of the Rhode Island Survey. Much of his work consisted of observational and analytical studies of soils and minerals. After 1850, he conducted numerous investigations and analyses for individuals and private companies. It was in the capacity of consultant that he performed the survey of the Deep River coalfield for his employer, F.W. Camman. See "Charles Thomas Jackson," in *The History of Science in the United States*, edited by Marc Rothenberg (New York: Garland Publishing Company, 2001), 287. During this time, the Greenpoint Gas Light Co., of Greenpoint, Brooklyn, New York, announced that it was no longer dependent on British coal and substituted North Carolina coal for fueling its gas works, finding that it was "more bituminous and purer than any imported." See the *Daily Eagle*, November 22, 1853. At nine dollars per ton for North Carolina coal compared to fourteen dollars per ton for its British counterpart, the gas company had found a better coal at a better price. See the *New-York Daily Tribune*, November 23, 1853.

11. Ibid. Jackson's report also included an inspection of the mines at Belmont and the Taylor property. J&L Haughton was the company owned by John H. Haughton and Lawrence J. Haughton.

12. Ibid. It was not uncommon for seams in the Deep River coalfield to dwindle from several feet to inches which frustrated mine owners' and investors' attempts to reach the veins of coal, often resulting in further drilling costs or abandoning the mine altogether.

13. Ibid.

14. Ibid. It was the bane of mines, both coal and mineral, to contain significant levels of water in the workings particularly if operations extended below the water table.

15. Ibid. Newcastle upon Tyne was one of the first and largest coalfields in England, spurring the idiom "carrying coal to Newcastle" to mean a pointless activity.

16. *Weekly Raleigh Register*, January 17, 1855. With navigation the only means for carriage improved river navigation was imperative for the survival of mining companies. McClane had made a name for himself as Wilkes's assistant during the latter's geological survey as well as serving in a similar capacity for Ebenezer Emmons.

17. *Private Laws of the State of North Carolina*,

Session of 1854–55, 150–151. John Hooker Haughton (1810–1876) was born in Chowan County and moved to Pittsboro, Chatham County, in 1837. A lawyer by profession, Hooker early on supported efforts in the internal-improvements push, including opening coal mines and improving water transportation to the mines. He is credited with bringing the first Chatham County coal downriver in 1856. He was especially involved in the Cape Fear and Deep River Navigation Company as one of its principal stockholders. He was the older brother of Lawrence J. Haughton.

18. Johnson, *Report on the Coal Lands*, 21. As an example, Johnson alluded to some mining districts in Pennsylvania with access to timber lands for supplying mine props. Johnson inserted a caveat into his study that the coal samples came from outcroppings, which did not necessarily represent the true value of a coal as impurities caused by the weather and other meteorological events, could affect the quality of the coal. Typically, coal taken from a more considerable depth represented a more accurate indicator as to quality but normally required great capital investment in adequate machinery, as well as experienced engineers.

19. *Ibid*. The company directors, in presenting to the stockholders Johnson's report, stressed the quality of the coal and its adaptation for steam purposes. Its low sulfur content proved to be an advantage for ocean-going vessels, which sometimes caught fire when highly sulfuric coal was on board ship. Citing the best market as Smithville, at the outlet of the Cape Fear River, the directors pointed out that vessels plying to Smithville also were in the track of steamers traveling from New York to Charleston, Savannah, New Orleans, Texas, the West Indies, and Mexico but also giving the company an advantage in southern markets.

20. *Asheville News*, February 28, 1856. This article originally appeared in the *Fayetteville Observer* two weeks earlier.

21. The company was incorporated as the Governor's Creek Steam Transportation and Mining Company on January 28, 1851, Lemuel Williams and M.Q. Waddill incorporators, with $100,000 in capital stock. See *Laws of the State of North Carolina, Passed by the General Assembly, at the Session of 1850-'51*, 603–604.

22. *Western Democrat* (Charlotte, NC), March 3, 1857. Fire-damp, an extremely flammable gas, was largely prevalent in mines, often exploding with a tremendous force when ignited by a flame. One month later, on April 6, another explosion also attributed to fire damp resulted in the deaths of six men, including the mine superintendent, at the Governor's Creek Mine at Egypt. See the *Western Democrat*, April 21, 1857. The tragedies at the Egypt shaft were the first in a series of explosions that frequented the mine throughout its history.

23. *Weekly Standard*, December 1, 1858.

24. *Ibid*., Pig iron was an intermediate product resulting from the smelting of iron ore in a blast furnace. The product could then be used in applications such as the manufacture of wrought iron or steel. Gammell may be referring to "mine pig," made from ores only.

25. *Ibid*.

26. *New York Times*, November 5, 1860.

27. *The Mining Journal, and Journal of Geology, Mineralogy, Metallurgy, Chemistry, and the Arts in Their Applications to Mining and Working Useful Ores and Metals*, January 1861, 90. The article appeared first in December 1860 in the *New York Times* and considered mostly the Egypt beds, which the Deep River Coal and Iron Company were to develop for the "the mining of coal for the local and New York market." Oil production from shales was carried out on a much larger scale in the early 1900s. Indeed, H.M. Chance, who conducted an investigation into the Deep River coalfield from 1884 to 1885, reported that before the Civil War a slope was sunk on the upper bed at the Farmville property near Egypt, and soon works were established for the "manufacture of illuminating oil from the bituminous slates overlying the coal." See H.M. Chance, *Report on an Exploration of the Coalfields of North Carolina, Made for the State Board of Agriculture. By H.M. Chance. Published by Order of the Board* (Raleigh: P.M. Hale, 1885), 27.

28. *Ibid*., 92. Black band ore must go through this process before it can be smelted.

29. *Ibid*.

30. *Ibid*., 93.

31. *Ibid*., 94. The new structures and machinery were estimated at over $300,000.

32. Northern speculators had provided a much-needed infusion of capital to organize the railroad. In an article entitled "Northern Enterprise," appearing in the *North Carolinian* for March 12, 1853, the writer referred to "Wall Street allies," and, particularly, a "mysterious stranger from Wall Street" who subscribed all the remaining stock of the Western Railroad. Quoting from the *Newbern News*, the writer noted that an agent from the Wall Street firm of Colby & Smith, stopped in Newbern for the purpose of selecting a point for the terminus of a railroad between Fayetteville and Beaufort. No doubt banking on the completion of the rail connection from Fayetteville to the Deep River coalfields as the first step, the agent's function involved securing the connection from Fayetteville to Beaufort harbor. In its articles of incorporation, the Western Railroad could "be continued to and connected with the North Carolina Railroad at any point," thus making the Western Railroad accessible to even newer and greater markets. Charleston interests noticed these developments when the possibility of connecting to the Deep River coalfields via the Western Railroad and the North Carolina Railroad became more evident.

33. In an article appearing in the November 29, 1858, issue of the *Fayetteville Observer*, reprinted from the *Raleigh Standard*, the writer, signed "Clinton," made a compelling case for completing the line to the coalfields by assuming a business approach based on: the amount of $40,000 the state had already expended in conducting geological surveys; the considerable amount of tonnage carried by coal traffic in serving market needs; an infusion of Northern and foreign capital into the area; increased value of Deep River land; par rates on state and railroad bonds for financing the project; and mortgaging the line to the state in order to indemnify it. See "The Geological and Mineralogical Survey—What Was Its Object?" *Fayetteville Observer*, November 29, 1858.

34. *Speech of John T. Gilmer, Senator from Cum-*

berland and Harnett, on the Bill to Aid in the Construction and Equipment of the Western Railroad from Fayetteville to the Coalfields, Delivered in the Senate of North-Carolina, December 2, 1858* (Raleigh: Holden and Wilson, 1859), 1. The legislation passed on January 19, 1859, with an option on the part of the state of subscribing $400,000 of stock in the company.

35. *Ibid.*, 2–3. Almost inextricably linked to the coalfields was the great wealth represented by the nearby iron ore deposits, which Gilmer astutely pointed out that if smelted and manufactured on location could add considerable wealth to the state's treasury. Gilmer's vision included profiting from the sale of Deep River coal in direct competition with that of Northern markets as well as providing ingress to the timber and coal areas needed in the day-to-day functioning of the U.S. Arsenal at Fayetteville. Anticipating the possible dissolution of the Union, he urged that an independent, readily accessible supply of coal and iron would be tantamount to not only the state's survival but the South's as well.

36. Watson, *Internal Improvements*, 106.

37. *Ibid.*, 109. One action authorized a commission to open books in Petersburg, Virginia, "for the purpose of receiving subscriptions to the amount of $400,000 to constitute a joint capital stock for the purpose of making a Rail Road from some point within the corporation of Petersburg to some convenient point on the North Carolina line." See *Acts Passed by the General Assembly of the State of North Carolina, at the Session of 1830–31* (Raleigh: Lawrence & LeMay, 1831), 43. Some individuals, such as University of North Carolina president Joseph Caldwell, proposed combining the existing canal infrastructure with the railroad to effect shorter distances between places and be accessible year-round. Caldwell had traveled to Europe and returned convinced that although railroads held more potential for the state than canals, a joint canal-railroad project could serve a wider area than either mode on its own could. However, a canal-railroad plan he submitted in 1828 to link Beaufort Harbor with the Ohio and Mississippi rivers never went beyond the planning stage. See Shaw, *Canals for a Nation*, 121–22.

38. Charles Clinton Weaver, *Internal Improvements in North Carolina, Previous to 1860*. The Johns Hopkins University Studies in Historical and Political Science, Series XXI, Nos. 3–4 (Baltimore: The Johns Hopkins Press, 1903), 77.

39. *Acts Passed by the General Assembly of the State of North Carolina, at the Session of 1831–32* (Raleigh: Lawrence & Lemay, 1832), 39.

40. *Fayetteville Weekly Observer*, May 22, 1832.

41. *Ibid.*, May 7, 1833. Lack of cooperation between the western counties and Fayetteville had been cited as one reason for abandoning the project and another for failing to meet its subscription as specified in its charter. Responding to the failure of the Cape Fear and Yadkin Rail Road, the editor of the *People's Press and Wilmington Advertiser* took up the pen to address the editor of the *Fayetteville Observer* who claimed that local jealousy and the inability of its subscribers of the railroad's stock to pay their deposits lead to the railroad's failure. In a rather tongue-in-cheek manner, the editor of the *People's Press* claimed that such reasoning on the part of the *Fayetteville Observer* was "a mere matter of speculation." Further noting and concurring with the *Fayetteville Observer's* statement that the objective of improvements should be to prevent trade from passing to the markets of other states, the *People's Press* dissented from the *Fayetteville Observer's* view that in lieu of the railroad the river [Cape Fear] provided ease of movement between Fayetteville and Wilmington. The editor of the *People's Press* observed that for three or four months the river was non-navigable, resulting in the loss of markets and spoilage of goods left at the docks of Wilmington for want of transportation. While championing the vision of Wilmington residents to construct a road between Fayetteville and the Yadkin, the *People's Press* noted that Fayetteville would never give its assent to a project such as an extension of a railroad "beyond their town towards the sea-board," claiming that the river held as much promise for fostering commercial trade as the railroad. See the *People's Press and Wilmington Advertiser*, June 5, 1833.

42. The legislative session of 1831–32 proved to be a banner one for railroad incorporation as two additional lines, the North Carolina Central Rail Road Company and the Tarborough and Hamilton Rail Road, were chartered at this time.

43. Roy Parker, Jr., *Cumberland County, A Brief History* (Raleigh: Division of Archives and History, North Carolina Department of Cultural Resources, 1993), 57.

44. *Fayetteville Weekly Observer*, April 3, 1839. In addition to railroads and steam navigation, plank roads, and canals were among the modes of transportation requiring accessible routes within the state for importing and exporting goods. Early in Great Britain's storied history of coal mining, the physical characteristics of coal—namely its bulk and weight—greatly influenced the means by which it could be transported. Only small amounts of coal could be carried long distances by land and due to its sheer mass to be moved from the larger mines, rudimentary "railways" were conceived and constructed to accommodate the shorter trips from the mine to the sea or river. Local customers came to the company's mine with their own means of transportation such as wagons, carts, and packhorses, to obtain small amounts for their own consumption. With the gentry and yeoman classes assuming the burden of the cost of coal mining, including internal improvements and transportation, partnerships similar in character to joint-stock companies also were formed to share the enormous cost involved in coal mining. See Hatcher, *The History of the British Coal Industry*, 459.

45. H.S. Tanner, *A Description of the Canals and Rail Roads of the United States, Comprehending Notices of All the Works of Internal Improvement Throughout the Several States* (New York: T.R. Tanner & J. Disturnell, 1840), 169. The author considered the Wilmington and Raleigh and the Raleigh and Gaston railroads, as well as the Dismal Swamp, Lake Drummond, North-West, Weldon, and the Clubfoot and Harlow canals. He also noted that the Cape Fear, Yadkin, Tar, and Catawba rivers had been greatly improved by joint-stock companies.

46. *Laws of the State of North Carolina, Passed by the General Assembly, at the Session of 1848–49. Published Agreeably to the Ninety-fifth Chapter of the*

Revised Statutes (Raleigh: Thos. J. Lemay, 1849), 211. The Cape Fear River is formed at Haywood near the Chatham and Lee county line by the confluence of the Deep and Haw rivers and empties into the Atlantic Ocean near Cape Fear, from which it takes its name. As early as colonial times, it served as a major transportation route into the interior of the state.

47. *Wilmington Journal*, January 12, 1849. Thompson's report was printed in this issue. A graduate of the U.S. Military Academy at West Point, he was hired as a consultant to report his findings on an examination of the Cape Fear and Deep rivers. His credentials up to that time included assistant engineer, Charleston and Hamburg Railroad, South Carolina; engineer, Norfolk and Portsmouth Railroad, Virginia; and contractor for the completion of the Chesapeake and Ohio Canal, Maryland. Thompson's completed report was submitted to Walter Gwynn, on behalf of the members of the Pittsboro Convention, who had funded the survey. See *Register of the Officers and Graduates of the U.S. Military Academy, at West Point, N.Y., from March 16, 1802, to January 1, 1850, Compiled by Captain George W. Cullum* (New York: J.F. Trow, 1850), 125, for Thompson's employment history.

48. *Ibid*. An article appearing in the December 26, 1849, issue of the *North-Carolina Star* noted that Solomon McCullough and James Hunter, previously involved in the construction of the Chesapeake and Ohio Canal, had contracted to build the slack-water improvement of the Cape Fear and Deep River Navigation Company, and during their preliminary investigation had located a large specimen of semi-anthracite coal at Deep River which they compared favorably with the same type found in the Cumberland Maryland coalfield. The C&O Canal, completed in 1850, became an active waterway for the transportation of coal and involved a leg of railway to complete its journey. Coal was mined ten-to-twelve miles above Cumberland and loaded onto railroad cars before being transferred to canal boat for its final destination of Alexandria, Virginia, which took as long as seven days. A writer named "Chatham" submitted an observation that during the journey from Cumberland to Alexandria, the coal became friable and crumbled causing its value to decrease compared to the lump variety the Deep River specimen represented. Furthermore, compared to Virginia's higher cost of canal and railroad haulage the transportation of Deep River coal to Wilmington via slack water was far cheaper, and could be exported on a large scale in the millions of dollars. One champion of canals was Dr. M.P. Williams, who, on January 14, 1850, gave a lecture to the state legislature in Raleigh about his findings on the value of the coalfields of Deep River. Williams, who had given months of attention to the subject, identified three kinds of coals: highly bituminous, semi-bituminous, and pure anthracite. Analyses of the coals had shown them to be "unsurpassed in variety, unequalled in quantity." Transportation would be accomplished through the slack-water improvement of the Cape Fear and Deep Rivers. See "Mr. Williams's Lecture on the Coal Formation of North Carolina," in the *Greensborough Patriot*, November 22, 1851.

49. *Ibid*.

50. *Acts of Assembly of Virginia*, 1857–58, Chapter 113, 90.

51. *Ibid*.

52. *Public Laws of the State of North-Carolina, Passed by the General Assembly, at its Session of 1858-'9: Together with the Comptroller's Statement of Public Revenue and Expenditure* (Raleigh: Holden and Wilson, 1859), 184. The Dan River and Coalfield Railroad Company would be re-chartered by the legislature in 1868.

53. Fairfax Harrison, *A History of the Legal Development of the Railroad System of the Southern Railway Company* (Washington: Southern Railway Company, 1901), 562. C.S.A. President Jefferson Davis had advocated constructing a railroad from Greensboro to Danville, Virginia, and plans were put into place to connect the two places. Leaksville, a coal-mining district on the Dan River in Rockingham County, ten miles from the two points, was seen as an embarkation point for hauling the semi-anthracite coal to Danville. At the time, coal from Leaksville was shipped by boat to Danville. See the *Weekly Standard*, November 27, 1861. Also, *Coal Men of America: A Biographical and Historical Review of the World's Greatest Industry*, Arthur M. Hull, Sydney A. Hale, eds. (Chicago: The Retail Coalman, 1918), 34.

54. *Laws of the State of North Carolina, Passed by the General Assembly, at the Session of 1852, Published Agreeably to the Fifty-ninth chapter of the Revised Statutes* (Raleigh: Wesley Whitaker, 1853), 560. The following day, on December 25, the legislature passed an act to incorporate the North Carolina Steam Carriage and Plank Road Company, giving the company authority to build a plank road from the Cape Fear River at Fayetteville to the Deep River coal mines. Clearly, plans were in the making to tap the coalfield by any means available even if it meant without the help of the railroad. See *Laws*, 352–53. There was no clear-cut immediate "winner" as to which would better serve transportation demands. Plank roads and canals continued to be constructed at the same time railroad track was laid, and railroads still had their fair share of detractors who hurled accusations of mismanagement and lack of experience in dealing with complex issues. Railroad track was also exceedingly expensive as costs for surveying, engineering, grading, land reclamation, rights of way, bridges, repair shops, warehouses, and depots constituted a significant investment by private interests as well as by the state. See also Watson, *Internal Improvements*, 120–21.

55. *North-Carolinian*, January 22, 1853.

56. *Ibid*., February 5, 1853.

57. *Report of the Committee on Int. Improvements on the Cape Fear and Deep River Navigation Company* (Raleigh: W.W. Holden, 1855), 2. While serving as principal engineer of the Lehigh Coal and Navigation Company, Douglass was engaged in slack water navigation of the Lehigh River. His experience and reputation earned in this capacity served him well in assessing internal improvements on the Cape Fear and Deep rivers especially in regard to accessing the coal deposits as he had successfully accomplished through the Lehigh and Delaware improvement. The *Wilmington Journal* had reported in its May 26, 1854, issue that U.S. Senator George E. Badger of North Carolina had introduced a bill in the Senate appropriating $200,000 for the purpose of "opening and deepening the mouth of the Cape Fear River, below Wilmington."

The appropriation, approved and signed by President Franklin Pierce on July 22, grew out of plans to construct locks and dams on the upper Cape Fear and Deep rivers to the coalfields in Chatham and Moore counties. See *Journal of the House of Representatives of the United States: Being the First Session of the Thirty-second Congress, Begun and Held at the City of Washington, December 5, 1853, and in the Seventy-eighth Year of the Independence of the United States. United States Congressional Serial Set, Issue 709.* (Washington: Robert Armstrong, 1853), 1196.

58. *Ibid.*, 3.

59. *Ibid.*, 5. Fisher alluded to the number of geologists who had already visited and reported on the Deep River coalfield as evidence that it was extensive and valuable once the coal was mined and prepared for shipment.

60. *Ibid.*, 6. McLane, representing the interests of investors, made a bold pronouncement that getting out the coal for market would begin the following March, presumably when the internal-improvement projects had been completed. He estimated 350,000 tons of coal being extracted the first year with the expectation that between 400,000 and 500,000 tons would be extracted in succeeding years. Proceeds from the sale of coal would go towards paying expenses and the interest on the bonds when it fell due with a consideration of $50,000 in annual revenue. To bolster his argument for expenditures, Fisher pointed out the "very finest iron ore near Deep River, and in immediate proximity with the coal." The 1856–1857 session laws passed by the legislature called for the Cape Fear and Deep River Navigation Company to pay a tax of two cents per ton, to be collected by the sheriff of Chatham County, on all coal shipped and conveyed from the mines on Deep River. See *Public Laws of the State of North-Carolina, Passed by the General Assembly, at Its Session of 1856–57* (Raleigh: Holden and Wilson, 1857), 61.

61. *Memorial of the Stockholders of the Cape Fear and Deep River Navigation Company to the General Assembly* (Raleigh: W.W. Holden, 1854), 51. The Lehigh and Delaware improvement consisted of a canal and slack water navigation to reach the coal mines in the interior of Pennsylvania which was roughly the same distance between Fayetteville and the Deep River coalfield.

62. *Report of the President and Directors of the Cape Fear & Deep River Navigation Company to the General Assembly* (Raleigh: W.W. Holden, 1854), 8. The report pointed out that approximately $400,000 already had been invested in the coalfield, with $50,000 for machinery and housing for miners, to sway even the most conservative capitalists of the potential of the coalfield. Providing an example of the estimated output from the mines, the report mentioned the W.W. Lane workings at Egypt whose company was projected to get out the first year 250,000 tons of coal; at Lagrange, 100,000 tons; and, at the Taylor mine, 600,000 tons. An estimated 500,000 tons of coal would pass down the rivers in a single year and at $.25 toll on the ton would generate for the company an annual revenue of $125,000 from coal alone which would be nearly 18 percent on a capital of $700,000 and would pay 6 percent interest on a loan of $300,000, including furnishing a sinking fund to bolster investor confidence, increasing at the rate of at least 4 percent to discharge the principal of the loan when due, saving the state and stockholders money already invested and turning the stock into a profitable enterprise.

63. *North Carolinian* (Fayetteville, NC), January 22, 1853.

64. *Private Laws of the State of North-Carolina, Passed by the General Assembly at its Session of 1854–55* (Raleigh: Holden and Wilson, 1855), 257. During its annual meeting held in August 1859 at Salisbury, stockholders of the Western North Carolina Railroad passed a resolution to notify the Board of Internal Improvements that it be requested "to make immediate provision for paying the expenses of the surveys, directed by said act out of the Public Treasury, as part of the $4,000,000 already appropriated by the State, to the end that a corps of Engineers may be organised [sic] and the surveys commenced at the earliest practicable moment." *Proceedings. Annual Meeting of Stockholders. Held in Salisbury, August 25–26, 1859* (Salisbury: Carolina Watchman, 1859), 4.

65. Contributing to this dormancy, especially during the 1820s, were partisanship, lack of financial backing, and indifference. By the 1830s, the state began to have high expectations for its transportation infrastructure that included railroads, canals, and plank roads. In 1848, Gov. William A. Graham remarked in his address to the General Assembly that many residents of the Upper Cape Fear pushed for opening the river for navigation to or above the confluence of the Haw and Deep Rivers near the coal deposits. See Watson, *Internal Improvements*, p. 68 for a more detailed analysis.

66. *Private Laws of the State of North-Carolina, Passed by the General Assembly, at Its Session of 1854-'55*, 280. See also Watson, *Internal Improvements*, 121. Another Chatham Railroad Company, incorporated by the General Assembly on February 15, 1861, was authorized to build and construct a railroad with one or more tracks to be used with steam or other motive power from the coalfields in Chatham County to Raleigh or some point west of Raleigh. See *Private Laws of the State of North-Carolina, Passed by the General Assembly at its Session of 1860-'61* (Raleigh: John Spelman, 1861), 116.

67. Taylor, *Internal Improvements*, 121. The state of South Carolina chartered the Cheraw and Coalfields Railroad in 1857.

68. *Weekly Raleigh Register*, March 28, 1855.

69. *Ibid.*, 69–70. In Northeast Pennsylvania, between 1815 and 1834, nearly $10,000,000 was allocated to construct canals from the already tapped anthracite coal deposits to supply an ever-growing market for cheap fuel. The Lehigh Coal and Navigation Company, which was created with the 1820 merger of the Lehigh Navigation Company and the Lehigh Coal Company, was at the forefront of these efforts and became one of the largest coal producers with a ready means to markets in the northeast and beyond. With the advent of the railroad, entrepreneurs and state governments began to put their money into this more promising and reliable form of transportation. See H. Benjamin Powell, *Philadelphia's First Fuel Crisis: Jacob Cist and the Developing Market for Pennsylvania Anthracite* (University Park: Pennsylvania State University Press, 1978). The afore-

mentioned steamer was named for John Hooker Haughton (1810–1876), lawyer, politician, entrepreneur, advocate of internal improvements, and one of the largest stockholders in the company. Iron ore from the same Deep River region was another commodity investors and board members of the company believed could be mined with the nearby coal supply used in iron production and transported as well.

70. *Fayetteville Observer*, July 21, 1856.
71. *Ibid.*
72. *Ibid.*
73. *Ibid.*

Chapter 5

1. Whisonant, *Arming the Confederacy*, 2.
2. *United States. Nonpopulation Census Schedules for North Carolina, 1850–1880, Mortality and Manufacturing, 1860, Schedule 5, Products of Industry in Western Division in the County of Chatham*, 1. Also, "Art. XIV.—Description of the Coal and Iron Works of the Deep River Coal and Iron Company," *The Mining Magazine and Journal of Geology: Mineralogy, Metallurgy, Chemistry, and the Arts in Their Applications to Mining and Working Useful Ores and Metals*, January, 1861, 90–94. Goldston is thirteen miles southwest of Pittsboro. In 1859, North Carolina possessed twenty-six bloomery forges, including three in Stokes County, one of which, the Tunnel Bloomery Forge, was situated on the Dan River. See Lesley, *The Iron Manufacturer's Guide*, 186–92.
3. Whisonant, *Arming the Confederacy*, 137. The North's dominant hold on anthracite markets had its beginnings in the late eighteenth century, when coal companies began to tap the enormous deposits and incorporated themselves as both mining and transportation enterprises. Despite the North's monopoly on anthracite coal, there was a real or perceived threat that if General Robert E. Lee's troops pushed into Pennsylvania and held Harrisburg, the enemy would soon sever all railroad connections to the anthracite coalfields, destroy valuable machinery, and set fire to the pits. Of course, this scenario never played out and the North continued to enjoy its hold on anthracite throughout the war. See "What the Rebels Propose to do with Our Coal Mines, *Scientific American*, Vol. IX, No. 3 (July 18, 1863), 35. The article first appeared in the July 2nd issue of the *Richmond Whig*, a semiweekly newspaper.
4. *Ibid.*, 16. Charcoal production began to decrease after the end of the war as coal and coke became the main fuel sources for industry and manufacturing particularly in the iron and steel centers. So critical was anthracite coal to the Union's war effort that by July 1861 Secretary of the Navy, Gideon Welles, had placed announcements in East Coast newspapers soliciting bids for furnishing anthracite coal for the naval department. *New York Times*, July 25, 1861. To keep anthracite coal from falling into the hands of Confederate blockade runners, Welles sent a dispatch to President Lincoln on April 12 1862 stating that exportation of anthracite from U.S. ports to "any and all foreign ports should be absolutely prohibited." *Official Records of the Union and Confederate Navies in the War of the Rebellion. Published Under the Direction of The Hon. H.A. Herbert, Secretary of the Navy, by Lieut. Richard Rush, U.S. Navy, Superintendent Naval War Records, and Mr. Robert H. Woods. By Authority of an Act of Congress Approved July 31, 1894. Series I—Volume I. The Operations of the Cruisers, from January 19, 1861, to December 31, 1862* (Washington: Government Printing Office, 1894), 229. Though anthracite coal was extremely reliable compared to other varieties, modifications to marine boilers and grates were sometimes required because of its intense heat. Holds of ships could carry less coal and more cargo because of anthracite's longer burning capability. See Frederick M. Binder, "Pennsylvania Coal and the Beginnings of American Steam Navigation," *Pennsylvania Magazine of History and Biography*, 83 (1959), No. 4, 420. In a dispatch dated July 2, 1864, a Capt. Morris, C.S. Navy, commanding the *C.S.S. Florida*, reported that in reference to the anthracite coal he captured as part of the cargo on board the Union bark *Greenland*, "it was of no use for our furnace." *Official Records of the Union and Confederate Navies in the War of the Rebellion. Published Under the Direction of The Hon. H.A. Herbert, Series I—Volume I*, 623.
5. *Ibid.*, 150.
6. Whisonant, *Arming the Confederacy*, 139.
7. *Ibid.*, 141. The author also noted that in 1864 seven iron furnaces were operated by the Confederate government while forty-five were privately owned. Fifty-two forges produced iron in the same year, twenty-five each in Virginia and North Carolina. (See 129–30.) Before the war in 1859, the Tredegar Iron Works imported 1,500 tons of anthracite for use in its foundries and furnaces and by the next year only the Midlothian and Clover Hill coal companies actively mined coal on a large scale. See Sean Patrick Adams, *Old Dominion Industrial Commonwealth: Coal, Politics, and Economy in Antebellum America* (Baltimore: Johns Hopkins University Press), 208–9.
8. "An Act to Establish a Nitre and Mining Bureau," in *Public Laws of the Confederate States of America, Passed at the Third Session of the First Congress; 1863. Carefully Collated with the Originals at Richmond. Edited by James M. Matthews* (Richmond: R.M. Smith, 1863), 114. Additionally, the bureau allocated desperately needed slave and free labor to points in the South. See Adams, *Old Dominion Industrial Commonwealth*, 209.
9. Whisonant, *Arming the Confederacy*, 17–18, 54.
10. Adams, *Old Dominion Industrial Commonwealth*, 209. Despite disruptions, the Richmond basin mines still managed to ship a significant amount of coal to many needy areas in the Confederacy until Union forces late in the war captured and shut down existing coal and iron operations. Virginia and North Carolina coal became even more critical for the South's war effort when the occupation of Chattanooga, Tennessee, by Union forces in August 1863 resulted in the demise of coal supplies from mines in that region. Public works in Georgia and South Carolina, along with Confederate naval vessels, soon came to rely on coal mined near Richmond and from the Egypt mine in North Carolina. Output from the Egypt mine had been greatly increased under the direction of the Nitre and Mining Bureau, with an allotment of 290 tons monthly earmarked for the Confederate Navy and its steamers. Supplies of North Carolina coal shipped

from Fayetteville also reached the naval workshops in Charlotte but delays and disrupted service, along with unfulfilled contracts by individual coal merchants to supply coal to the Confederacy, meant that delays affected the flow of industry, causing some to resort to wood and charcoal as their primary fuel sources. During the war, the Silver Hill Mine in nearby Davidson County came under the control of the Nitre and Mining Bureau, producing much needed lead, whose deposits Ebenezer Emmons noted and analyzed for his *Geological Report of the Midland Counties of North Carolina* (1854). See *Official Records of the Union and Confederate Navies in the War of the Rebellion. Published under the Direction of The Hon. Josephus Daniels, Secretary of the Navy, by C.C. Marsh, Captain, U.S.N. Retired, Series II, Volume 2, Navy Department Correspondence, 1861–1865, with Agents Abroad* (Washington: n.p., 1921), 543–44. Demand for coal was considerable at the beginning of the "steam age" and especially during the Civil War as railroads and ships vied for much-needed supplies. Most steamers, i.e., steamships, required a minimum of ten tons daily to fire their boilers, quickly depleting their onboard hold of around one hundred tons. See Dawson Carr, *Gray Phantoms of the Cape Fear: Running the Civil War Blockade* (Winston-Salem: John F. Blair, 1998), 24.

11. "Remarks of Mr. Manning, of Chatham," in the *Weekly Standard*, February 12, 1862. The son of a well-regarded navy commander, who later served the Confederacy as chief of bureau, ordnance, and clothing, Manning was well acquainted with previous studies of the coal and iron deposits at Deep River and had a practical knowledge of the mining works in Chatham County. Manning acknowledged that the coal in the region made an excellent gas coal, which was an illuminating gas distilled from coal. He had visited the deep shaft at Egypt and testified to its output of one ton of coal every two minutes. Other individuals had observed at the outset of the war the proximity of the coal and iron deposits to each other at Deep River, including the potential for significant iron production at Ore Hill, near Egypt in Chatham County. For an assessment of those deposits, see especially the June 12, 1861, issue of the *Weekly Standard* in which an individual named "Practical" submitted a letter to the editor noting the potential for ironmaking. Appearing in the October 14, 1861, issue of the *Fayetteville Weekly Observer*, the Chatham Ore Hill Company placed an advertisement announcing that their operations were producing twelve to fifteen tons of pig iron per week for manufacturing pots, ovens, plow castings, wagon boxes, and andirons.

12. *Ibid*. In January 1862, the *Fayetteville Weekly Observer* printed an article that first appeared in the *Greensborough Patriot* noting that the Deep River coalfields contained deposits of more annual yield and value to the state and the Confederacy than "half the whole of the cotton fields." The newspaper chastised the Confederate government for its reluctance to open up the fields to provide work for laborers in support of the Confederate war effort as well as promoting a railroad connection to the coalfields. See *Fayetteville Weekly Observer*, January 6, 1862.

13. *Public Laws of the State of North Carolina, Passed by the General Assembly, at its Session of 1854–55. Together with the Comptroller's Statement of Public Revenue and Expenditure* (Raleigh: Holden & Wilson, 1855), 280.

14. *Public Laws of the State of North Carolina, Passed by the General Assembly, at its Session of 1860–61. Together with the Comptroller's Statement of Public Revenue and Expenditure* (Raleigh: John Spelman, 1861), 124.

15. *History of the Raleigh & Augusta Air-Line R.R. Co. Known originally as the Chatham Railroad Company, Including all the Acts of the General Assemblies of North and South Carolina Relating Thereto. Compiled by Walter Clark, Esq.* (Raleigh: The Raleigh News Steam Job Print, 1877), 53.

16. *Ibid.*, 55. At about this time the Cheraw and Coalfields Railroad Company, chartered in 1857 by the South Carolina legislature, was poised in 1863 to run a line from Cheraw, South Carolina, to the coalfield at Gulf. The Chatham Railroad was intended to be an extension of the Cheraw line from Gulf to Raleigh by which the Chatham Railroad Company was allowed to extend its line from Gulf to Cheraw, providing a more direct line to Cheraw. However, the road never was completed during the war and its intended destination of Gulf was changed to Salisbury, North Carolina, thus becoming the Cheraw and Salisbury Railroad Company. See "Report of the President of the Cheraw and Coalfields Railroad Company, Office of the President of the Cheraw and Coalfields Railroad Company, Society Hill, S.C., November 17, 1868," in *Report of the Comptroller General to the General Assembly of the State of South Carolina, November 1868/69* (Columbia: John W. Denny, 1868), 40–41.

17. *Ibid.*

18. *Private Laws of the State of North-Carolina, Passed by the General Assembly at Its Session of 1860–'61* (Raleigh: John Spelman), 116.

19. *Private Laws of the State of North-Carolina, Passed by the General Assembly at Its Session of 1862–'63* (Raleigh: W.W. Holden), 29.

20. *Weekly Standard*, October 8, 1862. Actually the newspaper reported that the railroad would be located "from Raleigh; thence on the ridge to Haw River, crossing near Haywood; thence up the north bank of Deep River, by way of Lockville, to the coalfields."

21. Kemp P. Battle, *Reminiscences: The Chatham Railroad Company*, unpublished typescript, Civil War Collection. Military Collection. State Archives of North Carolina. MilColl Civil War Box 76, Folder 63, 1. Battle had been appointed to the Secession Convention in 1861 and signed the Ordinance of Secession for the state. Acting in his capacity as president of the Chatham Railroad Co., Battle, on January 21, 1863, wrote Gov. Zebulon Vance, seeking a subscription from the state of the capital stock of the railroad "to facilitate a speedy access to, and development of, the rich resources of the coal and mine region of Deep River" for the purpose of supplying the Confederacy with much needed iron as well as coal for domestic use. Battle continued to plead that the coal from the mines of the Deep River region was hauled by wagons over treacherous terrain and had been found in the "Railroad Shops and Foundries at Raleigh, Charlotte, and the Company shops and other places, by actual trial, superior to any in the Confederacy." Attributing the need of iron to the Union blockade of Confederate

ports and the continued failure to realize shipments from Europe, the presidents of the railroad companies of Virginia, North Carolina, and South Carolina met to recommend the establishment of foundries and rolling mills at the combined expense of those private companies that had been awarded large contracts. See *Doc. No. 21, Office of the Chatham R.R. Co., Raleigh, Jan. 21st, 1863, To His Excellency, Z.B. Vance, Governor and President of the Board of Internal Improvements. Ordered to be Printed. W.W. Holden*, 1–4.

22. *Ibid.*, 3. Battle claimed that he appeared before the directors of the North Carolina Railroad Company in 1862 to secure an arrangement for grading but the Confederate government took no interest in the matter for two years. Realizing the need in completing the grading and laying the track, Battle, in August 1863, advertised in local newspapers, such as the *Semi-Weekly Standard*, for "300 Negroes to work on the grading from eight miles west of Raleigh to Deep River in Chatham County." However, by the time grading had been completed, Sherman's army came through the area in March 1865, effectively halting and dismantling additional efforts toward completion of the railroad, which was not completed until 1870 and took a southerly route to tap the timber reserves in Lee County.

23. "Charles Beatty Mallett," ncpedia.org/biography/mallet-charles-beatty. Accessed 3 May 2015. On behalf of the Western Railroad Company, Mallett and George McNeil in 1856 were engaged to construct a railroad from Fayetteville to the coalfields and "opening communication from the mines to Wilmington as well as Fayetteville." See *Wilmington Tri-Weekly Commercial*, March 1, 1856.

24. *Private Laws of the State of North-Carolina, Passed by the General Assembly at Its Session of 1860–61*, 151. The legislature ratified the act on February 16, 1861. Insufficient funding of the railroad on the part of the state was the impetus behind this additional legislation to purchase rolling stock and equipment, build workshops, construct the necessary wharves at Fayetteville, and erect a permanent bridge across Deep River.

25. *Fayetteville Weekly Observer*, December 5, 1861.

26. *Private Laws of the State of North-Carolina, Passed by the General Assembly at Its Session of 1862–63*, 82.

27. *Fayetteville Weekly Observer*, January 12, 1863. Horne claimed that approximately eighty acres of his land contained a six-foot vein of coal running underneath the outcroppings on his property. See also, Emmons, *Report of Professor Emmons, on His Geological Survey of North Carolina*.

28. *Ibid.*, May 4, 1863. Continuation of the critical line to the Egypt coalfields was completed in September 1863, allowing trains to haul coal to Fayetteville for the war effort or ship by barge via the Cape Fear River, which was navigable by locks, to Wilmington for local use or shipped to other markets on the Eastern Seaboard. See also "Western Railroad Company," in *North Carolina Business History*, www.historync.org/railroad-WRR.htm. Accessed 3 May 2015.

29. *Ibid.*, June 1, 1863.

30. Sales receipt dated June 13, 1861, in the *C.B. Mallett Papers*. A letter in the Mallett papers dated March 23, 1863, from Asa Biggs, one of Mallett's more prominent champions, stated that he was pleased to learn that Mallett was "getting on so well at the mines" and that he hoped Mallett would be able to supply the coal that was so urgently needed. Indeed, an advertisement appearing in the September 25, 1861, issue of the *Weekly Raleigh Register* announced that a "constant supply of coal from Egypt" was available for delivery to customers. The merchant, P. Ferrell, no doubt had a business relationship with Mallett and Browne to sell their coal on a retail basis. Mallett already planned to supply the need in South Carolina as well. The *Columbia South Carolinian* in its July 11, 1861, issue carried a notice that an agent representing the Egypt Coal Company was in the area offering North Carolina coal at about twelve dollars a ton, while the *Charleston Mercury* favorably announced that it was using Deep River coal for running its engines since the northern supply was cut off. The unnamed agent in regard to the availability of North Carolina coal in South Carolina was most likely James Browne, a resident of Charleston, who, while associated with Mallett, realized a business opportunity to supply his native state.

31. *Fayetteville Observer*, August 12, 1861.

32. Sales receipts dated March 12 and March 16, 1864, in the C.B. Mallett Papers #3165, Folder No. 26., Southern Historical Collection, The Wilson Library, University of North Carolina at Chapel Hill. Hereafter cited as the *CBM Papers*.

33. Sales receipts dated March 24, 1864, in the *CBM Papers*.

34. Sales receipt dated July 25, 1864, in the *CBM Papers*. For more information about the *Florie*, see James Sprunt, *Derelicts: an Account of Ships Lost at Sea in General Commercial Traffic and a Brief History of Blockade Runners Stranded Along the North Carolina Coast 1861–1865* (Wilmington, NC: n.p., 1920), 109. Also, *Official Records of the Union and Confederate Navies. Published Under the Direction of the Hon. John D. Long, Secretary of the Navy, by Prof. Edward K. Rowson, U.S. Navy, Series I—Volume 10. North Atlantic Blockading Squadron* (Washington: Government Printing Office, 1900), 504.

35. Sales receipt dated July 27, 1864, in the *CBM Papers*. A fire on April 2, 1863, attributed to non-war related causes, destroyed most of the mill, which represented a great economic loss for the Southern economy as paper had to be scrounged from other locations to make up for the loss. The mill was rebuilt and production resumed in the summer of 1864. See Donald B. Ball, "The Paper Mills in the Confederate South: Industrial Archaeology of a Forgotten Industry," in *Ohio Valley Historical Archaeology*, Vol. 17, 2002, 10.

36. Sales receipt dated July 27, 1864, in the *CBM Papers*.

37. *Ibid.*

38. *Fayetteville Observer*, Nov. 16, 1863. The November 17 edition of the *Semi-Weekly Standard* also covered the event. One source of labor used in the Egypt mine consisted of captured Union Army deserters. According to the March 1, 1864, edition of the *Fayetteville Intelligencer*, forty-three Union Army deserters who had been interned at Castle Thunder Prison in Richmond had accepted an arrangement in which they would exchange prison life for work in the

coal mines at Egypt. Obviously, for some, the benefits of mining outweighed life in a Confederate prison, where malnutrition and disease were rampant. See the *Weekly Intelligencer*, March 1, 1864.

39. *Daily Progress*, November 19, 1863, 2. When the lease for the Egypt property, including the mine, came up for consideration before the Confederate Court in November 1864, the two lessees, C.B. Mallett and James Browne, applied for renewal of their lease which would end January 31, 1865. The application was resisted by the Lockville Mining and Manufacturing Company, of Chatham County, which wished to dispose of the property at Egypt by public auction. The presiding judge declined to order a sale because of the need for coal by the Confederate government and its citizens, while receiving rent for the property. See the *Weekly Conservative* (Raleigh, NC), November 9, 1864. In addition to coal, Lockville contained gold, silver, iron, and copper deposits worth pursuing.

40. *Fayetteville Observer*, February 8, 1864. Having readily available coal to furnish the home front was important, and Deep River coal served this need especially when other markets were unavailable on account of Union control of Southern ports.

41. Mrs. John.H. Anderson, "Confederate Arsenal at Fayetteville, N.C.," in the *Confederate Veteran*, Vol. XXXVI, No. 6, June 1928, Nashville, 224. Rpt. *The Confederate Veteran Magazine*, Volume XXXVI, January 1928–December 1928 (Wilmington, NC: Broadfoot, 1988). Spartanburg, South Carolina, had once been considered for the manufacture of Confederate weaponry but a lack of sufficient coal resources shifted production to Fayetteville. See the *Semi-Weekly Standard*, July 20, 1861. The article originally appeared in the *Charleston Mercury*.

42. *The War of the Rebellion: A Compilation of the Official Records of the Union and Confederate Armies. Published under the Direction of The Hon. Daniel S. Lamont, Secretary of War, by Maj. George B. Davis, U.S.A., Mr. Leslie J. Perry, and Mr. Joseph W. Kirkley, Board of Publications*, Series I—Volume XLVII—in Three Parts, Part II—Correspondence, Etc., 1264.

43. John G. Barrett, *The Civil War in North Carolina* (Chapel Hill: University of North Carolina Press, 1963), 313.

44. *Fayetteville Observer*, January 26, 1863.

45. James Maglenn, "The Steamer Ad-Vance," in *Histories of the Several Regiments and Battalions from North Carolina in the Great War 1861–65, Written by Members of the Respective Commands, Edited by Walter Clark, Vol. V, with Index, Published by the State* (Goldsboro: Nash Brothers, 1901), 337–338.

46. S.R. Mallory, *Communication of Secretary of the Navy ... Jan. 31st, 1864 [i.e., 1865]* (Richmond: s.n., 1865), 1–12. Welsh steam coal, a superior variety of bituminous and closely approximating the best North Carolina Deep River examples, had properties of burning cleaner because of its higher carbon content, and its ability to generate speed quickly often meant the difference between escape and capture. In his study of blockade running at Cape Fear, historian Dawson Carr noted that Vance was infuriated over the "good coal" being allocated to the Confederate-owned vessels, while privately built and state-owned blockade runners, such as the *Advance*, the pride of North Carolina's blockade-running squadron and of which Vance was a part owner, were left with inferior fuel supplies of North Carolina coal. A similar incident with purportedly inferior North Carolina coal involved the Confederate blockade runner *Robert E. Lee*, whose captain, John Wilkinson, resorted to soaking wads of cotton in turpentine before flinging them into the ship's furnace to provide a surge in speed. See Carr, *Gray Phantoms of the Cape*. There is no statistical data as to how many Confederate vessels suffered from poor performance as the result of use of allegedly inferior coal. Only those whose notoriety made them stand out, as in the cases of the *Advance* and the *Robert E. Lee*, was there an issue. Much later, in his defense of the use of North Carolina coal in steam vessels, Jasper Leonidis Stuckey, former state geologist who had assisted J.H. Pratt in the preparation of the North Carolina Geologic and Economic Survey in the 1920s, claimed that the quality of Deep River coal had tested very satisfactorily in trials conducted by Emmons and other prominent geologists of the day even in comparison to coals from other regions of the country as well as Europe. However, improper mining methods, such as failure to extract the purer coals and lack of screening and cleaning, had more to do with this negative assessment. The presence of slate, bone matter, dirt, and sulfur, as well as other impurities, in coal was a contributing factor leading to unacceptable levels of emissions according to Stuckey. Stockpiles of Confederate coal used primarily by blockade runners most likely were rushed into service without the benefit of the removal of these impurities and lack of proper storage, as in cases where coal was left openly on wharves for extended periods. Mining operations in the North, however, especially in Ohio and Pennsylvania, benefitted from the latest mining technology to produce not only cleaner coal but to tap large seams at greater depths, where deposits were more likely to be better quality. Anthracite coal could also be stored more advantageously than bituminous because of its harder, stone-like properties. Stuckey's assessment seemed more accurate given that numerous tests of Deep River coal found it to be among the best in quality and that nothing was inherently wrong with it, only in how it was mined and stored. See the *High Point Enterprise*, January 8, 1961. Also, Stuckey, *North Carolina: Its Geology and Mineral Resources* (Raleigh: Dept. of Conservation and Development, 1965), 510–516. Perhaps the main difference in terms of overall quality, the North, with its vast deposits of anthracite coal, held a decisive advantage over the South's supply of bituminous. As early as 1826, the Lehigh Coal and Navigation Company began trials using anthracite in the steam boilers of tow boats. By the 1840s the Northeast had successfully adapted marine boilers and grates to burn anthracite coal. See Binder, "Pennsylvania Coal and the Beginnings of American Steam Navigation," 420. The U.S. Navy, in a succession of trials conducted by Senator James Cooper of Pennsylvania in the early 1850s, concluded that anthracite coal and its ability to produce steam exceeded its bituminous counterpart thirty-six percent, a considerable difference given the need for ships to evade and pursue enemy vessels. See Eli Bowen, *The Pictorial Sketch-Book of Pennsylvania, or Its Scenery, Internal Improvements, Resources, and Agriculture, Popularly Described, Illustrated with over Two Hundred Engravings, and a Colored Map* (Philadelphia: William Bromwell, 1853), 221. Also, *Official Records of the Union*

and Confederate Navies in the War of the Rebellion. Published under the Direction of The Hon. Josephus Daniels, Series II, Vol. 1, 600, 729, where "want of coal" appears to be a predicament for the Confederacy leading to their reliance on whatever supplies they could find. Vance biographer Joe A. Mobley noted this controversy in his study of Vance's tenure as governor during the Civil War and sides with historian Stephen R. Wise's conclusion that the *Advance* was lost two months before the Navy Department's impressment of its coal supplies. See *"War Governor of the South": North Carolina's Zeb Vance in the Confederacy* (Gainesville: University Press of Florida, 2005), 143–44.

47. *Ibid.*, 623.

48. *War of the Rebellion: A Compilation of the Official Records of the Union and Confederate Armies. Published under the Direction of The Hon. Russell A. Alger, Secretary of War. By Brig. Gen. Fred C. Ainsworth, Chief of the Record and Pension Office, War Department, and Mr. Joseph W. Kirkley. Series II—Volume VII. Section 2.* (Washington: Government Printing Office, 1899), 1057. This represented Special Orders No. 257, Richmond, October 28, 1864. See also, "Jonathan McGee Heck," in NCpedia, ncpedia.org. http://ncpedia.org/biography/heck-jonathan-mcgee. Accessed December 27, 2015.

49. *Ibid.*, 1058. Lincoln's directive to end the policy of prisoner exchange midway in the war, which otherwise would have given the South more manpower militarily and on the home front, resulted in the Confederacy's lack of able-bodied men available to serve as factory workers, mill hands, and laborers. Also, prison labor in the aid of the enemy's war effort typically was unacceptable to the opposing side, leading some officials to enforce the same conditions on their captured enemy combatants.

50. *PrivateLaws of the State of North-Carolina, Passed by the General Assembly, at Its RegularSession of 1864–65, by Order of the State Convention* (Raleigh: Cannon and Holden, 1865), 16. The legislature incorporated the Gorgas Mining and Manufacturing Company on December 20, 1864, with a capital stock of $2,000,000 in shares of $100 each. The officers of the company also included R.H. Butler, superintendent, and P.T. Norwood, treasurer.

51. *Daily Confederate*, January 4, 1865. In 1866, the company continued to operate in Chatham County under the presidency of J.M. Heck. See *Branson and Farrar's North Carolina Business Directory for 1866–67, Containing Facts, Figures, Names and Locations, Revised and Corrected Annually* (Raleigh: Branson & Farrar, 1866), 29.

52. Ruth Little-Stokes, "Logan Manufacturing Company/Oakdale Cotton Mill Village," Inventory/Nomination Form. Division of Archives and History, Raleigh, NC, August 31, 1975. The business was dissolved on April 4, 1865, and its equipment sold at auction. For a more detailed analysis of water power on Deep River, see Geo. F. Swain, et al., *Papers on the Waterpower of North Carolina: A Preliminary Report. North Carolina Geological Survey, Bulletin No. 8* (Raleigh: Guy V. Barnes, 1899).

53. Barrett, *The Civil War in North Carolina*, 27–28.

54. *Official Records of the Union and Confederate Navies in the War of the Rebellion. Published under the Direction of The Hon. Josephus Daniels*, Series 2, Vol. 2, 241, 547. A sufficient number of coke ovens was needed in the manufacturing process along with a dedicated labor pool for this purpose alone. However, Deep River bituminous coal was acknowledged as a high-quality coking coal in previous tests of its ability to perform in iron production.

55. *Ibid.*, 543–544. The navy department drew its coal supplies mainly from three points: Richmond, the Egypt mines in North Carolina, and the mines near Montevallo, Alabama. Supplies from the Chattanooga coal mines were shut off in September 1863 when Federal troops occupied the city. See *Official Records of the Union and Confederate Navies in the War of the Rebellion. Published under the Direction of The Hon. Josephus Daniels*, Series 2, Vol. 2, 534. In spring of 1864, production at the Alabama mines had increased sufficiently to meet the demands of the navy at "points immediately dependent upon them." In North Carolina, South Carolina, and Georgia, transportation problems caused inadequate supplies of coal. The naval stations in the three states were supplied mainly by the Egypt mines and from those in the vicinity of Richmond. See *Official Records of the Union and Confederate Navies in the War of the Rebellion. Published under the Direction of The Hon. Josephus Daniels*, Series 2, Vol. 2, 640. Two such transactions involved the Mallett and Browne Coal Company whose Egypt Mine in March 1864 supplied thirty-eight tons of coal valued at $1,570.70 to the Nitre and Mining Bureau. See Sales receipts dated March 12, 1864, and March 16, 1864, in the *Charles Beatty Mallett Papers.*

56. The Tredgar Iron Works in Richmond represented the largest and most important heavy manufacturing plant in the South, but the Endor Iron Furnace at Deep River was important for manufacturing implements and iron plating for the Confederacy particularly the Confederate Navy Yard at Charlotte. A foundry located at Buckhorn in Harnett County near the Cape Fear made iron in cold blast for the Confederate Arsenal at Fayetteville. See Malcolm Fowler, "Iron Mining in Upper Cape Fear: 3. The Blast Furnace at Buckhorn," in *The State: A Weekly Survey of North Carolina*, April 26, 1941, 11.

57. *Fayetteville Observer*, October 20, 2014.

58. J. Daniel Pezzoni, "Historic and Architectural Resources of Lee County, North Carolina, ca. 1800–1942," National Register of Historic Places/Multiple Property Documentation Form. Preservation Technologies, March 1993.

59. Endor Iron Works Ledger, #2279-z, Southern Historical Collection, The Wilson Library, University of North Carolina at Chapel Hill. The company's ledger consists of one volume containing entries dating from 1864 to 1865, together with a number of checks and receipts loosely inserted inside. See appendix for additional expenditures by the company for coal.

60. *Ibid.*

61. *Ibid.* The source is unnamed but most likely the Farmville Coal Company and the Mallett and Browne Coal Company, as they were the largest producers in the area.

62. *Wilmington Daily Dispatch*, October 23, 1865.

63. Endor Iron Works Ledger, #2279-z, entry dated June 6, 1866. The state legislature had chartered the

Lockville Mining and Manufacturing Company on December 12, 1863, with commissioners George W. Mordecai, C.P. Mendenhall, Joseph S. Jones, Thomas D. Hogg, Kemp B. Battle, and J.M. Heck, Thomas B. Harris, William J. Hawkins, and David A. Barnes, for the purpose of mining, foundry work, and metal manufacturing. See *Public Laws of the State of North-Carolina, Passed by the General Assembly at its Adjourned Session of 1863* (Raleigh: W.W. Holden, 1863), 6–7. Superintendents of the Lockville company included Capt. E. Bryan, J.N. Craig, and John W. Scott. See also the *Daily Progress (Raleigh)*, December 24, 1863. The Endor Iron Works, located near Gulf in present-day Lee County, experienced a succession of ownership beginning in 1872 when it was announced that George B. Lobdell and associates of Wilmington, Delaware, had bought the Endor facility to manufacture pig iron for making railroad car wheels. Interestingly, the current owners relied mostly on water power generated at Deep River to operate their works. Due to transportation problems, lack of mineral deposits, and a depression in the iron business, the enterprise ceased operations by 1876. See the *Greensboro Patriot*, November 13, 1872, and Swain, Holmes, and Myers, *Papers on the Water Power in North Carolina*, 129.

Chapter 6

1. *Raleigh Daily Standard*, January 24, 1866. See also the *Greensboro Patriot*, December 21, 1866. At this time plans were being discussed to widen the Western Railroad from Fayetteville to Salisbury to "open up a wider market for Deep River coal." The *Fayetteville News*, November 6, 1866.
2. *North Carolina Advertiser*, July 8, 1865. The article first appeared in the *Richmond Republic*. One of the concerns of promoting land sales in Southern states had to do with the U.S. government's real or perceived confiscation and disposition of Southern property immediately after the war.
3. *Ibid.*, July 29, 1865. In March 1867, the state legislature incorporated the North Carolina Land Agency for the Encouragement of Immigration, with R.C. Badger; C.B. Root; James Litchford, Jr.; F.G. Foster; and J.G. Hester serving as directors. See *Private Laws of the State of North Carolina, Passed by the General Assembly at the Session of 1866–67* (Raleigh: Wm. E. Pell, 1867), 298–300. After the conclusion of the Civil War, other states in the nation hopped on the band wagon of self-promotion to welcome immigrants and investors to exploit their regions' abundant natural resources such as timber, minerals, agriculture, and wildlife. West Virginia, another state seeking investment in its nascent coalfields, published its own handbook in 1870 extolling its state's opportunities and salubrious climate. For North Carolina, as well as West Virginia and other states, internal improvements played a large part in opening up the region for transporting goods and commodities to market.
4. *North Carolina Advertiser*, July 8, 1865. This issue of the newspaper featured no less than thirty-seven advertisements taken out by the North Carolina Land Agency. One contemporary observer noted that with the exception of Missouri, North Carolina was "foremost, since the close of the war, in encouraging immigration ... establishing offices in the Northern cities for the purpose of representing the advantages that North Carolina possesses." See Thomas Wallace Knox, *Camp-fire and Cotton-field: Southern Adventure in Time of War. Life with the Union Armies, and Residence on a Louisiana Plantation* (1865; repr. Bedford, MA: Applewood, n.d.), 513.
5. *Fayetteville News*, May 15, 1866.
6. *Raleigh Weekly Sentinel*, October 8, 1866.
7. *Branson's North Carolina Business Directory, for 1867–68, Containing Facts, Figures, Names and Locations. Revised and Corrected Annually* (Raleigh: Branson & Jones, c. 1867), 28.
8. *Wilmington Daily Dispatch*, May 29, 1867.
9. The Egypt Mine's output continued to decline beginning in1870, realizing virtually no production from 1874 until about 1890 on account of insufficient capital needed to open or continue operations, as well as prolonged flooding. Ore Hill, located near the Wilcox Iron Works in western Chatham County, was a site of significant iron-production potential, later becoming known for its mineral springs. In 1907, the community was incorporated as Ore Hill with local entrepreneurs poised to develop the area to accommodate the tourist trade. Mount Vernon Springs, located near Ore Hill, represented a new business orientation from mining to therapeutic water treatments on a large scale. One local businessman, J.M. Foust, built the Mt. Vernon Springs Hotel for that purpose. See the *Greensboro Patriot*, July 24, 1907, for a discussion of plans for investment in the area.
10. *Ibid*. Kerr drew on reports conducted by earlier scientists attesting to the superior quality of Deep River coal and mentioned that the bituminous shales associated with the coal yielded a high percentage of kerosene oil, which promised considerable sales on the domestic front.
11. *Ibid.*
12. *North Carolina Advertiser*, July 8, 1865. One such champion of Deep River coal was Salmon Adams of the Tredegar Iron Works in Richmond, Virginia, who in 1868 addressed a letter to Silas Burns, a director of the North Carolina Iron and Steel Rail Company, paying tribute to the natural resources of the state for the manufacture of iron. Noting especially Egypt coal, Adams claimed that it surpassed local Virginia coals used in iron production. The only criticism he offered was that a reduction in price of Egypt coal was necessary to be competitive in the market but that it represented a great enterprise ready for North Carolina to seize upon. See Letter from S. Adams, Tredegar Iron Works, Richmond, Nov. 12, 1868, Document No. 7, Session 1868–69.
13. *Private Laws of the State of North Carolina, Passed by the General Assembly at Its Session 1868-'69* (Raleigh: M.S. Littlefield, 1869), 37.
14. North Carolina Land Company, *A Guide to Capitalists and Emigrants: Being a Statistical and Descriptive Account of the State of North Carolina, United States of America; Together with Letters of Prominent Citizens of the State in Relation to the Soil, Climate, Productions, Mine-rails, AC., and an Account of the Swamp Lands of the State.* (Raleigh: Nichols & Gorman), 3. In a letter to the *Brooklyn (NY) Eagle*, appearing in its July 5, 1869 issue, a contributing

writer described his recent visit to the South, noting that North Carolina was desirous of welcoming the emigrant "from whatever part, and an eager desire to embrace every good suggestion and improvement." He added that trips for prospective emigrants had been arranged to places in the South showing the greatest promise and that cheap return tickets could be obtained from the North Carolina Land Company as well as the Virginia State Land Company. In a further example of enticement to potential investors, a northern director of the North Carolina Land Company offered certificates for roundtrip travel from Boston or New York for twenty-four dollars.

15. *Ibid.*

16. *Ibid*, 4. During its legislation passed during the 1870–71 session, the General Assembly appointed Little as commissioner of immigration for the state charged with selecting one or more individuals to act as assistant commissioner "resident in Great Britain, France, and Germany." See *Public Laws of the State of North Carolina, Passed by the General Assembly at Its Session 1870-'71, Begun and Held in the City of Raleigh on the Sixteenth of November, 1870, to Which Are Prefixed the Constitution of the State and a Register of State Officers, Members of the General Assembly and Judiciary, with the Auditor's Statement of the Public Revenue and Expenditure* (Raleigh: James H. Moore, 1871), 171.

17. *A Guide to Capitalists and Emigrants*, 91.

18. *Ibid.* As seen earlier, the need for a statewide railroad network linking the Deep River coalfield with important manufacturing and financial centers was a constant discussion point that never seemed to be completed soon enough.

19. *Ibid.*, 94.

20. *North Carolina Advertiser* (Raleigh), October 28, 1865. Although the South's economy lay mostly tattered in ruins, in some areas of the old Confederacy sectarian pride and enthusiasm never abated and were the order of the day for promoting opportunities requiring commitment and investment.

21. *Branson's North Carolina Business Directory, for 1867–8*, 28.

22. *Daily Standard*, December 24, 1868. In 1870, Burns placed notices in local newspapers for his Lockville Foundry and Machine Shop, which utilized water power instead of wood or coal, soliciting jobs for castings, mill work, and engine repairs all in the name of prosperity.

23. *Ibid.*, January 13, 1870.

24. *Ibid.*, June 3, 1870.

25. *Raleigh Daily Telegram*, April 23, 1871. The *Fayetteville Eagle* announced in its October 19 issue that the mines were being operated and large quantities of coal produced.

26. *The Raleigh News*, March 22, 1872. The standard, or short, ton weighed 2,000 pounds while the long, or British, ton was 2,240 pounds.

27. *Daily Journal*, May 25, 1872.

28. *Public Laws of the State of North Carolina, Passed by the General Assembly at Its Session 1868-'69, Begun and Held in the City of Raleigh on the Sixteenth of November 1868, to Which Are Prefixed the Constitution of the State and a Register of State Officers, Members of the General Assembly and Judiciary. With the Auditor's Statement of the Public Revenue and Expenditure. Published by Authority* (Raleigh: M.S. Littlefield, 1869), 548–49.

29. *Ibid.* At this time there was no one standard gauge, leaving some areas with different gauges to contend with, causing problematic interchanges and linkages.

30. Emmons, *Geological Report of the Midland Counties*, 1856, 254.

31. R.W. Stone, "Coal on Dan River North Carolina," in *Contributions to Economic Geology, 1910, Part II, Mineral Fuels—Coal on Dan River, North Carolina, Bulletin 471-B*, 137–69. The negative assessments of Dan River coal did not deter some businessmen and experts from continuing to make forays into the area. Born in Laurens, South Carolina, in 1859, Joseph Austin Holmes earned a B.S. degree in agriculture from Cornell University. Upon graduation, the University of North Carolina hired him as professor of geology and natural history. Appointed state geologist to head the new geological survey, Holmes took the position of encouraging economic development and investment through long-range planning and principled conservation methods that sought to create less waste. In 1904, he became chief of the Department of Mines and Metallurgy at the Louisiana Purchase Exhibition in St. Louis. When Congress authorized an investigation into fuels and building materials under the United States Geological Survey, Holmes was appointed as a geologist in fuel investigation. He and his department investigated the relative values of coal, lignite, oil and other fuels that led to more economic and efficient methods of purchase and use. For his efforts, President Theodore Roosevelt, in 1908, selected Holmes as one of four secretaries of the National Conservation Commission, which was charged with inventorying the nation's natural resources. See *Dictionary of North Carolina Biography, Volume 3, H-K*, 177–78.

32. *Journal of the Constitutional Convention of the State of North-Carolina, at Its Session 1868* (Raleigh: Joseph W. Holden, Convention Printer, 1868), 326.

33. *Raleigh Daily Telegram*, April 23, 1871. Throughout its existence the Egypt Mine had a long history of flooding.

34. *Wilmington Journal*, June 23, 1871. Also included in the plan was a survey being conducted by the U.S. Corps of Engineers in anticipation of opening up Deep River for accessing the coal and iron deposits.

35. *Wilmington Daily Journal*, September 2, 1871.

36. *The Eagle*, November 30, 1871.

37. *Greensboro Patriot*, May 23, 1872. The legislature had incorporated the Governor's Creek Steam Transportation and Mining Company on January 28, 1851, with directors Lemuel Williams and M.Q. Waddill having authority for opening and working mines of the company. In 1866, an ordinance was passed to rename the company the Egypt Company for the purpose of establishing a foundry and machine shop, and to distill coal and other bituminous matter in the manufacture of oil. See *Ordinances Passed by the North Carolina State Convention at the Sessions of 1865-'66* (Raleigh: Wm. E. Pell, 1867), 48.

38. *Daily News*, November 10, 1872. Coal production at Deep River about 1868 began to fall off until 1873 when it reached 10,000 tons. From 1873 to 1880 no tonnage was reported except at Gulf. In 1880, the

following mines were noted by the U.S. Geological Survey in the Deep River area: (1) Gulf Mine. Operator, E.L. Houghton. Product, 200 tons. Thickness of coal, 6½ feet. Market: Raleigh, Charlotte, Fayetteville, and Lawrenceburg. (2) Egypt Mine. Not operated. See *Twenty-second Annual Report of the United States Geological Survey to the Secretary of the Interior, 1900–1901, in Four Parts, Part III—Coal, Oil, Cement* (Washington: Government Printing Office, 1902), 48.

39. "A.J. Derbyshire," in *Biographies of Successful Philadelphia Merchants* (Philadelphia: James K. Simon, 1864), 117–120.

40. *Daily Sentinel*, October 12, 1874. During its 1871–1872 session, the state legislature passed an act to change the name of the Chatham Railroad Company to the Raleigh and Augusta Air Line Railroad Company. The legislature granted authority to the Raleigh and Gaston Railroad Company or any other railroad company connecting with it to purchase stock of any kind of the Raleigh and Augusta Air Line Railroad Company. The legislature also approved the Raleigh and Augusta Air Line Railroad Company to discharge the bonds of the Chatham Railroad Company. See *Public Laws of the State of North Carolina, Passed by the General Assembly at its Session 1871-'72, Begun and Held in the City of Raleigh, on the Twentieth Day of November, 1871. To Which Are Prefixed, A Register of State Officers, Members of the General Assembly and Judiciary, and a List of Commissioners of Affidavits. Published by Authority* (Raleigh: Theo. N. Ramsay, 1872), 11–13.

41. *Evening Review* (Wilmington, NC) December 13, 1875.

42. *Daily News*, January 12, 1876.

43. *Chatham Record*, February 27, 1879.

44. Ibid. The completion of the bridge across Deep River and the laying of track fell to convict labor. Approximately 100 convicts from the state penitentiary were assigned to railroad construction and housed in a stockade near the work site. As will be seen later, the use of this controversial practice would extend to the actual mining operations at Deep River.

45. *Evening Visitor*, April 18, 1881.

46. *Chatham Record*, September 1, 1881. Francis Percival Dewees was born in 1837 at Pottsville, Schuylkill Co., Pennsylvania and admitted to the county bar in 1855. After service in the Union Army during the Civil War, he moved to Kentucky, where he entered the charcoal iron-making business, becoming managing director of F.P. Dewees & Co., and afterwards president of the Belmont and Nelson Iron Co. Having a proven track record with operations in anthracite coal, Dewees represented a new breed of northern entrepreneurs whose investment in North Carolina provided the necessary capital to open works and purchase the necessary machinery to adequately back a promising business venture. See *The Dewees Family: Genealogical Data, Biographical Facts and Historical Information. Collected by Mrs. Philip E. LaMunyan. Ellwood Roberts, Editor* (Norristown, PA.: William H. Roberts, 1905) 188.

47. Speculators sought land to lease or purchase for the minerals the land contained, but property ownership did not always mean possession of mineral rights. In a lease, the mineral owner granted the lessee access to the land's mineral content for a specified length of time in exchange for royalties and bonuses paid out by the mining company, which also assumed a calculated risk.

48. Stuart Noblin, "Washington Caruthers Kerr," http://ncpedia.org/biography/kerr-washington-caruthers. Accessed 29 March 2015.

49. *Raleigh Sentinel*, July 1, 1867. The paper appeared to be saying that stability of the state in the aftermath of the war would bode well for achieving wealth and prosperity for its citizenry. At the time, Kerr did recognize the value of the coal deposits on Deep River and spoke highly of them.

50. *Daily Journal*, October 23, 1867.

51. W.C. Kerr, *Report of the Geological Survey of North Carolina. Volume I. Physical Geography, Resume, Economical Geology. By Authority of the General Assembly*. (Raleigh: Josiah Turner, 1875), 293–95.

52. Ibid., 294–295. Drawing on Emmons's data, Kerr surmised that the oil-bearing strata could yield thirty per cent of their weight in kerosene oil. In 1863, H.E. Colton of the Kerosene Oil Works in Fayetteville successfully extracted oil from shale deposits at Deep River for use as a fuel primarily in domestic lighting to fill the need of depleted supplies typically imported from the North. It is not known to what extent Colton was able to manufacture coal oil while in the grip of the war. See the *Fayetteville Observer*, September 24, 1863. Gas extraction from coal had been successfully tested as early as 1853 when samples of the North Carolina variety were used to light the village of Greenport, New York, with an outcome more favorable, by some standards, than the best Newcastle, England, coal. Also, *Fayetteville Weekly-Observer*, December 5, 1853.

53. W.C. Kerr, *North Carolina as a Place for Investment, Manufactures, Mining, Stock Raising, Fruit and Farming: What Northern Residents in North Carolina Say of It as a Place to Live in* (Raleigh: The Observer, 1879). This curiously titled work no doubt beckoned other Northerners to come and invest in the state.

54. F.A. Genth papers, 1832–1901, The Pennsylvania State University, Penn State University Libraries. http://www.libraries.psu.edu/findingaids/392.htm#378ac0b8603ca1982669cc88b2761674. Accessed May 7, 2015.

55. Kerr, *Report of the Geological Survey of North Carolina, Vol. I*, 3.

56. Ibid.

57. P.M. Hale, *The Coal and Iron Industries of North Carolina* (Raleigh: P.M. Hale, 1883), iv. Although Hale was a newspaper man and publisher by trade, and not a trained scientist, his work nevertheless made accessible to readers earlier geological reports and new data with the intention of educating the populace of North Carolina as to their state's "inexhaustible supplies of coal and iron." Representing one in a series of "industrial books" published by Hale, *Coal and Iron Industries of North Carolina* intended to "place within reach of home people and people abroad the knowledge that North Carolina contains inexhaustible supplies of coal and iron of superior quality." Another work, *The Woods and Timbers of North Carolina*, published the same year as his study of the coal and iron industries, represented a compilation from the botanical and geological reports of M.A. Curtis, Ebenezer Emmons, and W.C. Kerr. A concluding chap-

ter on the railroads of the state promised easy, accessible travel on any one of the thirty roads spanning sixty-two counties, 2,040 miles in length.
 58. *Ibid.*
 59. In 1883, the Cape Fear and Yadkin Valley Railroad was reorganized as the Cape Fear and Yadkin Valley Railway, with a completed line from the Deep River coalfields to Greensboro.
 60. Dr. Henry Martyn Chance, born January 18, 1856, at Philadelphia, was a mining engineer and medical doctor, having received his education at the University of Pennsylvania and the Jefferson Medical College. Before his appointment by the North Carolina Board of Agriculture, he had been assistant geologist for ten years with the Pennsylvania Geological Survey. His work for the Board of Agriculture lasted only from July 4 to December 31, 1884. *Who's Who in America: A Biographical Dictionary of Notable Living Men and Women of the United States, 1908–1909. Edited by Albert Nelson Marquis* (Chicago: A.N. Marquis & Company, c. 1908), 327. See also the *Raleigh Register*, April 23, 1884.
 61. *Raleigh Register*, April 23, 1884. The state paid $1,800 for the exploration excluding Chance's compensation. See H.M. Chance, *Report on an Exploration of the Coalfields of North* Carolina, 8.
 62. *Ibid.* The board also approved a study of the Dan River coalfield but the conclusion that no significant seams of coal could be mined closed the door on developing this field. Not long afterwards, another geologist said of the Dan River field, "it is useless ... to spend money and energy in the region in the hope of developing a coal mine." See Stone, *Contributions to Economic Geology*, 169. In 1907, "a citizen" of Winston-Salem invested several thousand dollars in prospecting for coal at Walnut Cove in Stokes County, an area thought to contain significant deposits of anthracite. However, a diamond drill hole of about 300 feet and a slope driven on a thin bed of coal did not produce satisfactory results either.
 63. Chance, *Report on the Exploration of the Coalfields*, 26.
 64. *Ibid.*, 12. Created the same year that saw legislation to establish the North Carolina Department of Agriculture, the Experiment Station was an innovative department that was the only example of its kind in the South. Responsible for determining which fertilizers were best suited to the various crops in the state, experimenting on the nutrition and growth of plants, and determining whether other crops could be grown advantageously on its soils the station used an analytic and scientific approach to its experimentations. Working in conjunction with the state geologist, who was to furnish abstracts of the geological survey when needed, the Experiment Station provided free of charge at its laboratory analysis of marls, soils, and minerals. For the year 1885, the Experiment Station analyzed fifteen samples of coal for the benefit of the Geological Survey. See *Annual Report of the North Carolina Agricultural Experiment Station for 1879. Printed by Order and Expense of the Board of Agriculture* (Raleigh: The Observer, 1879), 8–9, 12. Also, *Annual Report of the North Carolina Agricultural Experiment Station for 1885*, 16.
 65. *Ibid.*, 36. Some properties Chance visited were in a better position where coal could actually be extracted for sampling, but in some instances mining efforts occurred below the outcroppings for more favorable samples.
 66. Chance, *Report on the Exploration of the Coalfields*, 12.
 67. *Ibid.*
 68. *Ibid.*, 55. Mining of narrow and pitched coal seams, which were characteristic of the Deep River field, could cut appreciably into a company's profit margin.
 69. *Ibid.*, 57.
 70. *Ibid.*, A trap dike, or dyke, is a long mass of igneous rock which cuts across the structure of adjacent rocks and in some cases extending for a considerable distance. Trap dikes were typically the main hindrances in extracting coal, making attempts at mining all the more complex and costly.
 71. *Ibid.*, 58.
 72. *Ibid.* Though Chance astutely recognized potential obstacles to mining coal at Deep River, other challenges, such as transportation problems, inclement weather, natural disasters, and labor problems could affect production as well.
 73. *Ibid.*, 59.
 74. *Ibid.* The amount of displacement caused by a fault can be measured in inches or less, or feet. Most earthquakes are the result of energy release associated with rapid movement on active faults.
 75. *Ibid.* The State Board of Agriculture invited Chance to supply North Carolina coal to the State Exposition held in 1884 at Raleigh. Using this venue as an opportunity to test the quality of the coal, Chance reopened some of the old pits and slopes at Farmville, mining approximately 160 tons of coal, of which 105 tons were shipped to Raleigh, where it was utilized as one of the primary fuel sources for the event. See *Report*, 29–30.
 76. *Ibid.*
 77. Fay, *A Glossary of the Mining and Mineral Industry*, 96.
 78. *Ibid.*, 59. The various types of mine gasses will be discussed later in this work, as firedamp represents only one type of mine gas. Some mines were equipped with machinery, such as large electric or steam-driven fans, that helped ventilate poisonous and explosive gasses away from the workings. The Felling mine disasters of 1812 and 1813 in England encouraged invention and experimentation of mine safety lamps by eminent scientists and engineers such as Sir Humphrey Davy and George Stephenson. These lamps could illuminate safely in the presence of methane as well as measure its concentration inside the mine. Davy's seminal treatise on firedamp, entitled "On the Firedamp of Coal Mines, and on Methods of Lighting the Mines so as to Prevent Its Explosion," appeared in the *Philosophical Transactions of the Royal Society of London*, 1816, pt. 1. From 1800 to 1815, alone, explosions in Great Britain coal mines accounted for approximately 475 fatalities of men and boys. See the Coalmining History Resource Centre, www.cmhrc.co.uk. Accessed February 5, 2016. Similar statistics for U.S. mines were not kept during this time.
 79. *Ibid.* Most deep mines, whether coal, gas, ore, or oil were susceptible to flooding especially when located parallel to a river or other large body of water in the vicinity. Significant loss of life and property were associated with severe mine flooding.

80. *Ibid.*, 60. Spontaneous combustion of bituminous coal was a consideration the Confederate government faced when coal was stored in the holds of blockade runners where if situated near a heat source, such as a boiler or furnace, could combust without warning. Because of its hardness, anthracite coal was not as susceptible to spontaneous combustion. After the Civil War, the navy department experimented with strategically placed coaling stations where ships could resupply their depleted coal supplies without having to take on large amounts at any one time.

81. *Ibid.*
82. *Ibid.*
83. *Ibid.*
84. *Ibid.* A proven but antiquated system of ventilation, furnaces were located at or near the bottom of the mine and heated the upcast air causing a partial vacuum over the furnace with the cold air flowing in through the mine. During its time of production up to the close of the Civil War, the Gulf mine was operated by a single shaft with one compartment acting as the "upcast shaft" in which a furnace located at the bottom of the shaft changed the air density, creating a ventilating pressure for the even distribution of air in the mine workings. After the war, mine owners were moving in the direction of mechanized fans to distribute air flow without the inherent danger of open-flame furnaces.
85. E.G. Tuttle, "The Deep River Coalfield of North Carolina and the Egypt Coal Company's Plant," *The Engineering and Mining Journal*, Volume LVIII, November 10, 1894 (New York: Scientific Publishing Company, 1894), 441. Tuttle was hired as a geologist by the Egypt Coal Company to investigate the coalfield and report his findings and recommendations to the company. His interest in the company's physical plant began with the sinking of the shaft in 1853 until about the time he was investigating the workings in the late 1880s and early 1890s. At the time he was hired, Tuttle had recently opened an office at Newark, New Jersey, as a consulting mining and railroad engineer. Previously, he had been mining engineer and superintendent of the Alamo and Coahuila coal companies at San Felipe, Coahuila, Mexico. See *Engineering and Mining Journal*, Vol. 58, August 25, 1894, 178.
86. *Ibid.*
87. *Ibid.*
88. *Ibid.* Openings were made primarily in the four-foot seam.

Chapter 7

1. *Farmer and Mechanic* (Raleigh, NC), October 1, 1884. As was discussed previously, Chance had extracted a sufficient amount of coal at Farmville to ship to the exposition for testing purposes.
2. *Chatham Record*, October 13, 1887.
3. *Ibid.*
4. *The South, an Immigration Journal, Devoted to the Interests of the Southern States, and North Carolina Particularly*, April 1, 1888.
5. *Ibid*, 3.
6. *Ibid*, 4. In *Branson's North Carolina Agricultural Almanac* for 1890, Harrell's enticement for new settlement in the state was printed prominently at the bottom of the page for Halifax County: "Buy a home, and let that home be in the South, especially in North Carolina. J.A. Harrell, Immigrant Agent." Self-promotion aside, Harrell represented the type of individual the state relied on to promote the advantages of North Carolina.
7. *News and Observer*, March 20, 1888. Other newspapers announcing the reopening included the *Chatham Record* and the *Raleigh Christian Advocate*. An article in the *Wilmington Messenger* for March 30, 1888, pointed to improved transportation outlets in the wake of the mine's reopening, specifically the extension of the Cape Fear & Yadkin Valley Railroad to Egypt. The hope was to transport coal from Egypt to Wilmington, which would serve as a coaling station for steamships.
8. *Montgomery Vidette*, January 3, 1889. The article originally appeared in the *Sanford Express*.
9. *Ibid.*
10. *Chatham Record*, November 7, 1889.
11. *The Cape Fear and Yadkin Valley Railway: from Mt. Airy, at the Base of the Blue Ridge, to Wilmington, NC. Its Origin, Construction, Connections, and Extensions. Embracing Descriptive and Statistical Notices of Cities, Towns, Villages, and Stations: Industries, Agricultural, Manufacturing, and Mineral Resources. Scenery of the Route: Transmontane Extensions, &c. Illustrated with Engravings Made from Photographs* (Philadelphia: Allen, Lane & Scott, 1889), 14. Interspersed with photographs, the publication was part history, travelogue, guide to natural resources, and advertising for the area served by the Cape Fear and Yadkin Valley Railway. Of course, the railroad beckoned travelers to ride the train as an excursion to see all the sites mentioned in its handbook.
12. *Ibid.*, 18. In January 1893, the legislature incorporated the Cape Fear Harbor and Coal Company with J.C. Bayles of New York as president for the purpose of constructing a coaling station at Southport (formerly Smithville) in Brunswick County. The company was authorized to build the Brunswick, Western and Southern Rail Road to connect with the Cape Fear and Yadkin Valley and the Norfolk and Western railroads in order to reach the Virginia coalfields as well as "any fields that may be developed in North Carolina," particularly to benefit the coal-shipping trade of the Cape Fear and Yadkin Valley road. Bayles anticipated that from 300,000 to 500,000 tons of coal per year would arrive at Southport to fuel steamships plying their cargo as distant as the West Indies. However, a severe storm in October caused extensive destruction on the North Carolina coast, washing away a number of wharves at Southport. In April 1894, the *Wilmington Star* announced that plans were still under way for building a coaling station there, but Deep River coal figured very little in the plan and Virginia-mined Pocahontas coal became the main staple. See *Private Laws of the State of North Carolina Passed by the General Assembly at its Session of 1893, Begun and Held in the City of Raleigh on Wednesday, the Fourth Day of January, A.D. 1893, Published by Authority* (Raleigh: Josephus Daniels, 1893), 60–62. Also, the *Wilmington Messenger*, January 15, 1893.
13. *Ibid.*, 69.

14. *Chatham Record*, May 23, 1889. The re-laying of the railroad track at Egypt to the mine also contributed to work delays.
15. *News and Observer*, January 10, 1890.
16. *Press and Carolinian*, October 24, 1889.
17. *People's Press*, December 5, 1889. The hotel, appropriately named the Egyptian Inn, was advertised as a winter retreat for northern visitors. See the *Chatham Record*, November 7, 1889.
18. *Wilmington Messenger*, March 7, 1890. Two other coal-mining enterprises at Egypt listed in the 1890 edition of *Branson's North Carolina Business Directory* included E. McIver's estate and Caldwell & Co., with J.P. Steadman, agent. It appeared that with the three active coal-mining operations at Egypt, new enthusiasm and capital could help support and grow the local economy and provide jobs for the area. But with most any new business there were always setbacks as in February when thirty-three miners became trapped in their cage at Egypt halfway between the top and bottom of the shaft. The men remained in the mine overnight when the next morning a hole was cut in the cage and the men, fortunately, were extricated safely from the mine. See the *Concord Standard*, February 28, 1890. Interestingly, the *Charlotte Chronicle*, which initially covered the event, reported that the mine superintendent warned the telegraph operator not to transmit any report of the accident.
19. "Norfolk Southern Railway (1942–82)," en.wikipedia.org/wiki/Norfolk_Southern_Railway_(194–82). Accessed 19 May 2015. One of the incorporators, Samuel A. Henszey, also served as president of the Egypt Railway. Born in 1854 in Philadelphia, he started at an early age employed by the railroad first in the capacity of freight accountant for the North Pennsylvania Railroad from 1870 to 1872. He served in various roles with a number of railroads, including a six-year association as vice president and general manager of the Prescott and Thirty-fifth Parallel and Arizona Central Railroads. At the time of his tenure as president and general manager of the Egypt Railway he was also president and manager of the Raleigh and Western Railway. See *The Biographical Directory of the Railway Officials of America. Edition of 1893. Edited and Compiled by T.A. Busbey* (N.p.: The Railway Age and Northwestern Railroader, n.d.), 169. The Egypt Railway Company entered receivership on April 8, 1908. See *Interstate Commerce Commission. Twenty-first Annual Report on the Statistics of Railways in the United States for the Year Ending June 30, 1908. Prepared by the Bureau of Statistics and Accounts* (Washington: Government Printing Office, 1909), 17.
20. *Sixth Annual Report of the Board of Railroad Commissioners of North Carolina, for the Year Ending December 31, 1896* (Winston: M.I. & J.C. Stewart, 1896), 379.
21. *Laws and Resolutions of the State of North Carolina Passed by the General Assembly at Its Session of 1891, Begun and Held in the City of Raleigh on Wednesday, the Eighth Day of January, A.D. 1891. To Which are Prefixed a Register of State Officers, Judiciary, a List of Commissioners of Affidavits, Members of the General Assembly, and State Constitution. Published by Authority* (Raleigh: Josephus Daniels, 1891), 908–909. Litigation ensued when the Glendon and Gulf Railroad laid track across the Raleigh and Western line, resulting in the latter having employees of the Glendon and Gulf indicted for trespass. Additionally, employees of the Raleigh and Western tore up the track running across their line. In November, the State Supreme Court ruled in favor of the Glendon and Gulf maintaining the right of way, which eventually lead to completion of track between Durham and Charlotte. On July 15, 1896, the Durham and Charlotte Railroad Company acquired the Glendon and Gulf Railroad Company, including "all real estate, franchises and property of every description." See *Sixth Annual Report of the Board of Railroad Commissioners of North Carolina, for the Year Ending December 31, 1896* (Winston: M.I. & J.C. Stewart, 1896), 379.
22. *Jonesboro Leader*, April 1, 1891. Census population schedules for 1900 show that men living in Chatham County whose occupation was coal miner mostly rented accommodations, and the majority of those lived in boardinghouses. In the more heavily populated areas, such as Gulf and Oakland townships, where coal mining was more prevalent, less than one percent of heads of households owned their own homes. In much larger mining communities, such as those in West Virginia and Pennsylvania, the coal companies could evict employees on such pretexts as union involvement, illness, and alleged conspiracy against the company.
23. *Chatham Record*, April 23, 1891. It appeared that the coal received little or no processing other than being brought to the surface as soon as it was mined and dumped into awaiting cars.
24. *Union Republican*, August 20, 1891. Earlier in the year, the State Geological Survey, under the directorship of J.A. Holmes, rented a diamond drill from the Pennsylvania Diamond Drill Company for the purpose of ascertaining the extent of coal at Walnut Cove in Stokes County. W.L. Spoon, a member of the Geological Survey staff serving in the capacity of road engineer, took bore-hole samples beginning in July, continuing until May 1892. Two bore holes were drilled, the deeper of which reached approximately 1,112 feet but with no discernible amount of coal present. See *First Biennial Report of the State Geologist, 1891-'92* (Raleigh: Josephus Daniels, 1893), 16–17.
25. *Ibid*. The legislature incorporated the Greensboro Coal and Mining Company in 1891 to develop coal land in Stokes County. Officers of the company included T.C. Worth, president; O.W. Carr, vice-president; and S.E. Broderick, secretary and treasurer. The company went into receivership in 1901. See the *Greensboro Telegram*, April 19, 1901. Before Maher submitted his report to the owners, the company had already begun to extract considerable quantities of coal from one vein Maher called the "E" seam. Another vein, Maher referred to as the "B" seam, had not yet been worked but based on the large thickness of the outcroppings and convenient location to the railroad he believed held great promise for development.
26. *Ibid*., February 25, 1892. According to the February 19 issue of the *State Chronicle* (Raleigh, NC), W.E. Worth, president of the company, had arranged with the Cape Fear and Yadkin Valley Railway Company to lay side tracks a half-mile into their coal property at Walnut Cove.

27. *Coal and Coal Trade Journal*, Volume 30, No. 29 (July 22, 1891), 361.

28. *Laws and Resolutions of State of North Carolina, Passed by the General Assembly at its Session of 1891*, 483. The geologist was to work under the supervision of Elias Carr and J. Turner Morehead.

29. John R. Ross, "The North Carolina Geological Survey, 1891–1920," in *Forest History*, Vol. 16, No. 4, January 1973, 21.

30. *Ibid*, 23–24. See also *Dictionary of North Carolina Biography*, Vol. 3, 177–78.

31. *Biennial Report of the State Geologist, 1893–1894, North Carolina Geological Survey* (Raleigh: E.M. Uzzell & Co., 1894), 10. In May 1894, Prof. Henry Fairfield Osborn of Columbia University, New York, and a noted geologist and paleontologist, arrived at Chatham County and without compensation worked with the state geologist in making an examination of the Egypt Coal Mine. The following June, W.D. Matthews, also of Columbia University, and a faculty member of the university's mineralogical department, came to the area to continue the work of Osborn. However, state geologist J.A. Holmes failed to mention their work in subsequent biennial reports of the state geologist.

32. *Evening Visitor*, July 26, 1894. Actually, the Seaboard Air Line ended up contracting with the coal company to furnish it with 100 tons of coal daily, a decline from the earlier anticipated 300 tons daily. See the *Evening Visitor*, June 14, 1895.

33. *Sixteenth Annual Report of the United States Geological Survey, to the Secretary of the Interior, 1894–95. Charles D. Walcott, Director. In Four Parts. Part IV.—Mineral Resources of the United States, 1894. Nonmetallic Products. David T. Day, Chief of Division* (Washington: Government Printing Office, 1895), 153–54. Carbonton was named by Mrs. Peter Evans, daughter of Gov. John Motley Morehead, for the amount of carbon deposits reported to be in the area. See William S. Powell, *The North Carolina Gazetteer* (Chapel Hill: University of North Carolina Press, 1968), 88.

34. *Wilmington Morning Star*, December 27, 1894. Samuel A. Henszey, a co-director of the firm, was already serving as president of the Egypt Coal Company and the Raleigh & Western Railway Company. The Egypt Railroad, chartered in 1890, was leased to the Raleigh & Western, which had a line from Cumnock to Harpers Crossroads, in Chatham County. The Taylor place property, consisting of about 1,575 acres of land on Deep River between Cumnock and Gulf, was owned by D.F. Caldwell, of Greensboro, who, in 1895, contracted it to Langdon-Henszey. The company intended to operate it in a connection with its mine at Cumnock, across Deep River, but because of the explosion shortly thereafter in the Egypt Mine, and the company's financial problems, the purchase money was not paid and Caldwell proceeded to sell the Taylor place to other interests. See *The Engineering and Mining Journal*, Vol. LXIV, No. 3, July 17, 1897, 78.

35. *Weekly Messenger*, January 31, 1895. Point Peter is a peninsula extending into the Cape Fear River opposite Wilmington. Before the Civil War it was a terminal for river traffic, and most likely was named for Peter Mallett, a Revolutionary War land owner and military leader. ncpedia.org/gazetteer/search/point%20peter/0. Accessed 24 May 2015.

36. *Ibid*. Pennsylvania was already a major coal-mining center, having extensive fields of anthracite in the east and bituminous in the west, while Tennessee was making itself known as a regional producer for southern markets.

37. *Ibid*. In his letter of transmittal to labor commissioner Benjamin R. Lacy for the 1894 Annual Report of the Bureau of Labor Statistics, Henszey outlined a number of provisions his company had made for its employees. The company maintained a community store that offered goods with a small margin of profit, unlike many other companies known for gouging their employees while keeping them in servile indebtedness. At Egypt, mine employees were not compelled to trade at the company store and during times when mining was suspended were allowed to overdraw their accounts to a modest extent. Housing built by the coal company provided space for garden purposes and lower fuel costs offered to employees helped many families make ends meet. However, Henszey's report took on larger meaning with a view toward developing the region by means of material inducements, low fuel costs, and transportation facilities. The establishment of cotton mills and other manufacturing enterprises represented his broader vision of industrial development leading to increased population and jobs. See B.R. Lacy, *Eighth Annual Report of the Bureau of Labor Statistics* (Raleigh: Josephus Daniels, 1894), 239. Coal miners working for Henszey that year earned between $1.50 to $2.00 a day while the national average was $.17 an hour. Based on a ten-hour day a bituminous coal miner made on average $1.70 a day, certainly within the range Henszey was paying his miners. However, Henszey paid his unskilled laborers wages commensurate with their work. "Trailers, hangers-on, and bottom men" earned between $.80 and $1.00 a day. See also "Earnings and Hours for Bituminous Coal-Lignite Mining and Class I Steam Railroads: 1890 to 1970," in *Historical Statistics of the United States: Colonial Times to 1970, Part 1* (Washington: U.S. Department of Commerce, Bureau of the Census, 1975), 170.

38. *Weekly Messenger*, March 14, 1895. Langdon already had made a number of trips to Cumnock to investigate the coal mines and on his trip to Wilmington he spent time to ascertain the exporting capabilities of its harbor.

39. *Ibid*. The article continued to report that the chutes already completed had a capacity of handling 1,000 tons of coal per day and more if the need demanded it.

40. *Ibid*.

41. *Ninth Annual Report of the Bureau of Labor Statistics of the State of North Carolina, for the Year 1895* (Winston: M.I. & J.C. Stewart, 1895), 363. In his recommendation, Beckham noted that coal mined in North Carolina would account for savings of $3,000 per month based on the reemergence of using local coal as a fuel source for the railroad.

42. *Weekly Messenger*, March 14, 1895.

43. *Weekly Star*, April 26, 1895. Pocahontas coal was mined in Mercer and McDowell counties in West Virginia and Tazewell County, Virginia. It was valued as a superior steam coal. Indeed, the coal chutes were

constructed on the same plan of those at Newport News, thus allowing ships to run under them so that a coal car could dump its contents into the ship's hold in the shortest amount of time possible. In the meantime, the Seaboard Air Line Railway was put in the position of expanding its coal chutes at Portsmouth in order to accommodate ship traffic from Wilmington.

44. *Seventeenth Annual Report of the United States Geological Survey, to the Secretary of the Interior, 1895–96. Charles D. Walcott, Director. In Three Parts. Part III.—Mineral Resources of the United States, 1895. Metallic Products and Coal. David T. Day, Chief of Division* (Washington: Government Printing Office, 1896), 462.

45. *Ibid.* Among the machinery installed at the plant was the Litchfield fast motion engine, which was designed to hoist men, coal, and material vertically in the mine shaft. Powered by steam supplied by boilers in the engine house, the engine was noted for its speed and strength. An engineer worked the engine, which had a foot brake and relief valve giving the engineer more control over his engine or engines. Typically, miners were lowered down the shaft in metal cages and could be extracted by the same means. Though hoisting engines typically were used in shaft mines, they also were well adept for use in slope mines though at a greater angle of entry.

46. *Charlotte Democrat*, July 19, 1895.

47. *Greensboro Patriot*, August 21, 1895.

48. *North Carolina and Its Resources. Illustrated.* State Board of Agriculture, Raleigh (Winston: M.I. & J.C. Stewart, 1896), xiii.

49. *Ibid.*, 104. Later, the community of Egypt would change its name to Cumnock, which would be annexed as part of Lee County in 1907. The *North Carolina Gazetteer* states it was named for an official of the mine, but most likely named for the coal-mining district of Cumnock, Scotland.

50. *Ibid.*, 105. The total size of the Cumnock property was 4,300 acres.

51. *Ibid.*

52. *Wilmington Messenger*, December 15, 1895.

53. H.B. Humphrey, *Historical Summary of Coalmine Explosions in the United States, 1810–1958, Bulletin 586, Bureau of Mines* (Washington: United States Government Printing Office, 1960), 17. According to Bureau of Mines criteria, a disaster was defined as having at least five fatalities.

54. *Chatham Record*, December 26, 1895. Coverage of the accident also appeared in trade journals, such as *The Engineering and Mining Journal* and *The Colliery Engineer and Metal Miner*, which attributed the accident to either a sudden exposure of gas or by an explosion of coal dust. The year 1895 saw thirty-eight coal mine explosions resulting in 1,006 fatalities. See *Historical Summary of Coal-mine Explosions*, p. 17. These were fatalities attributable to explosions only. Separate statistics were also kept for non-explosive accidents.

Chapter 8

1. *Press-Visitor*, December 19, 1895. Though other newspapers would provide coverage of the story, the *Press-Visitor* heard the story firsthand from passengers on the Seaboard Air Line arriving from Moncure; however, recollections could vary from person to person and were not always accurate. Some individuals on the train claimed to have heard the explosion and felt its aftereffects. Pence would move on later in his career to become Washington correspondent of the *News and Observer*.

2. *News and Observer*, December 20, 1895.

3. *Ibid.* As was often the case with mine explosions, lingering gases hovered in the air which could suffocate rescuers or cause secondary explosions.

4. *Ibid.*

5. *Charlotte Observer*, December 20, 1895. Another newspaper earlier reported that London was a passenger on the Seaboard Air Line at the time of the explosion, not in Pittsboro.

6. *Ibid.* Firedamp, or methane, is a tasteless, odorless, colorless gas whose properties are highly flammable and combustible. Methane is the main component of natural gas and occurs naturally as a by-product of coal. It is lethal as an asphyxiant in large amounts, as it displaces oxygen, but the great danger is its combustibility. Afterdamp, or blackdamp, is a toxic mixture of gases left in the mine after an explosion of firedamp. In many instances those inside the mine who survived firedamp succumbed to afterdamp, often making rescue efforts more dangerous and difficult as equipment to deal with such instances was virtually non-existent.

7. *Charlotte Observer*, December 21, 1895.

8. *Raleigh North Carolinian*, December 26, 1895. In its coverage, of January 1896, the prominent mining journal, *The Colliery Engineer and Metal Miner*, reported that firedamp was the probable cause of the explosion, and substantiated the claim that the fire-boss had inspected the mine and found it to be clear of gas. See *The Colliery Engineer and Metal Miner*, Vol. XVI, No. 6, January 1896, 126. The fire-boss was the first person to descend into the mine in order to check for gas in the mine's working areas. In the day, the fire-boss was issued a safety lamp and held it outstretched above his head so that gas at higher levels could be detected. If gas was found, the fire-boss closed off the section and work would stop until the fire-boss deemed the area safe again. Due to its high volatility and unpredictable nature, mine gas was checked for at the beginning of each shift, prior to the opening of a new working area, and after any explosion or roof fall. Sometimes pertinent information about the status of the mine was posted on a bulletin or chalk board outside the workings so that the mine worker would be apprised of the conditions before entering the area. Like miners, fire-bosses had to have experience as well as pass a rigorous course of instruction in order to be certified in that capacity.

9. *News and Observer*, January 8, 1896. It is unclear as to the living arrangements the miners and their families had with the Langdon-Henszey Coal Co., though many mining companies charged rent and deducted the amount from miners' monthly pay when they and their families lived in company housing. Few miners owned their own homes and relied on the coal company or other family members for room and board. Some unmarried miners lived in local boarding houses and paid rent to a landlord.

10. *Ibid.*, January 16, 1896. When the mayor toured the affected area, he visited the homes of the miners and workers who were killed or injured in the explosion. He and his entourage seemed to want to know what the company had done to help the families. Again, it was not uncommon for the coal company to not send aid or assume any liability when an accident occurred in their mines. Authorities often accepted the word of the coal companies as sacrosanct since the companies were rich, powerful, and usually the largest employer in the community.

11. "North Carolina, Estate Files, 1663–1979," index and images, *FamilySearch* https://familysearch.org/pal:/MM9.1.1/VH6X-1YT Accessed 27 Mar 2014, Charles Poe, 1895; citing Chatham, North Carolina, United States, State Archives, Raleigh; FHL Microfilm 001064919. Usually, it took a class-action suit brought against a company to have any chance of compensation.

12. *North Carolinian*, April 9, 1896. Charges against Langdon were eventually dropped due to lack of evidence. Bickering continued between the two owners well into the year when Henszey accused Langdon of trying to oust him from his appointment as receiver of the company.

13. *Ibid.* The association with the New York Gas Coal Company was a dubious arrangement on the part of Langdon and Henszey to shore up the financial standing of their coal mine. The Langdon-Henszey Coal Co. assumed the name of the New York Gas Coal Co., but the stockholders and directors were the same as the coal company's. The newspaper reported that Langdon-Henszey was delinquent in payment of its property taxes and also had not paid for the coffins furnished for the burial of its employees though the superintendent of the mine claimed otherwise. Doctors who attended to the victims had not been paid either and sued the company in court, winning a judgment against the owners, who immediately appealed the decision to Superior Court. One week later, in its April 16 edition, the *Chatham Record* printed a notice of a sheriff's sale of Langdon-Henszey's property at Egypt to satisfy the claims of three plaintiffs against the company. Langdon and Henszey, or their lawyers, knew how to play the judicial system, often dragging out court cases against them until the plaintiff either gave up or took a small pittance as settlement with no other future recourse. See also the *News and Observer*, January 29, 1897.

14. *Wilmington Morning Star*, May 30, 1896. This story was originally reported in the *Press-Visitor*.

15. *Annual Reports of the Department of the Interior for the Fiscal Year Ended June 30, 1897. Eighteenth Annual Report of the United States Geological Survey*, Charles D. Walcott, Director, Part V (Washington: Government Printing Office, 1897), 560. During the year, Langdon-Henszey continued to operate the mine on a much smaller scale mainly to keep water from submerging the workings.

16. *North Carolinian*, June 10, 1897. With no significant workforce employed at their mine and their finances tied up in the courts, Langdon and Henszey agreed to a decree of foreclosure of their property.

17. *Chatham Record*, June 10, 1897. Judge Purnell had only recently been nominated for the post by President McKinley and confirmed by the U.S. Senate on May 5, 1897. Notices began to appear in the local newspapers announcing that on August 16 claims against the company would be heard by Edward Chambers Smith, of Raleigh.

18. *Wilson Advance*, August 19, 1897.

19. *Press-Visitor*, September 22, 1897. To give an idea of the size of the mining operation at Cumnock, the physical plant consisted of an engine and boiler house containing hoisting, pumping, and other machinery; platform chutes and other structures used in the mining and transportation of coal; fan house with fan and engine for producing ventilation; office building; store houses; blacksmith and tool house; boarding house; and fifty tenements for the occupancy of employees.

20. *Ibid.*, October 9, 1897.

21. *Chatham Record*, November 11, 1897.

22. "History of Mine Safety and Health Legislation," United States Department of Labor, Mine Safety and Health Administration, www.msha.gov/MSHAINFO/MSHAINF2.HTM Accessed 23 May 2015. Some states by that time had passed their own mine safety regulations and laws, beginning with Pennsylvania in 1870, following the Avondale Mine disaster of 1869 in which a massive fire that had started at the bottom of the mine's only shaft resulted in the deaths of 110 men and boys trapped below ground with no means of escape. The 1870 mining legislation required that at least two shafts or entrances must be present in any given mining operation. Children often accounted for a significant amount of a mine's labor force, performing dangerous jobs such as door tender, mule driver, coal picker, and spragger, who was responsible for slowing or stopping mine cars by thrusting short lengths of wood (a sprag) between the spokes of the car. The danger of clothing and hands becoming entangled in the process could result in severe injuries and sometimes fatalities. Also referred to as "door trappers," these boys opened and closed the large, heavy ventilation doors in the mine before and after the passage of a trip of mine cars. Trappers could become crushed against the wall by the sheer weight of the door or run over by a mine car, amputating a limb or worse. As there was no enforcement behind the legislation, a number of mine operators continued to use children under the required age, as they were paid less and performed work no one else in the mine would do. I have found no references to underage workers in the mines at Cumnock, though it was not uncommon for children to misrepresent their ages in order to obtain employment.

23. In some mining states, such as Illinois, where both experienced and inexperienced workers were employed, the miner had to attest that his laborer had two years' experience working at the face before he could be left on his own. It was the miner who took the laborer under his tutelage, and not the boss, to teach him his trade. See Carter Goodrich, *The Miner's Freedom: A Study of the Working Life in a Changing Industry* (Boston: Marshall Jones Co., 1925), 37.

24. William Graebner, *Coal-Mining Safety in the Progressive Period: the Political Economy of Reform* (Lexington: University Press of Kentucky, 1976), 3. Though some mine owners believed in the importance of their workers' safety, as well as property,

many still ignored the same safety standards they were required by law to obey.

25. *Ibid.*, 113.

26. *Public Laws and Resolutions of the State of North Carolina Passed by the General Assembly at its Session of 1897, Begun and Held in the City of Raleigh, on Wednesday, the Sixth Day of January, A.D. 1897. Published by Authority* (Winston: M.I. & J.C. Stewart, 1897), 423. Because the U.S. government technically never owned mineral land in the thirteen original states, North Carolina was basically exempt from federal mining laws. However, state legislation provided for the inspection and regulation of mines. See Geo. P. Costigan, Jr., *Handbook on American Mining Law* (St. Paul, Minnesota: West Publishing Co., 1908), 32 f. U.S. mining law became codified in 1866 but it would be until 1873 that the law included coal lands as a mineral lands. Before 1866, coal land had been "disposed of under special statutes." See J.W. Thompson, *United States Mining Statutes Annotated, Part I.—Sections and Statutes Relating to Metalliferous and Coal Mining* (Washington: Government Printing Office, 1915), 655.

27. *Ibid.*

28. *Ibid.*, 424. The mine operator, or owner, was responsible for providing the mine inspector the means necessary for entry in the mine for inspection purposes.

29. *Ibid.*

30. *Ibid.*, 425.

31. *Ibid.*

32. *Ibid.*

33. *Ibid.*, 426

34. Strikers were almost always made by the company that manufactured the lamp and were specific to the operation of that particular brand of lamp. They were usually constructed of metal and flint. There was always the danger of workers smuggling matches or improvised strikers into the mine works. As was the case with the Farmville Mine of the Carolina Coal Company at Coal Glen, no system was in place for miners to check in before going into the mine or checking out after their shifts ended. Furthermore, at the time of the May 27, 1925, explosion, seventy-one lamps were missing from the lamp room which provided the basis for the fear of mounting casualties relative to the number of lamps checked in. See the *News and Observer*, May 28, 1925.

35. *Ibid.*, 427

36. *Ibid.*, 427–28. In mining parlance a squeeze could be the settling of the roof over a large part of the working area or the gradual upheaval of the mine floor due to the weight of the underlying strata. A crush was the settlement of the strata above the mine due to failure of the pillars supporting the roof.

37. *Ibid.*, 428

38. *Ibid.* Mine operators were also required to keep on hand and have delivered to the mine an adequate supply of timber for roof props. According to the new legislation, miners were not to be held responsible for roof falls caused by the mine operator's failure to have sufficient timber on hand.

39. *Coal Trade Journal*, February 13, 1897, 192.

40. *Morning Post*, February 6, 1898.

41. *Chatham Record*, March 3, 1898. The article noted that attempts had been made to sabotage the property with fire and dynamite but this claim has not been substantiated.

42. *County Union*, March 30, 1898.

43. *Ibid.*, June 27, 1898. The newspaper put the cost of the ventilation system and hoisting engine at $40,000.

44. *Asheville Citizen*, July 30, 1898. Production may have started before August 1 at a slower pace until the operation was brought up to speed. Still at odds were Samuel A. Henszey and his former partner Samuel P. Langdon this time over a cut through the grade of Henszey's railroad, the Raleigh and Western, by Langdon. Henszey was granted a restraining order against Langdon, who wanted to build his own railroad, in this continuing saga of verbal fisticuffs between the two former friends and partners.

45. *Chatham Record*, August 4, 1898. Though unnamed at the time, the newly created corporation had all the hallmarks of Samuel P. Langdon. Mention was even made of the ongoing dispute between Langdon and Henszey over the Raleigh & Western Railway property.

46. *Twelfth Annual Report of the Bureau of Labor Statistics of North Carolina, Including the Second Annual Report of the Inspector of Mines, for the Year 1898, by James Y. Hamrick and Warren V. Hall* (Raleigh: Guy V. Barnes, 1899), 350.

47. *Farmer and Mechanic* (Raleigh, NC), January 10, 1899.

48. *Twelfth Annual Report of the Bureau of Labor Statistics of North Carolina*, 350.

49. *Twenty-first Annual Report of the United States Geological Survey, to the Secretary of the Interior, 1899–1900. Charles D. Walcott, Director. In Seven Parts. Part VI.—Mineral Resources of the United States, 1899. Metallic Products, Coal and Coke. David T. Day, Chief of Division* (Washington: Government Printing Office, 1901), 328. Additional statistics included the average number of days mines were in operation, 210, and the average number of employees, 70. Interestingly, no coke production was reported for the state during that time.

50. *1900 U.S. Census, Twelfth Census of the United States, North Carolina, Chatham County, Comprising Albright, Baldwin, Bear Creek, Cape Fear, Centre, Coles Mill, Gulf, Hadley, Hickory Mountain, Matthews, New Hope, Oakland, Rock Rest, and William Townships*. FHL Microfilm 1241188, Roll 1188.

51. Migratory patterns of poor whites and blacks around the turn of the twentieth century reveal that sharecroppers from North Carolina and other southern states with an agricultural economy relocated with their families to the coalfields of West Virginia for better paying jobs. See Joe William Trotter, Jr., *Coal, Class, and Color: Blacks in Southern West Virginia, 1915–32* (Urbana: University of Illinois Press, 1990), 24–25.

52. Thomas B. Womack, Needham Y. Gulley, and William B. Rodman, *Revisal of 1905 of North Carolina Prepared Under Chapter Three Hundred and Fourteen of the Laws of One Thousand Nine Hundred and Three, and Enacted as a Law at the Session of the General Assembly of One Thousand Nine Hundred and Five, in Two Volumes, Volume Two* (Raleigh: E.M. Uzzell & Co., 1905), 342.

53. *Ibid.* 343. Some families sent their sons to work

in the mines and relied on their earnings to make ends meet.

54. *Ibid.*, 345.

55. Annual state mining reports reflected a number of accidents involving miners and laborers whose limbs or clothing became caught in machinery, often leading to amputations and in severe cases death.

56. Humphrey, *Historical Summary of Coal-mine Explosions in the United States, 1810–1958*, 27. Mine accidents in the United States were reported as early as 1825 but were not categorized whether at the state or federal levels. See Graebner, *Coal-Mining Safety in the Progressive Period*, 2.

57. Graebner, *Coal-Mining Safety in the Progressive Period*, 1. The evolution of mine safety is lengthy and not covered here. A thorough study of the Bureau of Mines up to the early 1920s can be found in Fred Wilbur Powell, *The Bureau of Mines: Its History, Activities and Organization. Institute for Government Research. Service Monographs of the United States Government, No. 3* (New York: D. Appleton and Company, 1922).

58. Humphrey, *Historical Summary of Coal-mine Explosions*, iii. See Graebner, *Coal-Mining Safety in the Progressive Period*, 35. Also, George S. Rice, L.M. Jones, J.K. Clement, and W.L. Egy. *First Series of Coal-Dust Explosion Tests in the Experimental Mine. Bulletin 56. Department of the Interior. Bureau of Mines. Joseph A. Holmes, Director* (Washington: Government Printing Office, 1913). *Dictionary of North Carolina Biography*, Vol. 3, 178–79.

59. *Biennial Report of the State Geologist, 1915–1916. North Carolina Geological and Economic Survey, Joseph Hyde Pratt, State Geologist* (Raleigh: Edwards & Broughton), 182. Holmes also recommended that North Carolina become the site of a regional mining experimental station under the auspices of the Bureau of Mines. Based on their British counterparts, rescue stations in some states, such as Illinois and West Virginia, maintained their own teams of trained professionals who could be deployed to any accident site in the state. Built in the aftermath of the Cherry Mine disaster of 1909 the Illinois station at Springfield was the first dedicated institution in the nation responsible for disaster response and mine safety.

60. *Ibid.* Mining equipment was required to be rated and labeled "permissible" by the Bureau of Mines in order to be used in the mine workings. This requirement included the introduction of certain types of explosives tested and approved by the bureau. Other restrictions might apply, including the use of non-conductive metals, which were less likely to spark when struck against another object. Manufacturers of mining equipment, such as Mine Safety Appliances of Pittsburgh, Pennsylvania, developed many types of gear to sell to mining companies as well as directly to miners, often replacing antiquated gear that no longer conformed to bureau standards.

61. Graebner, *Coal-Mining Safety in the Progressive Period*, 62–63. One of the first safety considerations the U.S. Bureau of Mines addressed when it was created in 1910 was the danger of coal-dust explosions in mines. Coal dust explosions can be caused by spontaneous combustion or the introduction of an open flame in the presence of highly inflammable coal dust particles suspended in the air resulting from blasting or any other activity in or around the mines where coal dust may be created. George S. Rice, chief mining engineer and physicist H.P. Greenwald, both of the U.S. Bureau of Mines, released a study in 1929 focusing on the inflammability and explosiveness of coal dust and its prevention under various conditions in investigations conducted from 1911 to 1929 at the bureau's experimental mine at Pittsburgh. Based on their extensive testing, the authors determined no other means of preventing coal-dust explosions than "efficient generalized rock-dusting," a precautionary measure to aid in preventing explosions by distributing pulverized inert material, such as limestone, throughout the mine which, when it combines with coal dust, lowers the temperature of the surrounding air below the ignition point. See G.S. Rice and H.P. Greenwald, *Coal-dust Explosibility: Factors Indicated by Experimental Mine Investigations 1911 to 1929, U.S. Bureau of Mines Technical Paper 464* (Washington: U.S. Government Printing Office, 1929).

Chapter 9

1. *Raleigh Times*, May 23, 1900. The bodies of Connelly and a contractor named James McCarthy were returned to their homes in Pennsylvania. Connelly, a native of Wales, had been foreman of the Laflin Coal Company and a resident of Taylor, Lackawanna Co., PA, before becoming superintendent of the Cumnock mine. The *Scranton Republican*, May 25, 1900. His surname appears also as Connolly.

2. *Evening Telegram*, May 24, 1900. The total number of fatalities in the explosion is unknown, as sources reported the death toll from twenty-three to twenty-eight. The Bureau of Mines arrived at a total of twenty-three men. See H.B. Humphrey, *Historical Summary of Coal-mine Explosions in the United States*, 17.

3. After a number of starts and stoppages, the Cumnock Mine was closed and flooded from 1905 to 1915. See John A. Reinemund, *Geology of the Deep River Coal field, North Carolina, Geological Survey Professional Paper 246* (Washington: U.S. Government Printing Office, 1955), 89. Reinemund's study of the Deep River coalfield was the most important and comprehensive of its type in the twentieth century. It is still useful today and is referred to as such in many other publications on coal mining in North Carolina.

4. *Fourteenth Annual Report of the Bureau of Labor and Printing of the State of North Carolina, for the Year 1900. B.R. Lacy, Commissioner, W.E. Faison, Asst. Commissioner* (Raleigh: Edwards & Broughton, and E.M. Uzzell, 1901), 357. Tally's report of events was in addition to the oral report given to Lacy by the company's counsel "as soon as the accident was heard of." See Appendix for transcript of the evidence taken at the inquest. After the creation of the state mining laws in 1897 there were no further amendments or changes until 1905.

5. *Caucasian*, July 19, 1900. The newspaper fell just short of calling Lacy a liar for claiming no appropriation available to perform his job as mine inspector. Lacy was condemned in the article for choosing to pursue his career in banking and insurance instead of saving lives in the state's mines. Viewed as pandering to railroad and factory interests to gain their confi-

Chapter Notes—10

dence and support in an upcoming statewide election, Lacy was taken to task for not using $3,500 in discretionary funds, paid by the state treasurer, that "he may need and call for." Indeed, the paper stated that Lacy failed to inspect the mines as part of his job and was more concerned with running his campaign for state treasurer. That Lacy could plead ignorance of the mining law he helped create made him guilty of criminal neglect, according to the article, and rightly so. Also, see the *Raleigh Times*, May 23, 1900.

6. *Fourteenth Annual Report of the Bureau of Labor and Printing of the State of North Carolina, for the Year 1900*, 363. A condenser is a vessel or chamber used in conjunction with a turbine having large tubes of cool water running through it. Steam condensed on the tubes, changing back to water, which could then be reused in the boiler or discharged from the mine. News of the explosion reached as far as Oregon, where it was reported in the May 24 issue of the *Morning Oregonian*.

7. *Ibid*.

8. *Ibid*., 366. A fine wire mesh inside the lamp kept the flame from interacting with the surrounding atmosphere. McNath testified that no open flames were allowed in the mine and that the safety lamps had been locked.

9. *Ibid*, 367. The jury members included T.W. Segroves, Thomas J. Johnson, Oren Dowdy, J.R. Burns, R.R. Segroves, and G.G. Lutterloh.

10. *Ibid*, 368. The mine contained two shafts, one for hoisting coal from the workings and the other an air shaft used solely for ventilation or in an emergency for ingress and egress. A large steam-driven suction fan, which was three times larger than required for the mine, allowed for fresh air to be distributed in all parts of the mine workings.

11. *Ibid*.

12. *Ibid*., 369. It was common for gas to exist in various "pockets" in the mine and to be absent in others. Only testing with a safety lamp would have been able to confirm either scenario.

13. *Ibid*.

14. The lack of enforcement of state mining law was again criticized for the legislature's unwillingness to set aside an appropriation for the office to do its job. See *Fourteenth Annual Report of the Bureau of Labor and Printing*, 357.

15. Humphrey, *Historical Summary of Coal-mine Explosions in the United States, 1810–1958*, 17. The high ratio of deaths to the number of miners working at Cumnock was very troubling and as has been shown, any tragedy involving loss of life greatly affected the small mining community.

16. *Semi-Weekly Messenger*, May 29, 1900.

17. *Ibid*., June 8, 1900.

18. *Chatham Record*, July 12, 1900.

19. *News and Observer*, October 15, 1901.

20. *Chatham Record*, October 31, 1901. The plaintiffs in the case were H.M. Hayes and W.P. Henszey of Philadelphia, who sold the Cumnock mining property to the new company and alleged that the defendant was diverting to other purposes the proceeds from the sale of its bonds which was agreed to be applied to the purchase of the property. Judge Purnell dissolved the injunction in November.

21. *News and Observer*, December 24, 1901.

22. *Chatham Record*, October 23, 1902.

23. *Union Republican*, October 30, 1902. Some years previously, the Seaboard Air Line had been using 150 tons of coal a day.

24. *Ibid*. A devastating strike in the anthracite coal field in northeast Pennsylvania lasting from May to October 1902 shut down anthracite usage on the East Coast and directed customers to bituminous coal instead, causing on-hand supplies to dwindle and necessitating the need to increase production significantly for that market.

25. *Greensboro Patriot*, May 27, 1903. MacGregor's remarks that the line from Cumnock to Winston-Salem had been surveyed except for a short distance within Greensboro's city limits were encouraging. A short line was planned in Winston-Salem separate from the main line for passenger service. Setbacks had caused delays in surveying efforts to Winston-Salem. In September, Winston-Salem's *Union Republican* stated that plans had been put in place for the line to extend from the Cumnock mines eastward to connect with the Atlantic & Western and Atlantic Coast Railway Lines. See the *Union Republican*, September 17, 1903.

26. *Ibid*.

27. *Ibid*. Cost of coal per ton fluctuated greatly as a commodity. In 1903, bituminous coal mined at Georges Creek, Maryland, sold on average for $2.40 per long ton (2,240 pounds). Also, prices were dictated by the inordinately high number of strikes, which were responsible for dwindling supplies and higher prices. In 1904, one ton of Georges Creek coal sold for $1.75. See *Wholesale Prices, 1890 to 1912, U.S. Department of Labor, Bureau of Labor Statistics, Wholesale Prices Series: No. 1* (Washington: Government Printing Office, 1913), 187.

Chapter 10

1. Joseph Hyde Pratt, *The North Carolina Geological Survey, Economic Papers, No. 4. The Mining Industry in North Carolina During 1900, by Joseph Hyde Pratt, Mineralogist* (Raleigh: E.M. Uzzell, 1901), 34.

2. Pratt, *The Mining Industry in North Carolina, During 1901*, 72.

3. Pratt, *The Mining Industry in North Carolina, During 1902*, 17.

4. Pratt, *The Mining Industry in North Carolina, During 1903*, 66.

5. Pratt, *The Mining Industry in North Carolina, During 1906*, 128. During times the Cumnock mine was not in operation in 1905, the mine's pump was not kept running, resulting in considerable flooding and inaccessibility to the underground workings. Accounting for the spike in local usage, most likely the coal normally shipped from the mine was diverted to use by the local population.

6. Pratt, *The Mining Industry in North Carolina, During 1903*, 66. Fines are typically "rejected" pieces of coal usually too small to be of any use or consequence.

7. *Public Laws and Resolutions of the State of North Carolina, Passed by the General Assembly, at Its Session of 1905, Begun and Held in the City of*

Raleigh, on Wednesday, the Fourth Day of January, A.D. 1905, Published by Authority (Raleigh: E.M. Uzzell & Co.), 548–50. During ratification, the act establishing the North Carolina Geological Survey, which was created in 1823, was repealed in favor of the legislation creating the new agency, which became the forerunner of the Department of Environment and Natural Resources. In 1906, Joseph Hyde Pratt was appointed state geologist and assumed the directorship of the North Carolina Geological and Economic Survey, after serving as field geologist for the U.S. Geological Survey. Pratt was born in Hartford, Connecticut, in 1870, and developed an early interest in chemistry and mineralogy while a student at Yale University. In 1892, together with S.L. Penfield, of Yale, Holmes came to North Carolina on a mineral-collecting trip for the state's exhibition at the World's Columbian Exposition held in Chicago in 1893. In 1897, Pratt relocated to North Carolina and began teaching economic geology at the University of North Carolina while holding the post of state mineralogist for the North Carolina Geological Survey until 1906. Beginning in 1905, Pratt also served as secretary of the North Carolina Good Roads Association and during World War I enlisted in an engineering battalion with the American Expeditionary Forces, compiling a distinguished service record and promotion to the rank of colonel.

8. *High Point Enterprise*, January 31, 1906.
9. *Daily Industrial News*, April 22, 1906. A great deal of speculation centered on the completion of the rail links and not as much for the future of the coal industry.
10. Pratt, *The Mining Industry in North Carolina, During 1906*, 127–28.
11. *Western Sentinel*, August 23, 1906.
12. *Chatham Record*, August 15, 1906. No mention was made of others going in with Shedd on the purchase but this newspaper referred to Shedd as "the syndicate." Hallison, in Moore County, was named by railroad man Maj. W.C. Petty for his deceased son Hall Jefferson. In 1953, the name of the community was changed to Parkwood. *North Carolina Gazetteer*, ncpedia.org/gazetteer/search/parkwood/0. Accessed 15 June 2015.
13. *Charlotte Daily Observer*, June 15, 1906.
14. *Goldsboro Daily Argus*, December 28, 1906. As this study will later show, interest in the oil-bearing shales of the Deep River district came about in the 1920s as a result of greater demand for petroleum products.
15. *News and Observer*, August 18, 1907. In November, the *Wilmington Morning Star*, picking up a story that first appeared in the *Sanford Express*, observed that a German named Alex Legler was interested in locating a farm near Cumnock in Lee County for twenty-five newly arrived German immigrants with an additional sixty soon to arrive and 200 planned early the following year. See the *Wilmington Morning Star*, November 19, 1907. Shedd, president of the Randolph and Cumberland Railroad, resigned from his position in July 1907 and was succeeded by Ira W. McCormack. See the *News and Observer*, July 12, 1907. The following year Shedd traveled to Lincoln County to work on Lincolnton's sewer system.
16. Joseph Hyde Pratt and H.M. Berry, *The North Carolina Geological and Economic Survey, Economic Paper, No. 23. The Mining Industry in North Carolina During 1908, 1909 and 1910* (Raleigh: Edwards & Broughton), 98.
17. *Chatham Record*, November 17, 1909. Ironically, in the same paper was a report of the Cherry Mine disaster in Illinois.
18. *Ibid.*, December 8, 1909.
19. *Ibid.*, January 17, 1910.
20. *Ibid.*, January 3, 1912. According to the 1910 U.S. Federal Census, John B. Lennig, age 59, was living in Philadelphia, PA. Born in Germany, he apparently owned a chemical manufacturing plant. U.S. Census. 1910. Philadelphia Ward 45, Philadelphia, Pennsylvania, Roll T624_1414, 6A. Enumeration District 1146, FHL Microfilm 1375427.
21. *Charlotte Observer*, March 10, 1912. This article first appeared in the *Sanford Express*. See also "Visit to a Talc Mine," in the *Chatham Record*, November 13, 1912. A mine owned by the Deep River Talc Company was located in Moore County near the Chatham line with the shipping point at Glendon.
22. Pratt, Berry, *The Mining Industry in North Carolina, During 1913–17*, 123.
23. *Ibid.*, December 14, 1916. The article's glowing report was based largely on previous examinations and surveys of the coalfield with added testimonials from local businesses. It was the type of information and presentation that befitted a chamber of commerce pitch.
24. *Ibid.* The writer observed that the value and proximity of the Cumnock Mine would warrant its reopening even if the armor plate facility located elsewhere. Of course, all hope rested on the lucrative government contract to infuse money and jobs into the local economy.
25. *News and Observer*, January 14, 1917.
26. *Wilmington Morning Star*, July 11, 1917. With a capitalization of $1,000,000, the new company was supplemental to the recently chartered Piedmont-Cumnock Coal Company with $500,000 capital, which was later increased to $1,000,000 by the incorporators of the Cumnock Coal Mining Company. See *Standard Corporation Service, January-April 1918* (New York: Standard Statistics Company, Inc., 1918), 136. The two companies eventually became known as the Cumnock Coal Mining Company. The main officers of the company included I.C. Millard, of Norfolk, president, and W.H. Hill, superintendent.
27. *Charlotte News*, January 17, 1912. The company sought to build infrastructure on the property such as a water system with a capacity of 10,000 gallons. See the *Alamance Gleaner*, October 24, 1912. Local entrepreneur John H. Kennedy, secretary and treasurer of the Egypt Improvement Company, opened a general store at Cumnock to sell goods and farm implements to the surrounding community. He issued his own scrip, a currency substitute with set denominations, allowing customers to purchase merchandise at his store. Scrip could also be exchanged for legal tender but rarely at face value with the intention of making customers more reliant on the store while driving them deeper into debt. One of the more newsworthy events involving Kennedy was his purchase of a 43,000-pound plow, manufactured by the International Harvester Company, representing the largest of

its kind in the state, which could cut seven-foot swaths of soil in one pass.

28. *Charlotte Observer*, February 16, 1918. The newspaper brought to the attention of its readers that similar success had been realized with the Seaboard Air Line Railway when its steam engines ran on Cumnock coal. The cost of repairing the mine to a working condition was considerable even if owned by a railroad with greater resources. One of the beneficiaries of the reopening of the Cumnock mine was the city of Wilmington with its emerging role as a port and industrial city. It was hoped that coal traffic would resume on a large scale with the prospects of the U.S. government using North Carolina coal at its port. See the *Wilmington Morning Star*, March 26, 1918.

29. *Chatham Record*, March 6, 1918. Experiencing a logistical challenge locating a cheap, reliable fuel source, the railroad saw this move as an opportunity to develop the area on a large scale to furnish fuel for its locomotives directly at the mines with the possibility that a portion of the coal could be produced for the market. See *The Coal Trade Journal, January 23, 1918* (New York: The Journal), 101.

30. *Wilmington Morning Star*, August 8, 1918.

31. *Chatham Record*, July 25, 1918.

32. *Biennial Report of the State Geologist, 1919–1920*, 30–32. The use of water power to fuel hydroelectric plants caused great interest because of its low cost and availability. Already, during the Civil War, some small manufacturing concerns had taken advantage of water power on Deep River as an energy source. Coal and timber farming in support of the war effort also put a strain on fuel supplies all over the country. Pratt, *The Mining Industry in North Carolina During 1900*, 34. Each year of the survey provided an overview and summary of mining activity. One report noted that the North Carolina Coal and Coke Company at Gulf had been sold and that preparations were being made to work the coal beds on a large scale.

33. *Ibid*. Of equal concern was the fact that some towns and communities were experiencing water shortages and the task of investigating the circumstances fell on the shoulders of the Survey. Data collection and analysis, reports, and recommendations were submitted to the Survey for consideration but because of the Survey's limited funding potential solutions could not be implemented and, as a result, the municipalities themselves were singled out for "financial cooperation."

34. *The Roanoke (NC) Beacon*, March 1, 1918. Indeed, J.N. Powell, proprietor of the hotel and secretary of the coal company, saw no worry about keeping his hotel properly heated with the steady arrival of Deep River coal.

35. *Ibid*. One of its board members, the influential Sandhills developer and newspaper man, Bion H. Butler, editor of the *Southern Pines Pilot* and associate editor of the *Sandhills Daily News*, interjected into his coverage what the new mine promised for Pinehurst and Southern Pines as premier vacation destinations.

36. *Greensboro Daily News*, January 23, 1921, 5. The article is probably referring to the Babcock & Wilcox boiler, used for generating steam for motive power. This would have been situated in the mine's power plant, a separate building from the mining operations.

37. Brent S. Drane and Jasper L. Stuckey, *North Carolina Geological and Economic Survey, Economic Paper, No. 55. The Mining Industry in North Carolina from 1918 to 1923* (Raleigh: Mitchell Printing Co., 1925), 60.

38. *News and Observer*, November 17, 1921.

39. *Moore County News*, January 5, 1922. Bion Huntley Butler (1857–1935) was born in western Pennsylvania and came to the Southern Pines region with his wife in the late 1890s. Newspaperman, entrepreneur, amateur geologist, and author, he soon took interest in the coal-mining operations near Cumnock, promoting the Deep River area for investment and coal production on a commercial scale. See "Bion H. Butler, Editor of Pilot, Dies, Aged 77," *The Pilot*, February 22, 1935.

40. *Coal Age*, November 23, 1922, Vol. 22, No. 21, 837.

41. At the time of the survey only two companies were working in the Deep River coalfield: the Cumnock Coal Company in Lee County and the Carolina Coal Company in Chatham County.

42. Campbell and Kimball, *Deep River Coal field*, 8–9.

43. *Ibid.*, 80.

44. *Ibid.*, 86–87.

45. *Ibid.*, 91.

46. *Ibid.*, 95. Joseph Hyde Pratt, State Geologist, in his preface to Campbell and Kimball's study was more optimistic about the potential success of mining operations at Deep River, citing, among other things, that there was now a demand for a readily available domestic fuel supply and that an improved transportation network had reigned in expensive haulage costs.

47. James Saxon Childers, *Erskine Ramsay: His Life and Achievements* (New York: Cartwright & Ewing, 1942), 5.

48. *Ibid.*, 75.

49. *Charlotte Observer*, September 20, 1922. Also the *News and Observer*, October 26, 1922.

50. *News and Observer*, October 26, 1922. Part of the new equipment included a Goodman electrically driven shortwall coal cutter, which was capable of cutting down three or four carloads of coal per day. See the *Albemarle Press*, November 23, 1922.

51. Drane and Stuckey, *The Mineral Industry in North Carolina from 1918 to 1923*, 60. In 1923, the Deep River Coal Company at Gulf began work in an attempt to open a mine but eventually operations ceased due to "unforeseen circumstances." The Carolina Coal Company and the Erskine Ramsay Coal Company had helped extract samples from the bed for analysis.

52. *The Pilot*, May 25, 1923. By early 1924, the newspaper announced that the auxiliary steam plant of the Sandhills Power Company was near completion and would supply 1,800 horsepower any time that that amount was called for. The plant, located near the stream of Deep River, from which water for steam purposes was taken, was fueled by coal transported directly from the nearby mines at Coal Glen to the boiler room on a siding of railroad owned by the coal company. With the completion of the steam plant, the company anticipated that it would have sufficient

water power to care for its customers while having the steam plant in reserve. See *The Pilot*, February 8, 1924. In April, the same newspaper reported that several members of its staff had spent two days examining the newly constructed plant as well as the company's dam near Carbonton. The plant featured the latest technology, including steam turbine engines, condensing plant, a stack one-hundred fifty feet high, and three large boilers. Commenting on the plant's fuel source, the newspaper believed that the coal of the Carolina Coal Company "can be regarded as a certain force in the development of Moore and the adjacent counties." See *The Pilot*, April 18, 1924.

53. Ibid.

54. Childers, *Erskine Ramsay*, 295.

55. *Ibid.*, 296. Ramsay had considered reports about the geology of the Deep River field and believed that poor management and lack of funds had curtailed sustained efforts to develop the area. Most likely, he relied too heavily on his own experts for recommendations.

56. *Ibid.*, 297.

57. The *Statesville Landmark* reported in its December 6, 1926, edition that after the explosion Grist had toured the mine with Frank Cash of the Bureau of Mines. Damage to the workings was minimal and operations resumed soon after. The two fatalities were Sylvester Murchison, age 21, and Charlie Shirley, age 52. (North Carolina Death Certificates, 1909–1975, State Board of Health, Bureau of Vital Statistics. Microfilm S.123. North Carolina State Archives, Raleigh, NC).

58. Childers, *Erskine Ramsay*, 297. During the same year, coal mining at Deep River suffered another blow when the Deep River Coal Company closed its mine at Gulf, citing "unforeseen difficulties." See Drane and Stuckey, *The Mineral Industry in North Carolina from 1918 to 1923*, 60. These "unforeseen difficulties" were not expanded on in the state report.

59. *Ibid.*, 299–300.

60. Herman J. Bryson, *The Mineral Industry in North Carolina for 1924 and 1925. By Herman J. Bryson, Acting State Geologist. Department of Conservation and Development, Wade H. Phillips, Director. Economic Paper No. 60.* (Raleigh: Commercial Printing Co.), 37–38. Vilbrandt in 1927 published a study of the oil-bearing shales of the Deep River coalfield. See *Oil-Bearing Shales of Deep River Valley, by Frank C. Vilbrandt, Industrial Chemist, University of North Carolina. Department of Conservation and Development, Wade H. Phillips, Director. Economic Paper No. 59* (Raleigh: Edwards & Broughton Co., 1927).

Chapter 11

1. The Farmville Mine's shaft was driven at a forty-five degree angle, extending 2,300 feet from that angle under the earth. The first seam of coal was reached at about 1,000 feet. From that point on, the mine branched out into individual passages leading into pockets or workings where the miners were extracting coal from the nearby seam. See Raleigh *New and Observer*, May 28, 1925.

2. *Report of Mine Explosion. Farmville Mine, Carolina Coal Company, Coal Glen, North Carolina, May 27, 1925, by J.J. Forbes and C.W. Owings. Department of Commerce. U.S. Bureau of Mines*, 1. An unpublished typescript of the findings of the disaster was made available to the author through the Mine Safety and Health Administration (MSHA) of the U.S. Department of Labor. The year 1925 saw 2,230 coal-mining related fatalities, down from the previous year's 2,402 but still indicative of the inherent danger of coal mining. Three hundred forty-five of the deaths in 1925 were attributed to gas- and coal-dust explosions. See *William W. Adams, Coal Mine Fatalities in the United States, 1925, Department of Commerce, Bureau of Mines* (Washington: Government Printing Office, 1926), 2, 4.

3. *Ibid.* At the time of the explosion, the Bureau of Mines maintained twelve rescue stations—six stationary and six mobile—all based in coal-mining areas. The mobile stations were serviced in railroad cars. See Graebner, *Coal-Mining Safety*, 48–49. Newspaper coverage of the explosion was reported in over 200 newspapers nationwide. Mining journals and periodicals also reported the event at some length.

4. *Durham Sun*, May 27, 1925. The newspaper was the first to print same-day coverage of the accident. News of the event was beginning to be reported in numerous papers throughout the country. The *News and Observer* had dispatched one of its staff writers and photographers, Ben Dixon MacNeill, to cover the event. Interestingly the sleeve containing some of McNeill's approximately one-hundred negatives of the events stored in the North Carolina Collection at the University of North Carolina is erroneously labeled "Cumnock Mine Disaster." The explosion occurred at the Farmville Mine of the Carolina Coal Company at Coal Glen, Chatham County. The Cumnock Mine, formerly called the Egypt Mine, was located across Deep River in Lee County, and owned by the Erskine Ramsay Coal Co. MacNeill was a prolific writer who counted among his associates fellow journalists H.L. Mencken and Josephus Daniels. His long-time association with Paul Green's *Lost Colony* held annually on Roanoke Island and a reputation for storytelling brought the musical drama international fame.

5. At the time of the explosion, Brigadier General Albert J. Bowley, U.S. Army, commanded troops at Fort Bragg in Fayetteville and within two hours responded to the crisis by ordering personnel to set up field kitchens at the site. For four days, the army provided food to miners' families and rescue crews. Bowley had distinguished himself in the Spanish-American War and World War I before being assigned command of Fort Bragg. However, a dispute arose when Gen. Bowley was accused of "manhandling" Red Cross relief workers who arrived at the scene wanting to take over the operations on the second day. Only the intervention of Governor McLean saved his command. Due to the chaotic scene in the aftermath of the explosion, it is unclear who actually directed rescue efforts as the National Guard, U.S. Army, state highway department, the Red Cross, and the American Legion all contributed to the relief efforts. (Typed label in the Ben Dixon MacNeill Collection #P0078, North Carolina Photographic Archives, The Wilson Library, University of North Carolina at Chapel Hill.)

6. *Durham Sun*, May 27, 1925. The Adjutant Gen-

eral was the de facto commander of the state National Guard.

7. *News and Observer*, May 28, 1925. McNeill heard three distinct explosions, the final one closing the opening of the shaft where mine superintendent Howard Butler had last seen a number of men alive. A raging underground fire caused much panic from the onlookers as already oxygen-depleted working areas inside the mine gave miners limited places to find air. Mine rescue is not without its risks as lethal gasses, cave-ins, and smoke inhalation, to name only a few dangers, can persist days if not weeks after an explosion.

8. *Ibid*. One of Butler's first actions was to telephone the men working at the 1,800-feet level. It appeared that the men at this location were safe and that the real concern was the upper level near the mine opening, where the shaft spread out into individual workings. A system of bulkheads and cloth brattices directed air flow into the workings but the force of the explosion had blown both out. Butler nearly succumbed to the gas, which at that point had not completely dissipated, and was immediately admitted to a Sanford hospital. It should be noted that most mine gasses are odorless and colorless, necessitating the need for rescuers to use caution in the airways and workings when entering the mine.

9. *Ibid*. Part of the rescue efforts were hampered by the only road leading to the mine becoming choked with automobiles of onlookers. By nightfall, five thousand people—many from outside the area—had congregated at the mine shaft. Also, the absence of gas masks had a deleterious effect on the ability of rescue workers to perform their duties.

10. *Ibid.*, May 29, 1925.

11. *Ibid*, May 30, 1925. In his reportage for May 29, MacNeill noted that fifty-one victims had now been identified with the possibility of six more unaccounted for and deemed dead. He acknowledged the heroic efforts of rescuers battling standing water, the overwhelming stench of decomposing bodies, and exhaustion to complete the search for victims. The *News and Observer* announced in its May 30 edition that it would accept donations on behalf of the victims' families and forward them to the proper authorities.

12. *Albemarle Press*, June 4, 1925. The explosion represented the worst industrial accident in terms of fatalities recorded in the state of North Carolina. The second highest total of deaths reported in the state was the Hamlet chicken processing plant fire in September 1991 which took the lives of twenty-five workers. The ranking of total number of deaths in an accident was taken mostly from newspaper coverage of events in the state while the records of the U.S. Bureau of Mines show that the explosion at Coal Glen resulted in the highest number of workers killed in a coal-mining accident.

13. *News and Observer*, June 1, 1925. As a journalist, MacNeill's description of deep inside the Farmville Mine paralleled that of another journalist writing thirty years earlier. The famed American author Stephen Crane, who penned the *Red Badge of Courage*, had contributed a piece in 1894 for *McClure's Magazine* in which his visit to a coal mine in Scranton, Pennsylvania, elicited the same vivid, impressionistic description MacNeill experienced and wrote about at Coal Glen.

14. Forbes and Owings, *Report of Mine Explosion, Farmville Mine*, 15.

15. *Ibid.*, 24.

16. In some mines, such as the Farmville Mine, the explosive charge was detonated by a blasting cap attached to a cable or wire, which was attached to a battery that the shot-firer operated to send an electrical current through the cable to initiate the charge, allowing the shot-firer to put greater distance between himself and the explosion. See Fay, *A Glossary of the Mining and Mineral Industry*, and Dennis Richard Preston, *Bituminous Coal Mining Vocabulary of the Eastern United States*. Publication of the American Dialect Society, Number 59, April 1973 (N.p.: University of Alabama Press, 1973).

17. Forbes and Owings, *Report of the Mine Explosion, Farmville Mine*, 22–23. The practice of "shooting off the solid" was a dangerous, often illegal, shortcut miners sometimes took to bring down coal from the face without first undercutting the seam, increasing the likelihood of igniting volatile mine gases. Gases could form in pockets in mines without being detected by safety lamps. Any open flame in a mine had the potential of igniting carbon monoxide, hydrogen, methane, hydrogen sulfide, and coal dust which were main causes of explosions. West Virginia and Pennsylvania were among the states that had regulations against "shooting off the solid." See Graebner, *Coal-Mining Safety*, 95.

18. *Ibid.*, 16. Sometimes referred to as a brattice, a check curtain was a partition designed to confine the air and force it into the working places. Cloth, boards, and planks are examples of materials used as brattices. Mine bosses sometimes posted prevailing conditions at the entry of the mine to inform miners before beginning their shift.

19. *Ibid.*, 6. According to Forbes' and Owings' report, the fire-boss, after making an inspection of the mine, should make a written report in a book used for that purpose only. The report should state where he made the tests and whether gas was found and how it was treated. In no instance should any attempt to remove gas be made while men were in the mine workings. See 31–32. In my talks with relatives of men who worked at the mine, safety measures were sometimes ignored, giving the mine the reputation of being gaseous. Some miners either quit or refused to work in the mine knowing the inherent dangers they had to face.

20. *1897 Mining Legislation*, 425.

21. *Thirty-fifth Annual Report of the Department of Labor and Printing*, 1925–1926, 248. Though no official inspection of the mine had been performed prior to the explosion, the Commissioner of Labor and Printing was still responsible for his report to the governor of the events.

22. *Ibid*.

23. *Ibid.*, 251.

24. *Ibid*.

25. The last victim retrieved from the mine was John Lauscher, whose body was found June 4. See *The Pilot*, June 5, 1925.

26. Rescuers found it difficult to ascertain the whereabouts of some miners, or have an accurate count of the victims, as twelve additional miners'

lamps were missing from the bookkeeper's office. One theory put forth was that the miners working the night shift may have taken the lamps home after coming out of the mine, or these men may never have come from the mine at all. Another theory was that contractors working for the company may have neglected to give the bookkeeper their names, adding yet more confusion. Preliminary figures placed the deaths between fifty-nine and seventy-one, both of which proved incorrect in the final tally. See the *Alamance Gleaner*, June 4, 1925.

27. In his biennial report to Governor McLean, State Geologist Herman J. Bryson's favorable comments about the state of coal mining at Deep River included acknowledgment of the increased training of miners and the prospects of a high yield of synthetic petroleum. Indeed, for the fiscal year ending in 1925 coal production had reached its highest level up to that date with 65,153 long tons. See *The Mineral Industry in North Carolina for 1924 and 1925*, 38.

28. *Durham Sun*, May 31, 1925. Future North Carolina governor and U.S. Senator William B. Umstead chaired the relief committee.

29. *The Pilot*, June 12, 1925.

30. *Robesonian*, July 9, 1925. A worker's compensation act was passed by the legislature in 1929.

Chapter 12

1. Humphrey, *Historical Summary of Coal-Mine Explosions*, p. 42. This section deals mostly with explosions attributed to gas. One day after the December 19, 1895, disaster at Cumnock, an explosion attributed to firedamp took the lives of twenty-nine miners at the Nelson Mine near Chattanooga, Tennessee, thus underscoring the potential for dangerous gasses in a mine's workings and the need to test for their presence. See the *Charlotte Observer*, December 21, 1895.

2. *1897 Mining Legislation*, 425.

3. *Ibid.*, 426.

4. *Mine Gases. U.S. Department of Labor. Mine Safety and Health Administration. National Mine Health and Safety Academy. Programmed Instruction Workbook No. 2*. (N.p.: n.p., reprinted 1991), 6.

5. *Ibid.*, 11–12.

6. *Ibid.*, 10.

7. *Ibid.*, 7. When extensive drilling commenced in 1921–22, officials at the Carolina Coal Company mine discovered that the areas where they found coal "gassy" and open-flame lamps were in use, the recommendation was made to replace the exposed flame with safety lamps when required. See the *Moore County News*, January 5, 1922.

8. *Ibid.*, 8–9.

9. *Ibid.*, 13.

10. George A. Burrell and Frank M. Seibert, *Gases Found in Coal Mines. Miners' Circular 14. Department of the Interior, Franklin K. Lane, Secretary. Bureau of Mines, Van. H. Manning, Director* (Washington: Government Printing Office, 1916), 20.

11. *Ibid.*

12. *Thirty-fifth Report of the Department of Labor and Printing, 1925–1926*, 252.

13. Forbes and Owings, *Report of the Mine Explosion, Farmville Mine*, 34.

Chapter 13

1. *Thirty-fifth Report of the Department of Labor and Printing, 1925–1926*, v–vi. Grist's observations concerned not only inspection of coal mines but the state's other types of mineral mines such as iron ore, feldspar, mica, etc. His recommendation that more comprehensive training of miners would help reduce the number of accidents followed on the work of former state geologist and director of the Bureau of Mines, Joseph A. Holmes. North Carolina's Mining Act of 1971 addressed mostly environmental and land reclamation issues but no provision was made for mine safety. Currently, the Mine and Quarry Bureau, within the NC Department of Labor, is charged with the responsibility of enforcing the 1975 Mine Safety and Health Act of North Carolina and conducting a program of inspections, education and training, technical assistance, and consultations to implement provisions of the act. North Carolina Department of Labor, http://www.nclabor.com. Accessed 15 August 2015.

2. *Albemarle Press*, June 18, 1925. Kennedy also had owned a mercantile store in Cumnock from 1913 to 1922 which served miners and the community. Coal production for North Carolina reached 59,936 tons in 1926; 53,377 tons in 1927; and, 60,860 tons in 1928. See *The Mining Industry in North Carolina During 1927 and 1928*, 72.

3. *Ibid.*, 65. The shales that were used for analysis came from the dump heaps of the Cumnock mine and Carolina Coal Company mine at Coal Glen.

4. Frank C. Vilbrandt, *Oil-Bearing Shales of Deep River Valley. Department of Conservation and Development, Wade H. Phillips, Director. Economic Paper No. 59* (Raleigh: Edwards & Broughton Co., 1927), 5. For data concerning the petroleum industry in relation to meeting demand, see 20–21.

5. *The Mining Industry in North Carolina, During 1927 and 1928*, 72. For some southern states, such as Alabama and Tennessee, farming out prison labor to private businesses often resulted in considerable financial returns. Historian Ronald L. Lewis noted that the system was advantageous to both parties, as it was a means for state governments to alleviate pressures on their depleted treasuries while company owners "secured cheap, tractable labor." However, it is not known to what extent inmates received the necessary training to cope with their new occupation that was wrought with injury and death. See Ronald L. Lewis, *Black Coal Miners in America: Race, Class and Community Conflict, 1780–1980* (Lexington: University Press of Kentucky, 1981), 15. Convict labor was already in place in North Carolina soon after the Civil War, for example, when over one hundred inmates from the state penitentiary between 1878 and 1879 helped construct the Western Railroad's extension from Egypt to the Gulf, including the bridge over Deep River which in 1863 had been abandoned due to lack of funding. See the *Chatham Record*, February 27, 1879. Similarly, a stockade was built on site to house the prisoners. Between 1891 and 1892 a sizeable contingency of free, mostly white, miners in the Tennessee coalfields armed themselves in protest over the state's convict leasing program, which was perceived as an unfair advantage given capitalists and producers.

Saddled with the costs of a protracted rebellion with the miners, the state terminated the practice of convict leasing in January 1896. See Karen A. Shapiro, *The Battle Against Convict Labor in the Tennessee Coalfields, 1871–1896* (Chapel Hill: University of North Carolina Press, 1998).

6. *The Pilot*, November 16, 1928. At a November 14, 1928, Kiwanis Club meeting held in Pinehurst, John R. McQueen, a former director of the Erskine Ramsay Coal Company of Cumnock, spoke to the members about the status of coal mining on Deep River, including the employment of prison inmates to work in the mines, and endorsing the continued mining of coal at Coal Glen for its nearby location as well as other valuable mineral products, such as fertilizer material and the oil-bearing shales, which could be distilled into oil and gasoline. Earlier explorations by the Carolina Coal Company between 1921 and 1922 included putting down drill holes at considerable depths to tap the oil reserves. The company accomplished this when extending a well 365 feet deep. See the *Moore County News*, January 5, 1922. Campbell and Kimball in their 1923 study of the Deep River field stated that the presence of oil offered more potential for wealth than the coal deposits but because of expensive testing recommended that efforts be applied to other uses. See Campbell and Kimball, *The Deep River Coalfield*, 95.

7. It is not known to what extent inmates received training in their new occupation of mining but the same safeguards that attend rudimentary coal mining would need to have been applied to shale mining as well.

8. *Biennial Report of the State's Prison, Raleigh, NC, 1927–1928*, 32. In fiscal year 1927–28, expenditures exceeded income at the Coal Glen camp by $1,423.51.

9. *Biennial Report of the State's Prison, Raleigh, NC, July 1, 1928–June 30, 1930*, 26.

10. *Robesonian*, July 18, 1929. The Department of Labor and Printing noted only one inspection of the mine in September 1929 by F.E. Cash of Birmingham, Alabama.

11. *Biennial Report of the State Prison Department, Raleigh, NC, July 1, 1930, June 30, 1932*, 32. Use of prison labor to mine coal at Coal Glen was mostly a failed proposition, as expenditures exceeded income at the Coal Glen camp in its first year of operation, and with the deaths of prisoners in the mines the continued risk of fatalities was unacceptable.

12. *Ibid.*, 33. The cost of purchasing the Coal Glen mine was estimated at $500,000. According to the *Handbook of Labor Statistics*, the average number of calendar days worked for all bituminous mines in the United States was 171. Coal mining, based on supply and demand, was very rarely fulltime work and as a result prisoners experienced considerable downtime. See *Handbook of Labor Statistics, 1924–1926* (Washington: United States Government Printing Office, 1927), 723. This figure is based on yearly reporting by the U.S.G.S. Though North Carolina did not figure into the overall statistics, the number gives an idea of the work load and downtime experienced in coal mines.

13. *The Mining Industry in North Carolina from 1929 to 1936*, 61. The year 1930 saw a sharp decrease in coal production from the previous year with a continued decline through 1935. In late March 1931, a cave-in at the Carolina Mine claimed the lives of L.A. Honeycutt and his thirteen-year-old son, who had been hired to descend 3,600 feet to operate a water pump. At a depth of about 2,600 feet, a pocket of mine gas exploded followed by an intense electrical storm that shut down the circulating fan, effectively cutting off any chance of air being forced into the shaft. The two workers were found the following day under a cave-in of rock and debris. See the *Statesville Landmark*, April 2, 1931.

14. *Ibid.*

15. L.D. Tracy, "Pulverized Coal Is Dangerous on the Surface as Well as Underground; Precautions to Be Taken in Handling It," in *Coal Age*, Vol. 22, No. 5, August 3, 1922, 164.

Chapter 14

1. North Carolina State Board of Health, Bureau of Vital Statistics. *North Carolina Death Certificates*. Microfilm S.123. Rolls 19–242, 280, 313–682, 1040–1297. North Carolina State Archives, Raleigh, North Carolina. Chapman, a resident of Sanford, was born in 1892 in Henry Co., Virginia.

2. *Thirty-fifth Report of the Department of Labor and Printing, 1925–1926*, 253–254. After the Cumnock coal mine owned by the Erskine Ramsay Coal Company ceased operations in 1927, the Carolina Coal By-Products Company acquired the property in 1928. However, the company declared bankruptcy in 1932, throwing the land into receivership. The parcel consisted of 1,000 acres in the fee simple and mineral rights on another nearly 4,000 acres. See *The Pilot*, December 23, 1932.

3. *The Pilot*, September 16, 1932. Some coal merchants continued to stock Pennsylvania and Virginia anthracite for their customers. Another argument as to which type of coal produced more smoke, anthracite or bituminous, played out in the *Asheville Citizen-Times* when enforcement of the city's smoke-abatement ordinance was challenged over the continued use of high volatile bituminous coal, which was noted for its proclivity to generate more smoke than anthracite. The discovery of a low-volatile variety of anthracite in Moore Co. bode well for those accustomed to the higher-grade anthracite which was also fifty-cents per ton cheaper than its bituminous counterpart, while discovery of a new vein of premium coal was a hedge against shortages as well as being a new industry for development. See the *Asheville Citizen-Times*, April 9, 1931.

4. Map: Coal-Bearing Areas of the United States. National Mining Association. www.nma.org/pdf/c_bearing_areas.pdf. Accessed 15 July 2015. North Carolina's anthracite field was quite the anomaly given the extensive beds of bituminous coal in the area. Anthracite was created when sedimentary rocks were subjected to higher pressures and temperatures. In other words, anthracitization was the transformation of bituminous coal into anthracite coal. Anthracite's semi-metallic sheen and hardness made it cleaner to handle though more difficult to ignite. One newspaper account referred to the anthracite in the area as a "freak of nature." An analysis of the anthracite coal at the site showed that its carbon content was eighty-five percent. Though deposits of anthracite were relatively

small compared to those of bituminous the owners nevertheless saw the potential for an industry that could supply a need for as long as twenty-five years.

5. *The Pilot*, September 16, 1932.

6. *Ibid.* McIver's father, John M., Sr., was one of the most well-known and prosperous citizens of Chatham County at the time of his death in 1923.

7. *Ibid.*

8. *Ibid.*, August 3, 1934. Many families in the Deep River basin sold the mineral rights to their property to pay taxes during the Great Depression, and significant attempts at mining occurred during the 1930s. See Jeffrey C. Reid and Kenneth B. Taylor, *Shale Gas Potential in Triassic Strata of the Deep River Basin, Lee and Chatham Counties, NC, with Pipeline and Infrastructure Data, North Carolina Geological Survey, 2009–01*, 1.

9. *Burlington Times-News*, August 3, 1936.

10. *The Pilot*, May 28, 1937.

11. *The Mining Industry in North Carolina, from 1937 to 1945*, 54.

12. *The Pilot*, June 27, 1941. Robinson and Robinson had offered their expertise in mining projects from Pennsylvania to Alabama. Carol Robinson had done considerable work under the Russian and British governments as a consulting engineer. The firm was to have complete oversight and management of the new operation at Coal Glen. Helen Chatfield, *nee* Butler, was the daughter of Bion H. Butler, newspaperman and land developer in the Sandhills region. Her father was vice-president of the Carolina Coal Company at the time of the 1925 explosion.

13. *Ibid.*

14. *Ibid.* The Deep River coalfield lay on both sides of the river as well as the Atlantic and Yadkin Railroad and the Norfolk Southern.

15. *Robesonian*, January 26, 1943. Sponge iron, also known as direct reduced iron (DRI), is produced by heating iron ore at a sufficiently high temperature to remove carbon and oxygen, turning the ore body porous, resembling the structure of a sponge.

16. *Monroe Inquirer*, January 14, 1943. One of the goals of the operation was to contribute to the country's war effort.

17. *Daily Times-News*, January 21, 1943.

18. *Ibid.*

19. Albert L. Toenges, et al., *Coal Deposits in the Deep River Field, Chatham, Lee, and Moore Counties*, 3. Congress, in the spring of 1946, made funds available for the Bureau of Mines to continue its drilling program during the fiscal year July 1, 1946 to June 30, 1947. See *Eleventh Biennial Report of the Department of Conservation and Development of the State of North Carolina, Biennium Ending June 30, 1946*, 119.

20. *The Mining Industry in North Carolina, from 1946 to 1953*, 83. Bledsoe's company provided coal for use in schools and hospitals during a nationwide coal "famine" because of strikes in the larger fields. See *The Daily Times-News* (Burlington, NC), February 25, 1958.

21. *Ibid.* The new firm was owned by the Walter A. Bledsoe and Company. John S. Marshall, of Scranton, Pennsylvania, vice president and treasurer of the corporation, forecast an ambitious production rate of 1,400 tons per day. See *The Pilot*, October 3, 1947.

22. *The Pilot*, November 18, 1949. The mine came to be known as the Carolina Slope Mine.

23. *The Mining Industry in North Carolina, from 1946 through 1953*, 4. The value was $104,000.

24. *Ibid.*, 81.

25. *Gastonia Gazette*, September 27, 1949. The strike represented the nation's first combined coal and steel strike, involving some 400,000 miners and mine workers together with 500,000 CIO-affiliated United Steel workers. One of the factors making the situation worse for North Carolinians was a failure on the part of customers to order coal earlier in the season. Many companies built up their coal reserves in the spring and summer to ensure an adequate supply of coal was on hand for the upcoming fall and winter seasons. Supplies were typically replenished by the railroads in instances where stock was being depleted. In July 1952, an official of the North Carolina Coal Merchants Association advised coal users to stock up during the summer, as supplies were ample to meet demand.

26. *Gastonia Gazette*, November 1, 1949. The state's entire annual production of coal for 1949 was a paltry 14,000 tons, not enough to make a dent in any area where coal was needed most. See *The Mining Industry in North Carolina from 1946 Through 1953*, 83.

27. *Daily Times*, December 28, 1949.

28. Reinemund had completed a survey of the region in 1952, appearing in print as "Future Coal Expectancy in North Carolina," in *Conservation and Development in North Carolina. Vol. 1. Compiled by C.S. Green* (Raleigh: North Carolina Dept. of Conservation and Development), 59–62. Here, he provided an explanatory text accompanying maps discussing mining conditions in the field especially in and near the Carolina Mine, the largest producing coal mine in the state. The author claimed that new structural information, obtained about coal deposits in the Deep River field, would aid in future mine planning and development.

29. *Robesonian*, October 27, 1952. There is no mention as to whether the coking coals were of the naturally occurring type as in higher grade anthracite or coals that were candidates for coking. Earlier studies of the Deep River coalfield showed that the coal had a strong coking ability as well as providing an intense heat for metal working, commercial, and domestic use.

30. *Daily-Times News*, May 12, 1953. Though Bidlack made this bold statement about coal's dominance, other fuel sources, such as gas, oil, and nuclear were poised to surpass coal's usage. Some experts at this time even referred to coal mining as a "sick industry."

31. North Carolina State Board of Health, Bureau of Vital Statistics. *North Carolina Death Certificates*. Microfilm S.123. Rolls 19–242, 280, 313–682, 1040–1297. North Carolina State Archives, Raleigh, North Carolina.

32. *Ibid.*

33. *The Mining Industry in North Carolina, from 1954 Through 1959*, 19. Interestingly, also in 1958, the *Burlington Daily Times-News* reported that the U.S. Bureau of Mines had been requested to make a study of the Deep River coalfield, February 25, 1958.

34. *Daily Times-News*, February 25, 1958.

35. *Biennial Report of the North Carolina Department of Conservation and Development, 1958–1960*, 38, 63.

36. Chatham County [NC] Planning Board, *Land*

Development Potential Study: Prepared for the County of Chatham, North Carolina (N.p.: n.p., 1970), 84.

37. *Charlotte Observer*, March 5, 1978. This condition had been noted a number of years before Duke's investigation, and had proved to be an almost insurmountable obstacle to mining coal. The Erskine Ramsay Coal Company had faced the same obstacle fifty years earlier, losing a considerable amount of investment. See also Toenges, et al., *Coal Deposits in the Deep River Field*, especially 11–18. In 1988, a study submitted by Textoris and Robbins found that reserves of coal in Chatham, Moore and Lee counties totaled 139,664,000 short tons (2,000 lbs.). At the time of active mining, more than two million tons of coal were extracted from the Deep River coalfield. See Daniel A. Textoris and Eleanora I. Robbins, *Coal Resources of the Triassic Deep River Basin, North Carolina*. Department of the Interior. U.S. Geological Survey (N.p.:n.p., 1988.). The North Carolina Geological Survey in 2009 compiled data regarding shale gas extraction in the Deep River basin and determined that shale deposits extended across 25,000 acres at depths less than 3,000 feet in the Sanford sub-basin of Lee and Chatham counties. Six of twenty-eight wells that had been drilled in the Cumnock formation indicated the presence of gas and oil, but controversial proposed methods of extraction resulting in potential harm to property and the environment lost favor among local residents. See Jeffrey C. Reid and Kenneth B. Taylor, *Shale Gas Potential in Triassic Strata of the Deep River Basin*, 1. (Textoris and Robbins' claim of two million tons is subject to question since the combined total tonnage from 1840, when the first amount of coal was recorded, until 1970 was approximately half the amount based on reported state and federal statistics. The authors may have included amounts later than 1970 which contributed to total tonnage.)

38. *North Carolina: Reasonably Foreseeable Development Scenario for Fluid Minerals, Prepared for U.S. Department of the Interior, Bureau of Land Management, Eastern States, Jackson Field Office* (N.p.: n.p., 2008), 5. In the early 1990s, consideration was given for generating hydrocarbons in the Dan River field but lake sediments were determined not to be conducive for such an enterprise.

39. Jeffrey C. Reid and Kenneth B. Taylor, *Shale Gas Potential in Triassic Strata of the Deep River Basin*, 1.

Conclusion

1. William E. Edmunds and Edwin F. Koppe, *Coal in Pennsylvania, Educational Series No. 7* (Harrisburg: Commonwealth of Pennsylvania, Department of Environmental Resources, Bureau of Topographical and Geological Survey, 1968), 2.

2. The North Carolina Workmen's Compensation Act, now known as the Workers' Compensation Act, was ratified by the General Assembly on March 11, 1929, and went into effect July 1 and was administered by the North Carolina Industrial Commission. The act initially covered all state employees, except elected officials, and public employments in which five or more people were regularly employed. Employers participating in the plan were required to carry group insurance and maintain proof of ability to pay claims submitted by employees. The new worker's compensation act helped to circumvent the stranglehold companies had over the justice system by requiring employers to pay out claims "promptly and directly." See *N.C. Session Laws, 1929*, pp. 117–147, for the full text of the legislation. In 1931, the legislature reorganized the Department of Labor and Printing, changing its name to the Department of Labor, which had three divisions, including the Workmen's Compensation Division. This division was created by the transfer of the Industrial Commission to the Department of Labor. See "Government Records Branch of North Carolina." Accessed August 1, 2016. http://www.stateschedules.ncdcr.gov/AgencyHistory.aspx?L1=Department%20of%20Labor.

3. The Deep River basin has been an important brick and common clay center in the United States, producing 2.6 million tons of common clay in 1985 for use in the brick industry. See P. Geoffrey Feiss et al., *The Geology of the Carolinas*, 338.

Appendix A

1. Emmons, *Geological Report of the Midland Counties*, 1856, 254.

2. Stone, *Contributions to Economic Geology*, 137–69.

3. *Acts of Assembly of Virginia*, 1857–58, Chapter 113, 90.

4. Ibid.

5. *Laws of North Carolina*, 1858–59, Chapter 161, 184.

6. Fairfax Harrison, *A History of the Legal Development of the Railroad System of the Southern Railway Company*, 562.

7. *Journal of the Constitutional Convention of the State of North-Carolina, at Its Session 1868*, 326.

8. *Public Laws of the State of North Carolina, Passed by the General Assembly at Its Session 1868–'69*, 549.

9. Ibid.

10. Stone, *Contributions to Economic Geology*, 137–69.

Appendix G

1. *Manufactures of the United States in 1860, Compiled from the Original Returns of the Eighth Census, Under the Direction of the Secretary of the Interior* (Washington: Government Printing House, 1865), clxiii.

Bibliography

Collections

Ben Dixon MacNeill Photograph Collection, North Carolina Collection, University of North Carolina at Chapel Hill

C.B. Mallett Papers, Southern Historical Collection, University of North Carolina at Chapel Hill

Endor Iron Works, Ledger Book, North Carolina Collection, University of North Carolina at Chapel Hill

Mallett-Brown Coal Company Papers, North Carolina Collection, University of North Carolina at Chapel Hill

Serials

Annual Report of the Board of Railroad Commissioners of North Carolina
Annual Report of the Bureau of Labor Statistics
Biennial Report of the State Geologist
Branson's Business Directory
Bureau of Labor Statistics of North Carolina
Coal Age
Coal and Coal Trade Journal
Coal Trade Journal
Colliery Engineer and Metal Miner
Colonial and State Records of North Carolina
Engineering and Mining Journal
Handbook of Labor Statistics
The Mineral Industry in North Carolina
Mining Journal, and Journal of Geology, Mineralogy, Metallurgy, Chemistry, and the Arts in Their Applications to Mining and Working Useful Ores and Metals
N.C. Bureau of Labor and Printing
N.C. Department of Labor and Printing
North Carolina Geological and Economic Survey
Official Records of the Union and Confederate Navies in the War of the Rebellion
Report of the Department of Conservation and Development of the State of North Carolina
Report of the State's Prison (Raleigh, NC)
Session Laws of North Carolina
The South, an Immigration Journal, Devoted to the Interests of the Southern States, and North Carolina Particularly
The State Records of North Carolina
Statistical Abstract of the United States
U.S. Census
U.S. Geological Survey
War of the Rebellion: A Compilation of the Official Records of the Union and Confederate Armies

Newspapers

Alamance Gleaner (Graham, NC)
Albemarle Press (Albemarle, NC)
Asheville Citizen-Times (Asheville, NC)
Asheville Daily Citizen (Asheville, NC)
Asheville News (Asheville, NC)
Burlington Daily Times-News (Burlington, NC)
Burlington Times (Burlington, NC)
Carolina Watchman (Salisbury, NC)
Caucasian (Raleigh, NC)
Charleston Mercury (Charleston, SC)
Charlotte Chronicle (Charlotte, NC)
Charlotte Daily Observer (Charlotte, NC)
Charlotte Democrat (Charlotte, NC)
Charlotte News (Charlotte, NC)
Charlotte Observer (Charlotte, NC)
Chatham Record (Pittsboro, NC)
Columbia South Carolinian (Columbia, SC)
County Union (Dunn, NC)
Daily Confederate (Raleigh, NC)
Daily Eagle (Raleigh, NC)
Daily Industrial News (Raleigh, NC)
Daily Journal (Wilmington, NC)
Daily News (Raleigh, NC)
Daily Progress (Raleigh, NC)
Daily Reporter (Wilmington, NC)
Daily Sentinel (Raleigh, NC)
Daily Standard (Raleigh, NC)
Durham Sun (Durham, NC)

Bibliography

The Eagle (Cherryville, NC)
Evening Journal (Albany, NY)
Evening Review (Wilmington, NC)
Evening Telegram (Rocky Mount, NC)
Evening Visitor (Raleigh, NC)
Farmer and Mechanic (Raleigh, NC)
Fayetteville Eagle (Fayetteville, NC)
Fayetteville Examiner (Fayetteville, NC0
Fayetteville Intelligencer (Fayetteville, NC)
Fayetteville News (Fayetteville, NC)
Fayetteville Observer (Fayetteville, NC)
Gastonia Gazette (Gastonia, NC)
Goldsboro Daily Argus (Goldsboro, NC)
Greensboro Telegram (Greensboro, NC)
Greensborough Patriot (Greensboro, NC)
High Point Enterprise (High Point, NC)
Jonesboro Leader (Jonesboro, NC)
Monroe Inquirer (Monroe, NC)
Montgomery Vidette (Montgomery, NC)
Moore County News (Carthage, NC)
Morning Post (Raleigh, NC)
New Bern News (New Bern, NC)
New-York Daily Tribune/New-York Tribune (New York, NY)
New York Times (New York, NY)
North Carolina Advertiser (Raleigh, NC)
North-Carolina Star (Raleigh, NC)
North Carolinian (Raleigh, NC)
People's Press (Winston-Salem, NC)
The Pilot (Vass, NC)
Press-Visitor (Raleigh, NC)
Press and Carolinian (Hickory, NC)
Raleigh Christian Advocate (Raleigh, NC)
Raleigh Daily Sentinel (Raleigh, NC)
Raleigh Daily Telegram (Raleigh, NC)
Raleigh News & Observer (Raleigh, NC)
Raleigh North Carolinian (Raleigh, NC)
Raleigh Register (Raleigh, NC)
Raleigh Semi-Weekly Standard (Raleigh, NC)
Raleigh Sentinel (Raleigh, NC)
Raleigh Times (Raleigh, NC)
Raleigh Weekly Sentinel (Raleigh, NC)
Richmond Dispatch (Richmond, VA)
Richmond Whig (Richmond, VA)
Roanoke Beacon (Plymouth, NC)
Robesonian (Lumberton, NC)
Salisbury Carolinian (Salisbury, NC)
Sanford Express (Sanford, NC)
Scranton Republican (Scranton, PA)
Scranton Times (Scranton, PA)
Semi-Weekly Raleigh Register (Raleigh, NC)
Southern Weekly Post (Raleigh, NC)
State Chronicle (Raleigh, NC)
State Gazette of North-Carolina (New Bern, NC)
Statesville Landmark (Statesville, NC)
Tarborough Southerner (Tarboro, NC)
Tri-Weekly Commercial (Wilmington, NC)
Union Republican (Winston-Salem, NC)
United States Daily (Washington, D.C.)
Weekly Conservative (Raleigh, NC)
Weekly Messenger (Wilmington, NC)
Weekly North Carolina Standard (Raleigh, NC)
Weekly Raleigh Register (Raleigh, NC)
Weekly Standard (Raleigh, NC)
Weekly Star (Wilmington, NC)
Western Carolinian (Salisbury, NC)
Western Democrat (Charlotte, NC)
Western Sentinel (Winston-Salem, NC)
Wilmington Daily Dispatch (Wilmington, NC)
Wilmington Messenger (Wilmington, NC)
Wilmington Star/Wilmington Morning Star (Wilmington, NC)
Wilmington Tri-Weekly Commercial (Wilmington, NC)
Wilson Advance (Wilson, NC)

Monographs

An Act to Encourage the Importation of Pig and Bar Iron from His Majesty's Colonies in America; and to Prevent the Erection of any Mill or other Engine for Slitting or Rolling of Iron; or any Plateing Forge to work with a Tilt Hammer; or any Furnace for making Steel in any of the said Colonies. London: Printed by Thomas Baskett, 1750.

Acts of Assembly of Virginia, 1857–58.

Adams, Sean Patrick. *Old Dominion Industrial Commonwealth: Coal, Politics, and Economy in Antebellum America*. Baltimore: Johns Hopkins University Press, 2010.

Annual Report of the North Carolina Agricultural Experiment Station for 1879. Printed by Order and Expense of the Board of Agriculture. Raleigh: The Observer, 1879.

Barrett, John G. *The Civil War in North Carolina*. Chapel Hill: University of North Carolina Press, 1963.

Battle, Kemp P. *Reminiscences: The Chatham Railroad Company*, unpublished typescript, Civil War Collection. Military Collection. State Archives of North Carolina. MilColl Civil War Box 76, Folder 63.

The Biographical Directory of the Railway Officials of America. Edition of 1893. Edited and Compiled by T.A. Busbey. N.p.: The Railway Age and Northwestern Railroader, n.d.

Burrell, George A. and Seibert, Frank M. *Gases Found in Coal Mines*. Miners' Circular 14. Department of the Interior, Franklin K. Lane, Secretary. Bureau of Mines, Van. H. Manning, Director. Washington: Government Printing Office, 1916.

Campbell, Marius and Kent W. Marshall. *The Deep River Coal field of North Carolina*. North Carolina Economic and Geological Survey. Joseph Hyde Pratt, Director and State Geologist. Bulletin No. 33. Prepared by United States Geological Survey

Bibliography

in Cooperation with the North Carolina Geological and Economic Survey. N.p.: n.p. Reprint 1995.

The Cape Fear and Yadkin Valley Railway: from Mt. Airy, at the Base of the Blue Ridge, to Wilmington, N.C. Its Origin, Construction, Connections, and Extensions. Embracing Descriptive and Statistical Notices of Cities, Towns, Villages, and Stations: Industries, Agricultural, Manufacturing, and Mineral Resources. Scenery of the Route: Transmontane Extensions, &c. Illustrated with Engravings Made from Photographs. Philadelphia: Allen, Lane & Scott, 1889.

Carr, Dawson. *Gray Phantoms of the Cape Fear: Running the Civil War Blockade.* Winston-Salem: John F. Blair, 1998.

Chance, H.M. *Report on an Exploration of the Coalfields of North Carolina, Made for the State Board of Agriculture. By H.M. Chance. Published by Order of the Board.* Raleigh: P.M. Hale, 1885.

Chatham County [NC] Planning Board. *Land Development Potential Study: Prepared for the County of Chatham, North Carolina.* N.p.: n.p., 1970.

Childers, James Saxon. *Erskine Ramsay: His Life and Achievements.* New York: Cartwright & Ewing, 1942.

Coal Men of America: A Biographical and Historical Review of the World's Greatest Industry, Arthur M. Hull, Sydney A. Hale, eds. Chicago: Retail Coalman, 1918.

Compendium of the Enumeration of the Inhabitants and Statistics of the United States, as Obtained at the Department of State, from the Returns of the Sixth Census, by Counties and Principal Towns Exhibiting the Population, Wealth, and Resources of the Country. Prepared at the Department of State. Washington, D.C.: Thomas Allen, 1841.

Coxe, Tench. *A View of the United States of America, In a Series of Papers, Written at Various Times, between the Years 1787 and 1794, By Tench Coxe of Philadelphia; Interspersed with Authentic Documents: The Whole Tending to Exhibit the Progress and Present State of Civil and Religious Liberty, Population, Agriculture, Exports, Imports, Fisheries, Navigation, Ship-Building, Manufactures, and General Improvement.* Philadelphia: Printed for William Hall and Wrigley & Berriman, 1794.

Debow, J.B.D. *The Industrial Resources, Etc., of the Southern and Western States: Embracing a View of Their Commerce, Agriculture, Manufactures, Internal Improvements, Slave and Free Labor, Slavery Institutions, Products, Etc., of the South, with an Appendix, in Three Volumes.* New Orleans: Office of De Bow's Review, 1852.

_____. *Report of the Coal Lands of the Deep River Mining and Transportation Company, in Chatham and Moore Counties, North Carolina, with Analyses of the Minerals.* Albany, NY: Weed, Parsons, and Company, 1851.

_____. *Statistics of Coal: the Geographical and Geological Distribution of Minerable Combustibles or Fossil Fuel, Including, also, Notices and Localities of the Various Mineral Bituminous Substances, Employed in Arts and Manufactures, from Official Reports of the Great Coal-Producing Countries, the Respective Amounts of their Production, Consumption and Commercial Distribution, in All Parts of the World, Together with Their Prices, Tariffs, Duties and International Regulations. Accompanied by Nearly Four Hundred Statistical Tables, and Eleven Hundred Analyses of Mineral Combustibles, with Incidental Statements of the Statistics of Iron Manufactures, Derived from Authentic Authorities.* Philadelphia: J.W. Moore, 1848.

De La Beche, Sir Henry, and Dr. Lyon Playfair. *Third Report on the Coals Suited to the Steam Navy, Presented to Both Houses of Parliament by Command of Her Majesty.* London: William Clowes and Sons, 1851.

Department of Commerce: Bureau of Foreign and Domestic Commerce. "Population—Occupations." *Statistical Abstract of the United States, 1920, Forty-third Number* Washington: Government Printing Office, 1921.

The Dewees Family: Genealogical Data, Biographical Facts and Historical Information. Collected by Mrs. Philip E. LaMunyan. Ellwood Roberts, Editor. Norristown, PA.: William H. Roberts, 1905.

Duane, William J. *The Internal Improvement of the Commonwealth, by Means of Roads and Canals.* Philadelphia: Jane Aitken, 1811.

Emmons, Ebenezer. *Geological Report of the Midland Counties of North Carolina.* New York: George P. Putnam; Raleigh: Henry D. Turner, 1856.

_____. *Report of Professor Emmons, on His Geological Survey of North Carolina.* Raleigh: Seaton Gales, 1852.

_____. *Special Report of E. Emmons, Geologist to the State of North-Carolina Concerning the Advantages of the Valley of the Deep River as a Site for the Establishment of a National Foundry. Made Pursuant to Instructions from Gov. Bragg, 29 December 1857.* Raleigh: Holden & Wilson, 1857.

Fay, Albert H. *A Glossary of the Mining and Mineral Industry.* Washington: U.S. Government Printing Office, 1947.

Fulton, Hamilton. *Annual Report of the Board of Public Improvements of North Carolina, to the General Assembly, December 10, 1822; Together with Mr. Fulton's Reports to the Board, On the Public Works Projected and Carrying on Throughout the State During the Present Year.* Raleigh: J. Gales & Son, 1822.

_____. *Report of Sundry Surveys Made by Hamilton Fulton, Esq., State Engineer, Agreeably to Certain Instructions, From Judge Murphey, Chairman, and Submitted to the General Assembly, at Their Session, in 1819.* Raleigh: Printed by Tho. Henderson, 1819.

Geological Essays; or an Inquiry into Some of the

Bibliography

Geological Phenomena to Be Found in Various Parts of America, and Elsewhere. Baltimore: J. Robinson, 1820.

Gilmer, John A. *Speech of Hon. John A. Gilmer, of North Carolina, on the Location of a National Foundry in the Deep River Valley, and the Mineral and Manufacturing Resources of North Carolina; Delivered in the House of Representatives, February 21, 1859.* Washington: Congressional Globe Office, 1859.

_____. *Speech of John A. Gilmer, Senator from Cumberland and Harnett, on the Bill to Aid in the Construction and Equipment of the Western Railroad from Fayetteville to the Coalfields, Delivered in the Senate of North-Carolina, December 2, 1858.* Raleigh: Holden and Wilson, 1859.

Goodrich, Carter. *The Miner's Freedom: A Study of the Working Life in a Changing Industry.* Boston: Marshall Jones Co., 1925.

Goodyear, W.A. *The Coal Mines of the Western Coast of the United States.* San Francisco: A.L. Bancroft, 1877.

Gordon, Robert B. *American Iron, 1607–1900.* Baltimore: Johns Hopkins University Press, 1996.

Hale, P.M. *The Coal and Iron Counties of North Carolina.* Raleigh: P.M. Hale, 1883.

Handbook of Labor Statistics, 1924–1926. Washington: United States Government Printing Office, 1927.

Harvey, Katherine A. *The Best Dressed Miners: Life and Labor in the Maryland Coal Region, 1835–1910.* Ithaca: Cornell University Press, 1969.

Hatcher, John. *The History of the British Coal Industry, Volume I, Before 1700: Towards the Age of Coal.* Oxford: Clarendon Press, 1993.

Hayden, Horace H. *Geological Essays; or an Inquiry into Some of the Geological Phenomena to Be Found in Various Parts of America, and Elsewhere.* Baltimore: J. Robinson, 1820.

Historical Statistics of the United States: Colonial Times to 1970, Part 1. Washington: U.S. Department of Commerce, Bureau of the Census, 1975.

Historical Statistics of the United States, Colonial Times to 1970, Part 2. Washington: U.S. Dept. of Commerce, Bureau of the Census, 1975.

History of the Raleigh & Augusta Air-Line R.R. Co. Known originally as the Chatham Railroad Company, Including all the Acts of the General Assemblies of North and South Carolina Relating Thereto. Compiled by Walter Clark, Esq. Raleigh: Raleigh News Steam Job Print, 1877.

Humphrey, H.B. *Historical Summary of Coal-mine Explosions in the United States, 1810–1958, Bulletin 586, Bureau of Mines.* Washington: United States Government Printing Office, 1960.

Interstate Commerce Commission. *Twenty-first Annual Report on the Statistics of Railways in the United States for the Year Ending June 30, 1908.* Prepared by the Bureau of Statistics and Accounts. Washington: Government Printing Office, 1909.

Johnson, Walter R. *The Coal Trade of British America, with Researches on the Characters and Practical Values of American and Foreign Coals.* Washington: Taylor & Maury; Philadelphia: A. Hart, 1850.

_____. *Report of the Coal Lands of the Deep River Mining and Transportation Company, in Chatham and Moore Counties, North Carolina, with Analyses of the Minerals.* Albany, NY: Weed, Parsons, and Company, 1851.

_____. *A Report to the Navy Department of the United States on American Coals Applicable to Steam Navigation, and Other Purposes.* Washington: Gales and Seaton, 1844.

Journal of the Constitutional Convention of the State of North-Carolina, at Its Session 1868. Raleigh: Joseph W. Holden, Convention Printer, 1868.

Journal of the House of Representatives of the United States: Being the First Session of the Thirty-second Congress, Begun and Held at the City of Washington, December 5, 1853, and in the Seventy-eighth Year of the Independence of the United States. United States Congressional Serial Set, Issue 709. Washington: Robert Armstrong, 1853.

The Journal of the Proceedings of the Provincial Congress of North Carolina, Held at Halifax, on the Fourth Day of April, 1776. New Bern: Printed by James Davis, 1776. Reprint. Raleigh: Lawrence and Lemay, 1831.

Keating, W.H. *Consideration upon the Art of Mining to Which Are Added Reflections on Its Actual State in Europe, and the Advantages Which Could Result from an Introduction of this Art into the United States.* Philadelphia: M. Carey and Sons, 1821.

Keating, William H. Accessed 26 February 2015. http://en.wikipedia.org/wiki/William_H._Keating.

Kerr, W.C. *North Carolina as a Place for Investment, Manufactures, Mining, Stock Raising, Fruit and Farming: What Northern Residents in North Carolina Say of It as a Place to Live in.* Raleigh: The Observer, 1879.

_____. *Report of the Geological Survey of North Carolina. Volume I. Physical Geography, Resume, Economical Geology. By Authority of the General Assembly.* Raleigh: Josiah Turner, 1875.

Knox, Thomas Wallace. *Camp-fire and Cotton-field: Southern Adventure in Time of War. Life with the Union Armies, and Residence on a Louisiana Plantation,* 1865. Reprint. Bedford, MA: Applewood Books, n.d.

Lesley, J.P. *The Iron Manufacturer's Guide to the Furnaces, Forges and Rolling Mills of the United States, with Discussions of Iron as a Chemical Element, an American Ore, and a Manufactured Article, in Commerce and in Industry, with Maps and Plates.* New York: John Wiley, 1859.

_____. *Manual of Coal and Its Typography, Illustrated by Original Drawings, Chiefly of Facts in the Geology of the Appalachian Region of the*

Bibliography

United States of North America. Philadelphia: J.B. Lippincott, 1856.

Lewis, Ronald. *Black Coal Miners in America: Race, Class, and Community Conflict, 1780–1980*. Lexington: University of Kentucky Press, 1987.

Little-Stokes, Ruth. "Logan Manufacturing Company/Oakdale Cotton Mill Village." *Inventory/Nomination Form*. Division of Archives and History, Raleigh, NC, August 31, 1975.

Long, Priscilla. *Where the Sun Never Shine: A History of America's Bloody Coal Industry*. New York: Paragon House, 1989.

Mallory, S.R. *Communication of Secretary of the Navy ... Jan. 31st, 1864 [i.e. 1865]*. Richmond: N.p.: n.p., 1865.

Manufactures of the United States in 1860, Compiled from the Original Returns of the Eighth Census, Under the Direction of the Secretary of the Interior. Washington: Government Printing House, 1865.

Memorial of the Stockholders of the Cape Fear and Deep River Navigation Company to the General Assembly. Raleigh: W.W. Holden, 1854.

Miller, Donald L., and Richard E. Sharples. *The Kingdom of Coal: Work, Enterprise, and Ethnic Communities in the Mine Fields*. Philadelphia: University of Pennsylvania Press, 1985.

Mine Gases. U.S. Department of Labor. Mine Safety and Health Administration. National Mine Health and Safety Academy. Programmed Instruction Workbook No. 2. N.p.: n.p., n.d. Reprinted. 1991.

Mobley, Joe A. *"War Governor of the South": North Carolina's Zeb Vance in the Confederacy*. Gainesville: University Press of Florida, 2005.

North Carolina and Its Resources. Illustrated. State Board of Agriculture, Raleigh. Winston: M.I. & J.C. Stewart, 1896.

North Carolina Death Certificates. North Carolina State Board of Health, Bureau of Vital Statistics. North Carolina State Archives. Raleigh, North Carolina.

North Carolina Land Company, A Guide to Capitalists and Emigrants: Being a Statistical and Descriptive Account of the State of North Carolina, United States of America; Together with Letters of Prominent Citizens of the State in Relation to the Soil, Climate, Productions, Mine-rails, AC., and an Account of the Swamp Lands of the State. Raleigh: Nichols & Gorman, 1869.

North Carolina: Reasonably Foreseeable Development Scenario for Fluid Minerals, Prepared for U.S. Department of the Interior, Bureau of Land Management, Eastern States, Jackson Field Office. N.p.: n.p., 2008.

Office of the Chatham R.R. Co., Raleigh, Jan. 21st, 1863, To His Excellency, Z.B. Vance, Governor and President of the Board of Internal Improvements. Ordered to be Printed. Doc. No. 21. W.W. Holden.

Oil-Bearing Shales of Deep River Valley, by Frank C. Vilbrandt, Industrial Chemist, University of North Carolina. Department of Conservation and Development, Wade H. Phillips, Director. Economic Paper No. 59. Raleigh: Edwards & Broughton Co., 1927.

Olmstead, Denison. *Report on the Geology of North Carolina, Conducted under the Direction of the Board of Agriculture. By Denison Olmstead. Part I*. November, 1824.

_____. *Report on the Geology of North Carolina. Conducted under the Direction of the Board of Agriculture. Part II. By Denison Olmstead*, November, 1825.

Parker, Roy, Jr. *Cumberland County, a Brief History*. Raleigh: Division of Archives and History, 1990.

Pezzoni, J. Daniel. "Historic and Architectural Resources of Lee County, North Carolina, ca. 1800–1942." *National Register of Historic Places/Multiple Property Documentation Form*. Preservation Technologies, March 1993.

Philosophical Transactions of the Royal Society of London, 1816, pt. 1. From 1800 to 1815.

Plans and Progress of Internal Improvements in South Carolina, with Observations on the Advantages Resulting there from to the Agricultural and Commercial Interests of the State. Columbia, SC: n.p., 1820.

Powell, Benjamin H. *Philadelphia's First Fuel Crisis: Jacob Cist and the Developing Market for Pennsylvania Anthracite*. University Park: Pennsylvania State University Press, 1978.

Powell, Fred Wilbur. *The Bureau of Mines: Its History, Activities and Organization*. Institute for Government Research. Service Monographs of the United States Government, No. 3. New York: D. Appleton and Company, 1922.

_____. *The North Carolina Gazetteer*. Chapel Hill: University of North Carolina Press, 1968.

Proceedings. Annual Meeting of Stockholders. Held in Salisbury, August 25–26, 1859. Salisbury: Carolina Watchman, 1859.

Register of the Officers and Graduates of the U.S. Military Academy, at West Point, N.Y., from March 16, 1802, to January 1, 1850, Compiled by Captain George W. Cullum. New York: J.F. Trow, 1850.

Reid, Jeffrey C., and Kenneth B. Taylor. *Shale Gas Potential in Triassic Strata of the Deep River Basin, Lee and Chatham Counties, N.C., with Pipeline and Infrastructure Data*." North Carolina Geological Survey.

Renemund, John. *Geology of the Deep River Coal field, North Carolina, Geological Survey Professional Paper 246*. Washington: U.S. Government Printing Office, 1955.

Report of Mine Explosion. Farmville Mine, Carolina Coal Company, Coal Glen, North Carolina, May 27, 1925, by J.J. Forbes and C.W. Owings. Department of Commerce. U.S. Bureau of Mines. (Unpublished typescript).

Report of the Committee on Int. Improvements on the Cape Fear and Deep River Navigation Company. Raleigh: W.W. Holden, 1855.

Report of the Comptroller General to the General

Bibliography

Assembly of the State of South Carolina, November 1868/69. Columbia: John W. Denny, 1868.

Report of the President and Directors of the Cape Fear & Deep River Navigation Company to the General Assembly. Raleigh: W.W. Holden, 1854.

Report of the Secretary of the Navy, Communicating the Report of Officers Appointed by Him to Make the Examination of the Iron, Coal, and Timber of the Deep River Country, in the State of North Carolina, Required by a Resolution of the Senate. 35th Congress, 2d Session, Ex. Doc. No. 26.

Report of the Situation of the Cape Fear Navigation Company. Raleigh: Lawrence & Lemay, 1832.

Rice, G.S., and H.P. Greenwald. *Coal-dust Explosibility: Factors Indicated by Experimental Mine Investigations 1911 to 1929, U.S. Bureau of Mines Technical Paper 464.* Washington: U.S. Government Printing Office, 1929.

____, L.M. Jones, J.K. Clement, and W.L. Egy. *First Series of Coal-Dust Explosion Tests in the Experimental Mine. Bulletin 56. Department of the Interior. Bureau of Mines. Joseph A. Holmes, Director.* Washington: U.S. Government Printing Office, 1913.

Shapiro, Karen A. *The Battle Against Convict Labor in the Tennessee Coalfields, 1871–1896.* Chapel Hill: University of North Carolina Press, 1998.

Shaw, Robert E. *Canals for a Nation: The Canal Era in the United States, 1790–1860.* Lexington: University of Kentucky, 1990.

Sixteenth Annual Report of the United States Geological Survey, to the Secretary of the Interior, 1894–95. Charles D. Walcott, Director. In Four Parts. Part IV.—Mineral Resources of the United States, 1894. Nonmetallic Products. David T. Day, Chief of Division. Washington: Government Printing Office, 1895.

Sprunt, James. *Derelicts: an Account of Ships Lost at Sea in General Commercial Traffic and a Brief History of Blockade Runners Stranded Along the North Carolina Coast 1861–1865.* Wilmington, NC: n.p., 1920.

Standard Corporation Service, January-April 1918. New York: Standard Statistics Company, Inc., 1918.

Stucky, Jasper Leonides. *North Carolina: Its Geology and Mineral Resources.* Raleigh: Department of Conservation and Development, 1965.

Tanner, H.S. *A Description of the Canals and Rail Roads of the United States, Comprehending Notices of All the Works of Internal Improvement Throughout the Several States.* New York: T.R. Tanner & J. Disturnell, 1840.

Taylor, Richard Cowling. *Statistics of Coal: the Geographical and Geological Distribution of Minerable Combustibles or Fossil Fuel, Including, also, Notices and Localities of the Various Mineral Bituminous Substances, Employed in Arts and Manufactures, from Official Reports of the Great Coal-Producing Countries, the Respective Amounts of their Production, Consumption and Commercial Distribution, in All Parts of the World, Together with Their Prices, Tariffs, Duties and International Regulations. Accompanied by Nearly Four Hundred Statistical Tables, and Eleven Hundred Analyses of Mineral Combustibles, with Incidental Statements of the Statistics of Iron Manufactures, Derived from Authentic Authorities.* Philadelphia: J.W. Moore, 1848.

Textoris, Daniel A., and Eleanora I. Robbins. *Coal Resources of the Triassic Deep River Basin, North Carolina.* Department of the Interior. U.S. Geological Survey. N.p.:n.p., 1988.

Thompson, J.W. *United States Mining Statutes Annotated, Part I. Sections and Statutes Relating to Metalliferous and Coal Mining.* Washington: Government Printing Office, 1915.

Tillman, David A. *Wood as an Energy Resource.* New York: Academic Press, 1978.

Trotter, Joe William, Jr. *Coal, Class, and Color: Blacks in Southern West Virginia, 1915–32.* Urbana: University of Illinois Press, 1990.

Twenty-second Annual Report of the United States Geological Survey to the Secretary of the Interior, 1900–1901, in Four Parts, Part III—Coal, Oil, Cement. Washington: Government Printing Office, 1902.

United States. *Nonpopulation Census Schedules for North Carolina, 1850–1880, Mortality and Manufacturing, 1860, Schedule 5, Products of Industry in Western Division in the County of Chatham.*

Vilbrandt, Frank C. *Oil-Bearing Shales of Deep River Valley. Department of Conservation and Development, Wade H. Phillips, Director. Economic Paper No. 59.* Raleigh: Edwards & Broughton Co., 1927.

Wakefield, Manville B. *Coal Boats to Tidewater: the Story of the Delaware & Hudson Canal.* Fleischmanns, NY: Purple Mountain Press, 1971.

Watson, Alan D. *Internal Improvements in Antebellum North Carolina.* Raleigh: Office of Archives and History: North Carolina Department of Cultural Resources, 2002.

Whisonant, Robert C. *Arming the Confederacy: How Virginia's Minerals Forged the Rebel War Machine.* Switzerland: Springer International Publishing AG, 2015.

Wholesale Prices, 1890 to 1912, U.S. Department of Labor, Bureau of Labor Statistics, Wholesale Prices Series: No. 1. Washington: Government Printing Office, 1913.

Who's Who in America: A Biographical Dictionary of Notable Living Men and Women of the United States, 1908–1909. Edited by Albert Nelson Marquis. Chicago: A.N. Marquis & Company, c. 1908.

Wood, Gordon H., Jr., Thomas M. Kehn, M. Devereux Carter, and William C. Culbertson. *Coal Resources Classification System of the U.S. Geological Survey .Geological Survey Circular 891.* Washington: U.S. Government Printing Office, 1983.

Bibliography

Articles

"An Act to Establish a Nitre and Mining Bureau." *Public Laws of the Confederate States of America, Passed at the Third Session of the First Congress; 1863. Carefully Collated with the Originals at Richmond.* Edited by James M. Matthews. Richmond: R.M. Smith, 1863.

"A.J. Derbyshire." *Biographies of Successful Philadelphia Merchants.* Philadelphia: James K. Simon, 1864.

Anderson, Mrs. John.H. "Confederate Arsenal at Fayetteville, N.C." *Confederate Veteran,* Vol. XXXVI, No. 6 (June 1928). Reprint. *The Confederate Veteran Magazine,* Volume XXXVI, January 1928—December 1928. Wilmington, NC: Broadfoot Publishing Company, 1988.

Ball, Donald B. "The Paper Mills in the Confederate South: Industrial Archaeology of a Forgotten Industry." *Ohio Valley Historical Archaeology,* Vol. 17 (2002).

Binder, Frederick M. "Pennsylvania Coal and the Beginnings of American Steam Navigation." *Pennsylvania Magazine of History and Biography,* 83 (1959).

Camman, F.W. "Charles Thomas Jackson." *The History of Science in the United States, Edited by Marc Rothenberg.* New York: Garland Publishing Company, 2001.

"Charles Thomas Jackson." *The History of Science in the United States,* edited by Marc Rothenberg. New York: Garland Publishing Company, 2001.

Clingman, Thomas Lanier. "North Carolina—Her Wealth, Resources, and History." *DeBow's Review and Industrial Resources, Statistics, etc. Devoted to Commerce, Agriculture, Manufactures, Internal Improvements, Political Economy, Education, General Literature, etc.* Edited by J.D.B. DeBow, Volume XXV. New Orleans and Washington City: N.p., 1858.

Drake, Francis S. "Walter Rogers Johnson." *Dictionary of American Biography, Including Men of the Time; Containing Nearly Ten Thousand Notices of Persons of Both Sexes, of Native and Foreign Birth, Who Have Been Remarkable, or Prominently Connected with the Arts, Sciences, Literature, Politics, or History, of the American Continent. Giving Also the Pronunciation of Many of the Foreign and Peculiar American Names, a Key to the Assumed Names of Writers, and a Supplement.* Boston: Houghton, Osgood & Company, 1879.

"Elisha Mitchell." *Dictionary of American Biography,* edited by Dumas Malone, Volume 13. New York: Charles Scribner's Sons, c. 1934.

Feiss, P. Geoffrey. "Mineral Resources of the Carolinas." J. Wright Horton and Victor A. Zullo, eds. *The Geology of the Carolinas: Carolina Geological Society, Fiftieth Anniversary Volume.* Knoxville: University of Tennessee Press, 1991.

Fowler, Malcolm. "Iron Mining in Upper Cape Fear: 3. The Blast Furnace at Buckhorn." *The State: A Weekly Survey of North Carolina,* April 26, 1941.

Hoyt, William Henry, ed. "Mr. Murphey's Report to the Legislature of North Carolina on Inland Navigation, December, 1816." *The Papers of Archibald D. Murphey,* edited by William Henry Hoyt, A.M., Volume II. Raleigh: E.M. Uzzell & Co., 1914.

"Internal Improvement: Extract of a Letter from a Member of the North-Carolina Catawba Navigation Company, to a Gentleman in Camden, S.C." *American Farmer: Rural Economy, Internal Improvements, News, Prices Current,* Vol. 1, No. 20, August 13, 1819.

"Jasper Leonidas Stuckey." *Dictionary of North Carolina Biography,* Volume 5, P-S. Edited by William S. Powell. Chapel Hill: University of North Carolina Press, 1994.

Lawrence, R.C. "The Deep River Coalfields." *The State Magazine.* May 24, 1941.

Maglenn, James. "The Steamer Ad-Vance." *Histories of the Several Regiments and Battalions from North Carolina in the Great War 1861–65, Written by Members of the Respective Commands.* Edited by Walter Clark, Vol. V, with Index, Published by the State. Goldsboro: Nash Brothers, 1901.

McClure, William. "Observations of the Geology of the United States." *American Philosophical Society, Trans.* 1809. Ser. I, v. 6.

"Mr. Murphey's Report to the Legislature of North Carolina on Inland Navigation, December, 1816." *The Papers of Archibald D. Murphey,* Volume II. Edited by William Henry Hoyt, A.M. Raleigh: E.M. Uzzell & Co., 1914.

"Mr. Williams's Lecture on the Coal Formation of North Carolina." *The Greensborough Patriot.* November 22, 1851.

Olmstead, Denison. "Red Sand Stone Formation of North Carolina, Extract of Letter from Professor D. Olmstead, of the College at Chapel Hill, North Carolina, dated Feb. 16, 1820." *American Journal of Science,* Vol. 2, 1820.

Olsen, Paul E., Albert J. Froelich, David L. Daniels, Joseph P. Smooth, and Pamela J.W. Gore. "Rift Basins of Early Mesozoic Age." *The Geology of the Carolinas: Carolina Geological Society Fiftieth Anniversary Volume,* edited by J. Wright Horton, Jr., and Victor A. Zullo. Knoxville: University of Tennessee Press, 1991.

Parks, B.C. "Petrography of Cumnock Coal." In Toenges, et al. *Coal Deposits in the Deep River Field, Chatham, Lee, and Moore Counties, N.C. Bulletin 515, Bureau of Mines.* Washington: Government Printing Office, 1952.

Powell, William S., ed. "Ebenezer Emmons." *Dictionary of North Carolina Biography.* Chapel Hill: University of North Carolina Press, 1986.

Reinemund, John. "Future Coal Expectancy in North

Bibliography

Carolina." *Conservation and Development in North Carolina.* Vol. 1. Compiled by C.S. Green. Raleigh: North Carolina Dept. of Conservation and Development.

"Report of the President of the Cheraw and Coalfields Railroad Company, Office of the President of the Cheraw and Coalfields Railroad Company, Society Hill, S.C. November 17, 1868." *Report of the Comptroller General to the General Assembly of the State of South Carolina, November 1868/69.* Columbia: John W. Denny, 1868.

Ross, John R. "The North Carolina Geological Survey, 1891–1920." *Forest History,* Vol. 16, No. 4 (January 1973).

Smith, Michael S. "The Conflict Between 'Practical Utility' and Geology: Denison Olmstead, Elisha Mitchell and the 1823 to 1828 Geologic Surveys of North Carolina." *Southeastern Geology,* Vol. 38, No. 5 (April 1999).

Stone, R.W. "Coal and Lignite: Coal on Dan River, North Carolina." In Marius R. Campbell, *Contributions to Economic Geology (Short Papers and Preliminary Reports), 1910. Part II. Mineral Fuels.* Department of the Interior. United States Geological Survey, Bulletin 471. Washington: Government Printing Office, 1912.

____. "Coal on Dan River North Carolina." *Contributions to Economic Geology, 1910, Part II, Mineral Fuels—Coal on Dan River, North Carolina,* Bulletin 471-B.

Sydnor, Charles S. "State Geological Surveys in the Old South." *American Studies in Honor of William Kenneth Boyd.* By Members of the Americana Club of Duke University. Edited by David Kelly Jackson. Durham, NC: Duke University Press, 1940.

Toenges, Albert L., Louis Turnbell, Joseph J. Shields, and Wilbur A. Haley. "Investigation of Field and Estimated Reserves of Coal." *Coal Deposits in the Deep River Field, Chatham, Lee, and Moore Counties, N.C. Bulletin 515, Bureau of Mines.* Washington: Government Printing Office, 1952.

Tracy, L.D. "Pulverized Coal Is Dangerous on the Surface as Well as Underground; Precautions to Be Taken in Handling It." *Coal Age.* Vol. 22, No. 5 (August 3, 1922).

Tuttle, E.G. "The Deep River Coalfield of North Carolina and the Egypt Coal Company's Plant." *The Engineering and Mining Journal,* Volume LVIII (November 10, 1894). New York: Scientific Publishing Company, 1894.

Internet Sources

"Charles Beatty Mallett." Accessed 3 May 2015. ncpedia.org/biography/mallet-charles-beatty.

"History of Mine Safety and Health Legislation." Accessed 23 May 2015. www.msha.gov/MSHA INFO/MSHAINF2.HTM.

Map: Coal-Bearing Areas of the United States. National Mining Association. Accessed 15 July 2015. www.nma.org/pdf/c_bearing_areas.pdf.

"Norfolk Southern Railway (1942–82)." Accessed 19 May 2015. en.wikipedia.org/wiki/Norfolk_Southern_Railway.

Olmstead, Denison. Accessed 16 February 2015. Ncpedia.org/biography/olmstead-denison.

"Thomas Lanier Clingman." Accessed 27 February 2015. http://ncpedia.org/biography/clingman-thomas-lanier.

"Washington Caruthers Kerr."Accessed 29 March 2015. http://ncpedia.org/biography/kerr-washington-caruthers.

"Western Railroad Company." Accessed 3 May 2015. www.historync.org/railroad-WRR.htm.

"William H. Keating." Accessed 26 February 2015. http://en.wikipedia.org/wiki/William_H._Keating.

Index

Numbers in ***bold italics*** indicate pages with illustrations.

Abernathy, R.F. 152
African-Americans ***130***, 178n1, 206n51
Albemarle Press 129–130
Alexander J. Derbyshire & Co. 66
Altoona and Phillipsburg Connecting Railway 8
American Legion 130
American Red Cross 136
American Retail Coal Dealers 153
American Revolution 12, 155, 179n2, 203n35; Continental Army 155
"Anaconda" Plan 55
Anderson, George 133
Anderson, W.E. 79
Andrews, Atlas 167
Andrews, Thomas 186n61
anthracite coal 7–10, 22–23, 40, 55–56, ***146***, 147–148, ***163***, 184n38, 192n3, 192n4, 195n46, 214n3; *see also* Deep River coal
Anthracite Mining Company 143
Arkansas 146
Arter, Glenn 153
Asheboro, North Carolina 85
Asheville, North Carolina 146
Asheville News 39, 188n20
Atkins, John 64
Atlantic and East Carolina Railway 149
Austin, W.L. 112
Avondale Coal Mine Disaster, 1869 (Plymouth, Pennsylvania) 100

Badger, George E. 190n57
Bagby, R.T. 125
Bailey, John Henry 145
Baldwin, Will 167
Baltimore, Maryland 48
Bath, South Carolina 54
Bath Paper Mill 54
Battle, Kent Plummer 52, 193n21, 193n22

Bayles, J.C. 201n12
Beaufort, North Carolina 45, 63
Beckham, C.H. 88, 203n41
Best, K.W. 63
Bible, D.P. 117
Bidlack, R.W. 153
Big Buffalo Creek 53
Big Stone Gap, Virginia 122
Biggs, Asa 194n30
Birdsong, John 179n4
Bishop, Julian T. 148
bituminous coal 7–10, 44, 195n46, 203n37; *see also* Deep River coal
Blacksburg, Virginia 49
blacksmiths 15, 20, 55, 58, 61, 74, 77, 92, 185n54, 186n57
Blair, Walter, Sheriff 129
Bledsoe, Walter A. ***150***, 215n21; Walter A. Bledsoe Company 150
blockade runners 54, 56, ***57***, 192n4, 194n34, 195n46; C.S. *Advance* 55–56, ***57***, 195n46; C.S. *Florida* 56; C.S. *Florie* 54; C.S. *Robert E. Lee* 195n46; C.S. *Tallahassee* 56; *see also* Vance, Zebulon B.
Board of Conservation and Development 153
Bowley, Gen. A.J. 129, 211n5
Branson and Farrar's North Carolina Business Directory 58, 62, 64, 67, 83, 176–177, 201n18
Branson's North Carolina Agricultural Almanac 201n6
Brassert, Herman A. 149
Brewer, W.W. 120
Broderick, S.E. 202n25
Brodie, William L. 57
Brooks, Harris 36
Brotherhood of Locomotive Engineers 88
Broughton, Joseph Melville 149
Brown and DeRossette 176
Browne, James 195n39
Browne, Peter 181n28

Bruceton, Pennsylvania 106
Brunswick County, NC 201n12
Brunswick, Western and Southern Rail Road 201n12
Bryson, Herman J. 142, 184n38, 213n27
Buckhill, North Carolina 66, 196n56
Buckhorn, North Carolina 45, 66; coal mine 176–177
Burr, Maj. W.A. 129
Butler, Bion H. 121, 210n28, 210n39
Butler, Howard N. 121, 127–129, 131, 133–134, 147–148, 212n7, 212n8
Butler, Reese W. 57

Caldwell, D.F. 203n34
Caldwell, Joseph 189n37
Caldwell & Company 176
Campbell, Marius R. 122–123, 152–153, 178n4, 187n3; *Contributions to Economic Geology, (Short Papers and Preliminary Reports) 1910. Part II. Mineral Fuels. Dept. of the Interior. United States Geological Survey* 178n4
canals 2, 13–16, 18, 25, 42–48, 181n21, 182n10, 189n37, 189n44, 189n45, 190n47, 190n48, 191n65, 191n69; Buckhorn 45; Fayetteville 181n28; Lumber River Canal Company 16
Cant, George T. 92, 96, 165
Cape Fear and Deep River Navigation Company 23, 36, 43, 45, 186n67, 188n17, 190n48, 190n57, 191n60, 191n61, 191n62
Cape Fear and Yadkin Rail Road Company 42–43, 69, 72, 76, 80, 82, 86, 90, 189n41
Cape Fear and Yadkin Valley Railway 85–86, 90, 200n61, 201n7, 201n12, 202n26

225

Index

Cape Fear Harbor and Coal Company 201n12
Cape Fear Railroad 77
Cape Fear River 28, 32, 43–44, 46–47; improvements 22; locks and dams 47, 53, 189n41, 190n46, 190n57, 194n28
Cape Fear River Navigation Company 36, 181n21, 186n67, 190n47
Cape Hatteras 88
Capps Hill Gold Mining Company 37
Carbonton, North Carolina 85, 143, 153, 203n33, 211n52
Carolina Central Railroad 67
Carolina Coal and Coke Company 104
Carolina Coal By-Products Company 214n2
Carolina Coal Company 3, 11, 120–121, 124–125, 127, 129, 133–134, **136**, 142–145, 147–148, 153, 173, 206n34, 210n41, 210n51, 211n2, 211n4, 211n52, 213n1, 213n3, 213n7, 213n13, 214n6, 215n12; *see also* Coal Glen, North Carolina; Farmville Mine
Carolina Fuel and Transportation Corporation 148
Carolina Mine 122, 149–151, **151**, 153, 215n28
Carolina Power and Light Company 151–152
Carr, Elias 203n28
Carr, O.W. 202n25
Cash, Frank 211n57
Carthage, North Carolina 117–118, 177
Cary, North Carolina 52
Caswell, Richard 179n4
Central Railroad 174–175
Chalk Level, North Carolina 176–177
Chambers Coal Company 176–177
Chance, Henry Martyn 72–76, 77, 188n27, 200n62, 200n67, 201n1
Chapin, H.T. 164
charcoal 48, 180n19, 184n38
Charleston, South Carolina 45, 53, 58; gas works 175
Charleston, West Virginia 149
Charleston Mercury 194n30
Charlotte, North Carolina 58, 70, 120
Charlotte Chronicle 202n18
Charlotte Democrat 204n46
Charlotte Observer 85, 95, 118, 204n5, 204n7, 209n21, 210n28, 210n49, 213n1, 216n37
Chatfield, H.B. 148
Chatham Coal and Coke Company 115

Chatham Coal and Iron Company 112–113, 118 *see also* Meyers, H.K.
Chatham County, NC 14, 30, 37, 43–44, 48, 63, 70, 104, 117, 120, 153–154, 184n25, 189n46, 202n22; *see also* Deep River coal
Chatham Mining and Transportation Company 36
Chatham Ore Hill Company 193n11
Chatham Railroad Company 51–52, 62–63, 66, 191n66, 193n16, 193n21, 199n42
Chatham Record 68, **79**, 80, 82–83, 94, 113, 117–118, 120, 164, 178n3, 199n45, 201n10, 202n14, 202n17, 202n23, 204n54, 205n17, 205n21, 206n41, 206n45, 208n18, 208n20, 208n22, 209n12, 209n17, 209n21, 210n29vis, 210n31, 213n5
Chattanooga Railroad 174–175
Cheraw and Coalfield Railroad 46, 193n16
Cherry Coal Mine Disaster, 1909 (Cherry, Illinois) 106, 207n59, 209n17
Chesapeake and Ohio Canal 44, 179n7, 190n47, 190n48
Cincinnati, Ohio 146
Civil War 3, 36, 39, 44, 48–59 passim, 64, 69, 84, 155, 174, 185n48, 186n62, 188n27, 192n10, 193n21, 197n3, 210n32
Clegg, Nathaniel 58
Clegg, Thomas W. 43
Clegg, Dennis & Co. 64
Clegg, Downer & Co. 59
Clingman, Thomas Lanier 25, **26**, 184n36, 184n37
Clover Hill, Virginia 49, 55, 192n7
Coal Age 121
coal "breaker" 147
Coal Deposits in the Deep River Field, Chatham, Lee, and Moore Counties 152, 216n37
coal dust 98, 107, 133–134, 138, 140, 149, 204n54, 207n61, 212n17
Coal Glen, North Carolina 3, 11, 66, 125, 127, **128**, 129, **130**, **131**, 133, **134**, 135–136, 138–139, 142–143, 145, 149, 155–156, 168–171, 178n4, 206n34, 211n2, 211n4, 211n52, 212n12, 212n13, 213n3, 214n6, 214n8, 214n11, 214n12, 215n12; convict labor 143–144; name changed from Farmville 178n3; segregated communities **130**; *see also* Farmville Mine
Coal, Iron, and Oil; or the Practi-

cal Miner. A Plain and Popular Work on Our Mines and Mineral Resources, and a Text-book or Guide to Their Economical Development. With Numerous Maps and Engravings by Samuel Harries Daddow, and Benjamin Bannan 163
coal strike of 1949 215n25
coal tipple 147
coking coal 30, 32, 41, 55, 58–59, 83, 92, 103–104, 114–115, 122–123, 125, 134, 175, 184n38, 196n54
Colon, North Carolina 90, 113
Colorado 146
Columbia, South Carolina 53
Concord Standard (NC) 202n18
Confederate States of America 49–50, 53, 155, 193n11, 193n12, 193n21, 195n46, 201n82; adjutant and inspector general 57; arsenal 195n41, 196n56; Castle Thunder Prison 194n38; Danville Prison 58; Gosport Navy Yard 58; locomotive shops 55; navy 55–56, 196n56; navy yard 155, 196n56; Nitre and Mining Corps 50, 55–56, 58, 175, 192n10; paper mill 194n35; Raleigh Bayonet Factory 57; secretary of the navy 56
Connellsville, Pennsylvania 123
Connelly, John 109–110; death 110; mine boss of Cumnock Mine 109–110, 207n1
Continental Army *see* American Revolution
Contra Costa, California, coal mining 178n2
convict leasing 43
copper 29–30, 50, 62, 132, 194n38, 195n39; Clegg Copper Mine (Chatham Co., NC) 71
H.M.S. *Corinthia* 88
Cox, Alfred 165
Coxe, Tench 14, 180n15
Cross Creek, North Carolina 47
Cumberland Co., North Carolina 42, 117, 189n34, 189n43
Cumnock Coal Mining Company 103, 119–121, 125, 209n26, 210n41
Cumnock, North Carolina 86; coal mines 88–90, 92, 94, 96–97, 103–104, 109–111, **110**, 112–115, 117–119, 121–124, 127, 138, 142, 144, 147–148, 187n2, 203n34, 204n49, 209n24; 1895 explosion **92**, **93**, 94–99, 164–167, 204n6; German immigrants 117; lamps, disrepair of 110; 1900 explosion 109, **110**, 111–114, 213n1; production 90, 120;

226

Index

transcript of 1895 explosion 164–167

Dabney, Charles W. 73
Daddow-Bannon map of the Deep River coalfield 162, **163**
Daily Eagle 187*n*10
Daily Journal (Wilmington, NC) 62, 65–66, 70, 190*n*57, 198*n*27, 198*n*36, 198*n*37
Daily News 199*n*40
Daily Progress 194*n*39
Daily Review (Wilmington, NC) 79
Daily Standard (Raleigh, NC) 64, 66, 197*n*1, 198*n*22
Daily Times (Burlington, NC) 148, 152
Dan River 28; coal field 7, 9, 29, 65–66, 71, 157–158, 177, 185*n*52
Dan River Coalfield Railroad 44, 65–66, 157–158, 190*n*52
Dante, Virginia 122
Danville, Virginia 44, 158, 190*n*53; prison 57
Davidson County, NC 30
Davis, E.H. 165
Davis, Jefferson 56, 190*n*53
Davy, Humphrey 111, 200*n*80; mine safety lamp **111**
De Bow, James Dunwoody Brownson 183*n*17; *De Bow's Review* 26
Deep and Haw River Navigation Company 181*n*21
Deep River 2–3, **4**, 30–32, 34, 36, 39–48, 51–54, 57–58, 62–63, 65–66, 72, 74–75, 78, 90, 113, 117, 121, 124, 146, 153, 179*n*6, 182*n*9, 184*n*35, 185*n*56, 187*n*71, 189*n*46, 190*n*57, 193*n*20, 194*n*22, 194*n*24, 196*n*52, 197*n*63, 199*n*46, 203*n*34, 209*n*21, 210*n*32, 210*n*34, 210*n*39, 210*n*41, 211*n*4, 211*n*52, 211*n*55, 213*n*5; basin 7; coalfield see Deep River coal; copper production 30, Daddow-Brannon map **163**; geological surveys 2, 15–16, 18–35 passim, 20–22, 24, **27**, 29–31, 34, 37–38, 40, 42, 44–45, 62, 69–72, 74, 76, 77, 81, 99, 117–118, 122, 153, 157, 159–161, 178*n*4, 178*n*5, 181*n*30, 185*n*42, 181*n*30, 182*n*10, 183*n*16, 184*n*39, 185*n*40, 185*n*51, 186*n*59, 186*n*61, 187*n*3, 187*n*10, 187*n*16, 187*n*71, 188*n*33, 190*n*47, 190*n*54, 191*n*64, 194*n*27, 198*n*31, 198*n*36, 199*n*40, 200*n*62, 200*n*66, 202*n*24, 203*n*29, 210*n*32, 210*n*33, 216*n*1; iron ore deposits see iron ore; locks and dams 62; navigation of 16, 24, 32, 38, 45; North Carolina Geological and Economic Survey 116, 118, 120, 187*n*3, 195*n*46; North Carolina Geological Survey 29, 84, 106, 115–118, 120, 194*n*27, 154, 182*n*1, 182*n*2, 182*n*5, 183*n*16, 185*n*40, 195*n*46, 196*n*52, 199*n*53, 199*n*57, 203*n*31, 207*n*3, 207*n*59, 208*n*1, 209*n*7, 209*n*16, 209*n*23, 210*n*37, 210*n*41, 215*n*8, 215*n*28; railroad link to coal fields 50, 52; resources 31; sandstone deposits 31;transportation 23, 32; U.S. Geological surveys 88, 97, 104, 106, 115–116, 122, 152–153, 158, 172, 178*n*4, 179*n*5, 182*n*1, 182*n*2, 182*n*5, 182*n*10, 182*n*12, 183*n*14, 184*n*7, 184*n*37, 184*n*39, 185*n*40, 185*n*41, 185*n*42, 185*n*51, 203*n*33, 204*n*44, 205*n*15, 206*n*49, 216*n*37; *see also* Deep River coal; North Carolina Geological and Economic Survey
Deep River coal 2, 7, 12, 14, 20, 22–24, **25**, 26–29, 31; cross section **32**, 36–39, 41–43, 45–46, 48, 50–53, 55–57, 62, 65–66, 69–76, **77**, 78, 81, 83–85, 88, 90, 92, 103, 113, 117, 121–127, 142, 144–145, 148, 150–163 passim, **159**, **163**, 175–177, 178*n*2, 178*n*3, 179*n*2, 180*n*19, 182*n*9, 182*n*11, 183*n*17, 183*n*19, 184*n*25, 184*n*33, 184*n*34, 185*n*52, 185*n*59, 185*n*61, 186*n*57, 186*n*62, 186*n*69, 187*n*12, 188*n*27, 188*n*32, 188*n*35, 189*n*35, 190*n*57, 191*n*59, 193*n*11, 193*n*12, 193*n*21, 194*n*30, 195*n*40, 196*n*54, 197*n*1, 197*n*10, 197*n*12, 198*n*18, 198*n*36, 199*n*51, 200*n*62, 209*n*21, 210*n*39, 210*n*41, 210*n*46, 210*n*51, 213*n*27, 214*n*6, 215*n*4, 215*n*8, 215*n*19, 215*n*28, 215*n*29, 216*n*33, 216*n*37; analysis 29; anthracite 22–24, 40, **148**; bituminous 8, 55, 71; capital 28; Chatham County 14, 63, 70, 86; Civil War 52; comparative analysis 10; controversy over use in blockade runners 195*n*46; faulted seams and veins 74–76, 118, 152, 200*n*70, 200*n*74, 201*n*12, 201*n*87; gas extraction 199*n*54; labor market 28; locations 29; mines 22; oil shale mining 117, 123, 126, 143, 154 199*n*54, 209*n*14, 211*n*60, 213*n*4, 216*n*3, 216*n*39; production 22, 36, 39, 49, 51, 55, 69, 77, 81, 88–90, 97, 104, 112–113, 115–126 passim, 141–144, 149, 152–153, 155, 172–173, 214*n*13; quality 28–29; railroads 30, semi-anthracite 8, 190*n*48; surveys 153; tonnage 172–173; transportation 23, 28; trap dikes 71, 74–76, 200*n*72; uses 12; value 22, **116**; *see also* Deep River
Deep River Coal and Iron Company 40–41, 48, 53, 174, 188*n*27, 192*n*2
Deep River Coal Company **148**, 175, 211*n*58
Deep River Mining and Manufacturing Company 23–24, 36–39, 184*n*25, 184*n*34
Deep River Mining and Transportation Company 24, 37, 186*n*69
Deep River Navigation Company 45, 191*n*60
Delaware and Hudson Canal 180*n*7
Derbyshire, Alexander J. 66
Devine, Arthur F. 153
Dewees, Francis Percival 68–69, 199*n*48
Dix, W.R. 58
Douglass, E.A. 45, 190*n*57
Duane, William J. 181*n*21
Duke Energy 154, 216*n*37
Duke Power Company see Duke Energy
Durham, North Carolina 113
Durham & Charlotte Railway Co., 118, 202*n*21
Durham Sun 178*n*1
Dye, John 104
Dye, W.W. 177

Eagle (Fayetteville, NC) 66, 198*n*38
East Tennessee and Georgia Rail Road 54
Edgecombe County, NC 180*n*9
Edison, Thomas Alva **140**
Egypt, North Carolina 29, 40, 43, 46, 48, 53–55, 58, **61**, 62, 68, 75, 81, 85; coal mine 29, 31, **33**, **34**, 36, 54–55, 58, 61–66, 68, 72–73, 75–77, 80–82, 85, 88, 97, **119**, 160, 165, 174, 176 187*n*2, 188*n*2, 188*n*27, 191*n*62, 192*n*10, 194*n*30, 194*n*38, 195*n*39, 196*n*55, 197*n*9, 198*n*18, 201*n*7, 202*n*18, 203*n*31, 203*n*37, 204*n*49; railroad line to mine 194*n*28; store **119**
Egypt Coal Company 39, 54, 62, 66, 77, 79–81, 83–84, 97, 149,

227

Index

176, 194*n*30, 198*n*39, 201*n*87, 203*n*34
Egypt Improvement Company 81, **119**
Egypt Mining Company 64
Egypt Railway Company 83, 202*n*19
electric cap lamp *see* Edison, Thomas Alva
Elkins, John 147
Emmons, Ebenezer 26–27, **27**, 28–30, 30, **30**; controversy with Elisha Mitchell 31, 34, 38, 45, 51, 53, 65, 69–72, 157, 183*n*18, 184*n*39, 185*n*40, 185*n*41, 185*n*42, 185*n*46, 185*n*51, 186*n*61
Endor Iron Works *see* iron ore
Engineering and Mining Journal 76
Enterprise Factory 117
Erie Canal 180*n*7
Erskine Ramsay Coal Company 121, 125, 127, 134, 145, 149, 209*n*51, 214*n*6, 216*n*37
Evans, Peter G. 36, 86, 186*n*61
Evening Review (Wilmington, NC) 67

Fair Haven, North Carolina 82
Faison, W.E. 111
Farmer and Mechanic (Raleigh, NC) 201*n*1
Farmersville, North Carolina *see* Farmville, North Carolina
Farmville, North Carolina 1, 28, 72–76, 78–79, 120, 136, 152, 173, 178*n*3, 188*n*27, 197*n*61, 200*n*77, 201*n*1; *see also* Farmville Mine
Farmville Coal Company 59, 175–176, 197*n*61
Farmville Mine 3, 28, 72–73, 78, 125, 127–137 passim, **128**, 138–139, **140**, 147, 152, 168–171, 206*n*34, 211*n*1, 211*n*4, 212*n*13, 212*n*14, 212*n*16, 212*n*17, 213*n*13, 221; 1925 explosion 3, 125, 28–134, **135**, 136, 142, 147, 178*Intro.n*1, 200*n*77, 206*n*34, 211*n*2, 211*n*4, 211*n*5, 211*n*57, 212*n*12, 212*n*21, 213*n*1, 213*n*13, 215*n*12, 221; relief fund 134; Report of Mine Explosion, Farmville Mine 212*n*14; shot firing 134; victims 168–171; *see also* Carolina Coal Company; Coal Glen, North Carolina; Farmville Coal Company
Fayetteville, North Carolina 31, 34, 38, 42–44, 46, 117–118; arsenal 55, 189*n*35, 195*n*41; canal 181*n*28, 189*n*41
Fayetteville and Coalfields Railroad 53

Fayetteville Examiner 68
Fayetteville Intelligencer 194*n*38
Fayetteville News 61, 197*n*5
Fayetteville Observer 24, 37, 43, 47, 184*n*34, 188*n*33, 193*n*11, 194*n*31, 194*n*38, 195*n*44, 196*n*57
Fayetteville Railroad Company 42
Fayetteville Standard 55
Federal Mining Bureau *see* U.S. Bureau of Mines
fire boss 111
Fisher, Charles F. 45
Fishing Creek, North Carolina 180*n*9
C.S. *Florida* 56, 192*n*4
C.S. *Florie* 54, 56, 175
Fooshee Place 72; mines 176
Forbes, J.J. 127, 131–132, 212*n*14
Fort Bragg, North Carolina 129
Foundry and Machine Shop 64
foundry, national 185*n*56
Fouse, F.S. 92
Fooshee 51, 118, 177
Frazier & Company 176
Frick, Henry Clay 124
Frog Town, Chatham Co., NC **130**
Fry, J.W. 86
Fulton, Hamilton 181*n*27

Gammell, William 39–40
Ganter, John A. 167
Gardner, O. Max 143
Gardner Place 72
Gaston and Raleigh Railroad 187*n*72
Genth, Frederick Augustus 69, 70–71, **73**
Geology of the Deep River Coal Field, North Carolina 152
Georgia Rail Road 54
Germanton, North Carolina 65, 158
Gilmer, Alexander 42, 185*n*56
Gilmer, John T. 42, 188*n*35, 189*n*34
Gilmer, North Carolina 176
Gilmore, John Alexander 185*n*56
Glendon, North Carolina 103, 118
Glendon and Gulf Mining and Manufacturing Company 82, 104
Glendon and Gulf Railroad 82, **83**, 202*n*21
Goddard, C.W. 36
gold mining 3, 19, 22, 29, 62, 105, 178*n*2, 185*n*40, 195*n*4
Goldsboro Daily Argus 117
Goldston, North Carolina 48, 192*n*2
Goodyear, W.A. 178*n*2
Gorgas, Josiah 55

Gorgas Mining and Manufacturing Company 57, 196*n*50
Gosport Navy Yard 58
Governor's Creek Coal and Iron Manufacturing and Transportation Company 32, 39
Governor's Creek Coal Company 65, 189*n*
Governor's Creek Steam Transportation and Mining Company 36, 188*n*21, 188*n*22, 198*n*39
Graf, E.H. 131
Graham, William A. 191*n*65
Greene, Nathanael 179*n*6
U.S.S. *Greenland* 56
Greensboro, North Carolina 44–45, 82, 113, 157
Greensboro Coal and Mining Company 83, 202*n*25
Greensboro Daily News 121
Greensboro Gas Company 92
Greensboro Patriot (NC) 38, 66, 88–89, 186*n*61, 187*n*10, 193*n*12, 198*n*39, 204*n*47
Greensboro Telegram 109
Greensborough Patriot see *Greensboro Patriot*
Grist, Frank D. 133, 142, 213*ch*13*n*1
Groves, George 131
A Guide to Capitalists and Emigrants 63, **64**
Guilford County, NC 117
Gulf, North Carolina 1, 3, 7, 16, 28, 30, 36–37, 39, 48, 51, 62, 66, 68–69, 72–77, 81–82, 85, 104, 109, 144, 147–148, 169–170, 173, 176–177, 186*n*63, 193*n*16, 197*n*63, 199*n*40, 201*n*86, 202*n*22, 206*n*50, 210*n*51, 211*n*58, 213*n*5; depot **85**
Gulf and Deep River Iron Manufacturing Company 36
Gulf and Glendon Mining and Manufacturing Company 85
Gulf Coal and Iron Company 176
Gulf Coal Mining Company 36, 39, 187*n*6
Guthrie, William A. 118–119
Gwynn, Walter 189*n*47

H.A. Brassert and Co. 149
Hairston, C. 177
Hale, Edward Joseph 71–72
Hale, Peter Mallett 71, 199*n*59; *In the Coal and Iron Counties of North Carolina* 72
Haley, Wilbur A. 152
Halifax County, NC 180*n*9
Hallison, North Carolina 117
Hamlet, North Carolina 113
Hancock's Mill, North Carolina 117

228

Index

Haninton, C. 104
Harpers Creek Crossroads, North Carolina 82–83, 117, 203n34
Harrell, J.A. 78–80
Harris, Clay 167
Hart, G.B. 166
Hart, J.D. 109–110
Haughton, John Hooker 36, 38–39, 187n6, 187n16, 191n69
Haughton, Lawrence J. 36, 38–39, 176
Haughton, Walter 167
Haw River 13–14 16, 46, 66, 181n21, 190n46, 191n65, 193n20
Hayes, James H.M. 112
Haywood, North Carolina 186n67, 189n45, 193n21
Haywood and Pittsboro Plank Road Company 185n42
Heck, Jonathan McGee 58, 196n51
Henderson, Thomas, Jr. 180n19
Hennessy, James A. 79
Henszey, Samuel A. 82–84, 86, 90, 97, 103–104, 117, 202n19, 203n34, 203n37, 204n34, 206n44; *see also* Langdon-Henszey Coal Company
Henszey, W.P. 208n20
Hewitt, Henry B. 36
Hickory Press and Carolinian 81
High Point, North Carolina 34, 117
High Point Enterprise 195n46
Hill, W.H. 134
Hill, William 104, 140–141
Hillsboro, North Carolina 36, 39
Hillsboro Coal Mining and Transportation Company 36
Hillsboro Mining and Transportation Company 39
Hines, Alex 109, 133, 142
Hinsdale, John W. 97
Hodges, Luther 153
Holden, W.W. 63
Holland, Lucian 167
Holmes, Joseph Austin 65, 84, *84*, 106–107, 157, 174–175, 197n63, 198n31, 202n24, 207n59, 209n7, 213ch13n1
Honeycutt, L.A. 214n13
Horne, W.T. 53, 194n27
Hornesville, North Carolina 28
Hornville Coal Company 176
Horton mine 28
Hot-Shot battery 131
Hunt, T.G. 131

internal improvements 2, 13–20, 22–26, 28, 34, 38, 40, 42–47, 52, 60, 63, 69, 72, 81, 86, *119*, 179n7, 180n13, 180n20, 180n21, 181n28, 181n31, 181n32, 182n10, 183n17, 183n19, 184n37, 186n67, 186n69, 187n6, 189n41, 189n44, 189n45, 190n48, 190n57, 191n61, 191n64, 191n65, 191n66, 191n67, 191n69, 193n21, 195n46, 197n3; Board of Internal Improvements 20; Committee on Internal Improvements 45; Egypt Improvement Company 81, *119*, 209n27
iron ore 3, 12, 26–31, 41, 66, 68, 82, 90, 149, 179n1, 184n38, 191n69, 192n2, 192n7; black band 90; Endor Iron Works 6, 66, 69, 196n56, 197n63; Iron Act of 1750 179n1; manufacturing 8, 39–40, 42, 51, 64, 179n1, 184n38, 185n56, 186n62, 186n63, 191n60, 191n69, 192n2, 193n11; Mecklenburg Iron Works 80–81; Novelty Iron Works 186n57; sponge iron 149; Wilcox Iron Works 197n9; *see also* Tredegar Iron Works

Jackson, Charles Thomas 37, 45; geological map of the Deep River coalfield 159, *160*, *161*, 186n61
Jacksonville, North Carolina 149
Jenkins, Will 167
John, A.H. 97
John, R.G. 143
John H. Haughton (ship) 46
Johnson, Walter Rogers 22–24, *25*, 45, 179n8, 183n17, 183n23, 185n54, 186n69, 188n18
Jones, A.J. 118
Jones, W.A. 121
Jonesboro Leader 202n22
Jordan, Ralph 143

Keating, William H. 182n4
Kelly, Marvin 125
Kennedy, John H. *119*, 142, 209n27
Kentucky 80, 117, 146, 150, 199n48; Geological Survey *8*
Kerosene Oil Works 199n54
Kerr, Washington Carruthers 62, 69–70, *71*, 83, 197n10, 199n51
Kerr, W.C. *see* Kerr, Washington Carruthers
Kessler, John 64
Kimball, Kent W. 122–123, 152–153, 187n3
Kohinoor Coal and Iron Company 85, 104, 177
Kyle, W.R. 86

Lacy, Benjamin Rice 109, 203n37, 208n5
Lacy, B.R. *see* Lacy, Benjamin Rice
Lagrange, North Carolina 36, 191n62
Lakeview Park, North Carolina *134*
Lamar, C.A.L. 54
Lambert, Jesse 167
Lambeth, Gaston 167
lamp house 100–101, *101*
Lane, F.A. 121
Lane, W.W. 191n62
Langdon, Samuel P. 84–86, 92, 97, 104, 203n38, 205n12, 205n13, 205n16, 206n44 206n45; *see also* Langdon-Henszey Coal Company
Langdon, North Carolina 85
Langdon-Henszey Coal Company 85–86, *87*, 88–89, 92, 96–97, 103–104, 203n34, 203n38, 205n9, 205n13, 205n15; *see also* Langdon, Samuel P.
Lagrange Mining and Transportation Company 36
Latrobe, Benjamin Harvey 180n15
Lawrence (steam tug) 81
Leaksville, North Carolina 190n53; mines 176
Lee County, NC 58, 117, *119*, 154; German immigrants 117
Leftwich, A.H. 79
Lehigh Canal *17*
Lennig, John 118
Lewis, John L. 151
lignite 70
Litchfield Foundry 89; Beach's balanced side valve 89; fast motion machine 89
Little, George 63
Lockville, North Carolina 46–47, 52, 58, 66, 193n20
Lockville Foundry and Machine Shop 198n22
Lockville Mining and Manufacturing Co. 59, 175, 195n39, 197n61
London, H.A. 94–95
Long, D.B. 146–147

MacGregor, George C. 113–114
MacNeill, Ben Dixon 128–130, *131*, 211n4, 212n7, 212n11, 212n13
Madison County, NC 65
Maglenn, James 55
Maher, Rody 83, 202n25
Mallett, Charles Beatty 52–54, 194n23, 194n30, 195n39; *see also* Mallett and Brown Coal Company
Mallett, Peter 203n35
Mallett and Brown Coal Com-

229

Index

pany 53–54; records 174–175, 194n30, 196n55, 197n61
Mallory, Stephen R. 56, 195n46
Manning, John 50–51, 193n11
Manufacturers Record of Baltimore 88
Marcy, William L. 27
Marie (steam tug) 81
Mason, W.B., Sr. **136**
Massachusetts 146
Matthews, W.D. 203n31
Matthews Township, North Carolina 104
McCarthy, James 110
McClanahan, Spencer "Spence" 28, 31
McClane, William H. 34, 38, 45, 187n16, 191n60
McClure, William 182n3
McDowell County, VA 204n43
McGehee, Montford 72
McIntyre, Sim 111
McIver, Alexander 176
McIver, Ed 145
McIver, Evander 53, 176; McIver's Depot 53
McIver, Jack 167
McIver, John McMillan 147, 215n6
McIver Coal Mining Company 36, 176
McIver's Depot *see* Evander McIver
McLean, Angus W., Governor 127, 223n13
McNath, George M. 109
McNeill, A.H. 177
McNeill, W.H. 121
McQueen, Rev. Angus 121
McQueen, John R. 121, 124, 214n6
McRae, Donald 58
McRae, John 58
Mecklenburg County, NC 3, 16, 37, 69
Mecklenburg Iron Works *see* iron ore
Mecklenburg Salt Company 69
Mendenhall, Jones, and Gardner 58
Mercer County, VA 204n43
Metts, John V.B. 128
Meyers, H.K. 112
Midlothian, Virginia 49, 192n7
Mills, Joe 96
Milton, North Carolina 44
mine gases 100, 107, 110–112, 133, 207n61; afterdamp 112; carbon dioxide 138; carbon monoxide 138; firedamp 110, 138–141; hydrogen 138; hydrogen sulfide 138; methane 139, 204n6, 204n8; nitrogen 139; nitrogen oxides 139; oxygen 139
Mine Hill Railroad 66

mine safety 98–100, 149; Federal Mining Bureau 98; 1897 legislation 99–103; Mine Safety and Health Administration 211n2; Mine Safety Appliances 207n60; reporting *102*
mine ventilation 40, 70, **98**, 138, 165–166; Fair Haven 82; Hancock's Mill 44
The Minerals and Mineral Localities of North Carolina 71
Miner's Journal 163
Mining Journal and Journal of Geology, Mineralogy, Metallurgy, Chemistry, and the Arts 40
mining legislation 98–100, 104, 133, 205n22, 206n26
Minter, James H. 153
Mitchell, Elisha 21, 23, **25**, 28–29, 31, 69, 71, 182n1, 183n15, 183n16, 184n37, 185n50, 185n53, 186n59, 186n60
Moncure, North Carolina 94, 152, 204n1
Monongah Mine Disaster, 1907 (Monongah, West Virginia) 106, 145
Monroe, George 167
Monroe, John 167
Montgomery Vidette 201n8
Moore County, NC 27, 40, 43–44, 70, 85, 115, 117, 125, 146–148; coal deposits 117, 214n3; strip mining 153
Morehead, J. Turner 203n25
Morrell, Maj. William H. 186n67
Morris, Daniel 167
Morris, Elwood 52
Morris, Henry 167
Mount Diablo, California coalfield 178n2
Mount Vernon Springs, North Carolina 197n9
MSHA *see* Mine Safety and Health Administration
Murchison 51, 72; mines 176–177
Murchison, Sylvester 145, 211n57
Murphey, Archibald 16, 181n22

Nashville and Chattanooga Railroad 54
Neuse River 28
New Bern, North Carolina 63
New Mexico 146
New York, NY 41, 48
New York Gas Coal Company 97, 205n13
New York Times 40
Newcastle coal 41
Newport News, Virginia 86
Norfolk, Virginia 30, 45, 58, 63, 86

Norfolk and Western Railway 92, 201n12
Norfolk Southern Railway 119–121, 124, 127, 146, 202n19
North Carolina 3, 5; Board of Agriculture 21–22, 72, 74, 89, 175, 183n14, 188n27, 200n62, 200n77, 204n38, 219, 221; brick and clay manufacturing 216n3; Bureau of Labor and Printing 105, 109, 133, 142; Bureau of Labor Statistics 99, 104; commissioner, 145; Commissioner of Labor Statistics 99, 109, 212n21; convict leasing in coal mines 143–144, 213ch13n5, 214n11; Department of Highway Commission 147; 1897 mining law 105, 141; National Guard 128; prison department 143; ranking in number of explosion-related mining deaths 112; State Board of Agriculture 89; State Constitutional Convention 65; Workmen's Compensation Law 216n2
North Carolina Advertiser 62, 197n2, 197n4, 198n20
North Carolina Agricultural Experiment Station 73
North Carolina and Its Resources 89
North Carolina Central Railroad Company 34–35, 44–45, 67, 157, 187n72, 189n42
North Carolina Coal and Coke Company 103–104, 209n28
North Carolina Coal Merchants Association 151, 215n25
North Carolina Department of Agriculture 200n62; Experiment Station 200n66
North Carolina General Assembly *see* North Carolina Legislature
North Carolina Hand Book 89
North Carolina Land Agency 61, **61**, 197n4
North Carolina Land Company 63, 198n14
North Carolina Legislature 36, 39, 42, 44–46, 51, 82, 84, 99, 158
North Carolina Mining Corporation 143
North Carolina Railroad Company 44, 51, 66, 157, 188n32, 194n22
North Carolina Star 14, 180n17, 180n19, 190n48
North Carolina State Board of Agriculture 200n77
North Carolina State Constitutional Convention 158
North Carolina State Exposition 200n77

230

Index

North Carolina Steam Carriage and Plank Road Company 190n54
North-Carolinian (Fayetteville, NC) 45, 97
Norwood, John 167
Novelty Iron Works *see* iron ore

Oakland Township, North Carolina 104, 202n22
Obie, John 167
Olmstead, Denison 65, 69, 182n1, 182n8
Olsen, Paul E. 178n1
Ore Hill, North Carolina 193n11, 197n9
Osborn, Henry Fairfield 203n31
Owings, C.W. 131–132, 212n14

Parish, Thomas 37
Parker, D.J. 131
Parks, B.C. 152
Parkton, North Carolina 128
Paton, Robert *see* Payton, Robert
Payton, Robert 61, 64, 176
Pence, Thomas 94
Pendleton, George 97
Pennsylvania 48, 94, 122, 155; geological survey 200n62
Pennsylvania Central Railroad 66
Pensacola Navy Yard 58
People's Press 43
Petersburg, Virginia 189n37
Philadelphia, Pennsylvania 48
Piedmont-Cumnock Coal Company 119, 121, 209n26
Piedmont Railroad 44
Pilot (Vass, NC) 124, 146–147, 150, 179n6
Pinehurst, North Carolina 120
Pinkney, R.F. 56
Pioneer Steam Ship Company: *Martin Mullen* **89**
Pitt County, NC 180n9
Pittsboro, North Carolina 66, 95, 176, 183n17, 192n2
Pittsboro Mining and Transportation Company 36
plank roads 46, 189n44, 190n54, 191n65, 212n18
Pocahontas, Virginia 122; coal 88
Poe, Charles 96, 166–167
Poe, Counts 167
Poe, James 166
Point Peter, North Carolina 86
Porter, J.A. **70**
Portsmouth, Virginia 30, 58
Potter, H.H. 65
Power, Law & Co. 56; *see also* C.S. *Advance*
Pratt, Joseph Hyde 120, 122, 195n46, 209n7, 209n46

Press Visitor 97, 204n1
Price Mountain Coal Company 49
pulverized coal *see* Deep River Coal
Purdie, Ed 121
Purdie, T.J. 121
Purdie, W.J. 121
Purnell, T.R. 113

Raeford, North Carolina 128
railroad shops and foundries 193n21
railroads 43; *see also* individual railroads
Raleigh, North Carolina 30, 51–52, 63, 76, 113, 120 128
Raleigh and Augusta Air Line Railroad 67, **67**, 80, 82, 199 42
Raleigh and Gaston Railroad 55, 66; shops 55, 155, 189n45, 199n42
Raleigh & Western Railroad 76, 85–86, **90**, 104, 113, 117, 202n21, 203n34
Raleigh Bayonet Factory 57
Raleigh, Charlotte & Southern Railway 118
Raleigh Daily News 66
Raleigh Daily Sentinel 66, 69
Raleigh Daily Telegram 64–65, 198n25, 198n35
Raleigh Evening Visitor 68, 84, 199n47, 203n32
Raleigh Gas Company 103
Raleigh Mining Company 150, **151**; steel tunnel 150
Raleigh Morning Post 103
Raleigh News 65, 68, 198n26; *see also* Raleigh News and Observer
Raleigh News and Observer 79, 94–95, 112–113, 117, 121, 124, 128–129, 178ch1n1, 201n7, 202n15, 204n2
Raleigh North Carolinian 95
Raleigh Press Visitor 94–95
Raleigh Register 28, 31, 184n33, 200n63
Raleigh State Exposition (1884) 74, 78
Raleigh Times 109, 164
Raleigh Weekly Sentinel 197n6
Ramsay, Ambrose 179n4
Ramsay, Charles 124–125
Ramsay, Erskine 121, *123*, 124–125, 127, 134, 145, 149, 210n47, 211n55, 214n6, 216n37, 219; *see also* Erskine Ramsay Coal Company
Ramseur, North Carolina 113
Randleman, North Carolina 90
Randolph and Cumberland Railroad 117–118
Randolph County, NC 117
Reconstruction 2

Red Springs, North Carolina 128
Red Town, Chatham Co., NC **130**
Reeves, C.M. 121, 124
Reid, Gov. David Settle 27, 30
Reid, Jeffrey C. 216n37
Reid, T.T. 131
Reinemund, John A. 152
Revolutionary War *see* American Revolution
Richardson, Joe 127
Richmond, Virginia 44, 53, 58, 148; Richmond basin 158
Richmond and Danville Railroad 44, 65, 157
Rives, Fisher 167
Rives, Jim 167
Robbins, Eleanor L. 216n37
Robert E. Lee (blockade runner) 195n46
Robinson and Robinson (West Virginia) 149; Carol Robinson, mine safety 149
rock dusting 108, 138, 207n61
Rockingham County, NC 27, 65
Rogers, B.M. 134
Russ, William M. 96

Salisbury, North Carolina 193n16; prison 57
San Francisco Call **95**
U.S.S. *San Jacinto* 56
Sandhills Power Company 124, 211n52
Sanford, North Carolina 94, 109, 121, 124, 130, 143, 148, 177; sub-basin 154
Saxon, C.E. 131
Schenck, Love 167
Scott, Kerr 152
Scurlock, Mial 179n4
Seaboard Air Line Railway 85, 90–92, 95, 103, 113, 203n32, 204n43, 210n28
Second Pennsylvania Geological Survey 71
Segroves, W.H. 78, **79**
Semi-Weekly Observer (Fayetteville) 54–55
Semi-Weekly Raleigh Register 185n49
Shale Gas Potential in Triassic Strata of the Deep River Basin 216n37
Shedd, Edward W. 117
Shields, Joseph J. 152
Shirley, Charlie 211n57
Siler City, North Carolina 104
Silliman, Benjamin 182n2
silver 22, 29, 41, 105, 195n39; silversmiths 41
Silver Hill, North Carolina 71, 192n10
Simmons, R.P. 146
slack water navigation 34

231

Index

Smith, John A. 59
Smithfield, North Carolina 188*n*19
The South, an Immigration Journal, Devoted to the Interests of the Southern States, and North Carolina Particularly 78, **80**
South Carolina 175
South Carolina Railroad 174–175
Southeastern Anthracite Company 146
Southern Pines, North Carolina 121, **146**; Southland Hotel 121
Southern Railway 90, 125
Southern Weekly Post (Raleigh, NC) 29, 185*n*56
Southport, North Carolina 201*n*12
Starbuck, M.C. 104
State Gazette of North Carolina 180*n*15
Statesville, North Carolina 65
Steadman, J.P. 176, 202*n*18
Stokes County, NC 27, 65, 71, 115–116, 158, 192*n*2, 202*n*25
Stone, R.W. 65, 158, 178*n*4, 198*n*31
Street Coal Company 177
Stuckey, Jasper Leonides 195*n*46
Sydnor, Charles S. 183*n*14

Taft, William H. 106
Tally, W.J. 109–110
Tanner, H.S. 189*n*45
Tar River 180*n*9
Tarborough and Hamilton Railroad Company 42
Taylor, Richard Cowling 183*n*17
Taylor Coal Company 176
Taylor Mine 28
Taylor Place 72–73; Taylor Place Coal Association 104
Tazewell Co., Virginia 204*n*43
Teusch, D.E. 165
Textoris, Daniel A. 216*n*37
Thompkins, D.A. 85
Thompson, Joe 167
Thompson, William Beverhant 44, 190*n*47
Tiffney, J.E. 134
Toenges, Albert L. 152, 215*n*19, 216*n*37
Toms Creek, Virginia 122
Tredegar Iron Works 49, 192*n*7, 196*n*56, 197*n*12
R.M.S. *Trent* 56
Trumbell, Louis L. 152
Tull, J.J. 19
Tuttle, Edgar G. 69, 76–77, 201*n*87
Tyler property 40
Tyson, G.W. 177
Tyson mines 176–177
Tysor, Wright 167

Union Republican 202*n*24
United Colliers Company 86
United Mine Workers of America 85, 151
U.S. Bureau of Mines 84, 102, 105, **106**, **107**, 122, 127, 131, 133–134, 139, **140**, 141, 145–146, 150, 152–153, 178*n*1, 207*n*59, 213*ch*13*n*2; experiments 207*n*61
U.S. Bureau of the Census 36, 178*n*3
U.S. Department of Housing and Urban Development 153
United States Geological Survey 88, 97, 106, 122, 153, 158, 198*n*31
University of North Carolina at Chapel Hill 126, 182*n*9
Upchurch, Charles D. 79
Utah 146

Vance, Gov. Zebulon B. 56, 69, 193*n*21, 195*n*46; *see also* C.S. *Advance*
Vildbrandt, Frank C. 126, 143
Virginia 44; coal 49, 55, 65, 146; iron ore deposits 49; Richmond Basin 65

Waddill, M.Q. 198*n*39
Walnut Cove, North Carolina 83, 104, 115, 177, 200*n*64, 202*n*24
Walnut Creek, North Carolina 52
Washington (state) 146
Washington, D.C. 35, 52, 60, 131, 146, 153
Weekly North Carolina Standard 186*n*57
Weekly Observer (Fayetteville) 43, 53, 193*n*11, 194*n*27
Weekly Raleigh Register 27, 38, 181*n*32, 185*n*40, 191*n*68
Weekly Standard (Raleigh, NC) 39, 55, 193*n*11
Weldon, North Carolina 30, 78, 189*n*45
Wells, Gideon 192*n*4
Welsh steam coal 55, 195*n*46
West Indies 180*n*15
West Virginia 2, 8, 11, 76, 106, 127, 146, 148, 150, 155, 202*n*22, 204*n*43, 206*n*51, 207*n*59, 212*n*17
Western and Coalfield Railroad 59
Western Carolinian 181*n*31
Western Democrat 188*n*22
Western North Carolina Rail Road 65, 68, 191*n*64
Western Railroad Company **41**, 42, 44, 46, 53, 62, 66, 188*n*32, 197*n*1
White, Arthur 167

White, Horace 36
White, Joe 167
White, Lewis 167
Wilcox, John 3, 12, 24, 29, 155, 176–177, 179*n*2, 179*n*6, 185*n*54, 197*n*9, 210*n*36; *see also* iron ore
Wilcox mines 176–177
Wilkes, Charles Denby (U.S.N.) 31–35, 186*n*62, 187*n*16, 187*n*72; advocate of national foundry 35; *Report of the Secretary of the Navy, Communicating the Report of Officers Appointed by Him to Make Examinations of the Iron, Coal and Timber of the Deep River* 32–33
Wilkinson, John 195*n*46
Williams, Emmitt C. 148
Williams, Lemuel 22–23, 184*n*33, 198*n*39
Williams, M.P. 190*n*48
Wilmington, North Carolina 24, 34, 45, 53–56, 58, 63, 76, 85–86, 88, 189*n*41, 210*n*28
Wilmington Advertiser 43
Wilmington and Manchester Railroad 174–175
Wilmington and Raleigh Railroad 189*n*45
Wilmington and Weldon Railroad 174–175
Wilmington Messenger 81, 112, 202*n*18, 204*n*52
Wilmington Morning Star 85, 120, 209*n*15, 209*n*26, 210*n*28
Wilmington Tri-Weekly Commercial 37, 194*n*23
Wilmington Weekly Messenger 86, 203*n*35, 203*n*38, 204*n*42
Wilmington Weekly Star 88, 204*n*43
Winston-Salem, North Carolina 113
Womack, Thomas B. 97, 111
Workers' Compensation Act of 1929 155, 213*n*30, 216*n*2
Workmen's Compensation Act of 1929 *see* Workers' Compensation Act of 1929
Worth, Jonathan 69
Worth, T.C. 202*n*25
Worth, W.E. 203*n*26
Worthy, Silas 145

Yadkin Valley, North Carolina 45–46
Yadkin Valley Railroad 79–80
Young, J.H. 119

www.ingramcontent.com/pod-product-compliance
Ingram Content Group UK Ltd.
Pitfield, Milton Keynes, MK11 3LW, UK
UKHW050533150426
5217IPUK00026B/1911